Premises Cabling
3rd Edition

Donald J. Sterling, Jr.

Les Baxter, P.E.

Australia • Brazil • Japan • Korea • Mexico • Singapore • Spain • United Kingdom • United States

DELMAR
CENGAGE Learning™

Premises Cabling, Third Edition
Donald J. Sterling Jr., Les Baxter, P. E.

Vice President, Technology and Trades SBU: David Garza

Director of Learning Solutions: Sandy Clark

Senior Acquisitions Editor: Stephen Helba

Senior Channel Manager: Dennis Williams

Production Director: Mary Ellen Black

Marketing Coordinator: Stacey Wiktorek

Senior Production Manager: Larry Main

Production Editor: Benj Gleeksman

Development: Dawn Daugherty

Library of Congress Control Number: 2005052904

ISBN-13: 978-1-4018-9820-5

ISBN-10: 1-4018-9820-3

Delmar
Executive Woods
5 Maxwell Drive
Clifton Park, NY 12065
USA

Cengage Learning is a leading provider of customized learning solutions with office locations around the globe, including Singapore, the United Kingdom, Australia, Mexico, Brazil, and Japan. Locate your local office at **www.cengage.com/global**

Cengage Learning products are represented in Canada by Nelson Education, Ltd.

To learn more about Delmar, visit **www.cengage.com/delmar**

Purchase any of our products at your local bookstore or at our preferred online store **www.cengagebrain.com**

Printed in the United States of America
2 3 4 5 6 20 19 18 17 16

Contents

About the Authors

Don Sterling has been with Godfrey Advertising for 15 years and is currently Director of Technical Strategic Marketing. He has produced technical marketing materials for several of the leading structured cabling manufacturers. He is the author of several widely read books, most recently, *Technicians Guide to Fiber Optics*.

Les Baxter was employed at Bell Laboratories for 24 years, serving as a Member of Technical Staff, Technical Manager, and Director in the Business Communication Systems, Optical Fiber Solutions, and Optical Networking business units. After retiring from Bell Labs in 2001, he founded Baxter Enterprises, a consulting firm specializing in the architecture and development of telecommunications products, systems, and applications.

Since founding Baxter Enterprises, Mr. Baxter has consulted with a variety of companies ranging from start-ups to industry leaders. His consulting projects have involved areas such as: product architecture (requirements, specifications, redundancy, reliability, etc.); product development (technological tradeoffs, design reviews, cost reductions, etc.); product line evolution (including phasing of functionality and adapting current products to address new markets); and industry standards (both complying with current standards and developing new standards.) More information on recent projects and publications is available at www.baxter-enterprises.com.

Mr. Baxter is a registered Professional Engineer in New Jersey and a Senior Member of the IEEE. His education includes an MSEE from the University of Delaware, a BSEE from Rochester Institute of Technology, and a mini-MBA in Global Business from Penn State. He has published more than 30 articles in technical and trade publications and is the author of ***Residential Networks.*** He has been a frequent speaker at industry conferences and has made technical presentations on 5 continents. He has contributed to a number of EIA/TIA, IEEE, and ISO/IEC standards, most recently serving as the secretary of the IEEE P1394c Working Group and participating in the definition of augmented Category 6 specifications in the TIA TR-42.7 committee. He holds 7 US patents, with 2 more pending.

Preface

Preface to the Third Edition

There have been many new developments in the cabling and networking industries in the five years since the second edition was published. The third edition incorporates discussions of many of the most important concepts, including:

- The development of Augmented Category 6 UTP cabling (Cat 6A), which is described in TIA-568-B.10.
- The IEEE 802.3 an standard, which specifies 10-gigabit Ethernet over UTP, also known as 10GBASE-T.
- The Power over Ethernet (PoE) specification (IEEE 802.3af).
- Several updates to the TIA cabling standards, including TIA-569B (Pathways and Spaces), TIA-758 (Customer-Owned Outside Plant), and TIA-606A (Administration).
- Several application-oriented cabling standards, including:
    - TIA-862 (Cabling for Building Automation Systems)
    - TIA-942 (Data Center Cabling)
    - TIA-1005 (Industrial Cabling Systems)
    - TSB-162 (Cabling for Wireless LANs)
- The chapter on Residential Networks has been updated to reflect TIA-570B and other developments.

Several other changes have also been made. Due to its increasing importance, IEEE 1394 (FireWire) is discussed in much more detail. A more complete review and tutorial on decibels has been added in Appendix F. Finally, about 100 new review questions have been added.

I would like to thank everyone who assisted with the preparation of this edition, including the following people and organizations who contributed new product photos:

- Chuck Grothaus and Stuart Reeves (ADC)
- Tim Adams and Joe McGuire (American Access Technologies)
- Dan Wright and Mark Johnston (Fluke Networks)
- Priscilla Gagnon (The Siemon Company)
- Jim Hulsey (SYSTIMAX Solutions)

Thanks to the people at Delmar, Cengage Learning for making the process fun. And finally, thanks to my family (Patti, Jenn, Cari, Ben, and Ziggy) for their support and encouragement.

Les Baxter
October 2005

The second edition of *Premises Cabling* follows the same general outline as the first edition. The text has been updated to include the many changes in technology that have occurred. These include Categories 5e, 6, and 7 cable, advances in fiber technology, and gigabit networking. Some topics in the first edition—such as FDDI and 100VG-AnyLAN—have been reduced or eliminated as their importance has declined or disappeared.

The first part of the book deals at the component level with cables and connecting hardware. Then to change pace, Chapter 5 deals mainly with networking. Because building cable evolved mainly to meet the needs of networks, we've tried to give some perspective on how networks fit into building cabling schemes—or, perhaps, how building cabling schemes fit into networks.

The remainder of the book considers the cabling system from a system-oriented viewpoint, with some guidelines on how to install it, test it, and administer it.

Two chapters have been added to cover residential cabling and building automation cabling. In both cases, we have concentrated on the cabling systems and have avoided large detours into automation and related topics. Similarly, although wireless systems are finding popularity, we have avoided them since they are, well, wireless.

Numerous people have helped make this book a reality. Of the many who helped, a few deserve special mention. Alyson Moore and Melanie Bridges of Siecor were models of efficiency, cheer, and responsiveness in supplying photos and illustrations. Bret Matz of Sun Conversion Technologies was equally helpful. Katherine Karter of The Siemon Company, Karen Curry of Lucent Technologies, and Holli Haswell of 3M were also helpful in providing photos.

Several people at Bell Labs lent their expertise in reviewing sections of the text. In particular, we would like to thank John Kee, Paul Kolesar, Bob Pritchard, Masood Shariff, Pete Tucker, and Bruce Wagner. The standard disclaimer applies: errors and shortcomings remain ours.

And, as always, thanks to our families for patience and understanding. And the best part of writing the book is the ability to dedicate it.

As before, this is for Megan.
DJS

For Patti, Jenn, Cari, and Ben.
LAB

ACKNOWLEDGMENTS

The authors and Delmar, Cengage Learning wish to acknowledge and thank the reviewers who participated in the development of the third edition. Thanks go to:

Bill Lacy
Northwest Kansas Technical College
Goodland, KS

Fred Lemke
Dunwoody Institute of Technology
Minneapolis, MN

CHAPTER 1 Introduction

OBJECTIVES

After reading this chapter, you should be able to:

- Define a structured cabling system.
- Describe the importance of standards in cabling systems.
- Distinguish between horizontal and backbone cable.
- Define the three-level hierarchy of a cabling system: main cross connect, intermediate cross connect, and horizontal cross connect.
- Discuss factors affecting the demand for bandwidth, including the Internet.
- Identify the main trend in local area networks: higher speeds.

Not too many years ago, wiring a building was a simple affair. Electricians wired the building for power. Telephone craftspeople wired the building for telephone service. If there were any large computers, they were usually wired together by the computer company that supplied the hardware. Lines of responsibility were clearly delineated, with little confusion or overlap.

The personal computer changed this orderly approach. Invented in the 1970s and widely adapted by business in the 1980s, the personal computer quickly became a fixture on the modern office desk. After the telephone, the PC is probably the most common piece of equipment in the office. People share copiers, printers, fax machines, and even coffeemakers. But the PC is an individual thing; like the telephone, it is installed on every desktop. What is shared is data over the network, not the computer hardware itself.

The PC by itself did not change the wiring of buildings: the networks, too, played a role. Indeed, networks predate the PC. Xerox invented Ethernet before the PC existed. IBM released its cabling system before it released either its PC or its Token Ring network. The earliest adopters of networks used workstations (essentially a very powerful and quite expensive desktop computer) and minicomputers.

If computers are neat and powerful tools allowing individuals to become more creative and productive, then linking them together brought the same benefits to the workgroup. People could share information, send electronic messages, and share resources like printers, fax machines, and mass storage.

The PC and network sowed confusion in how a building is wired. Different networks had different cabling requirements. Worse yet, computer networks tied together into larger networks (or large networks were subdivided into smaller networks), along with the need to link different types of networks. Remotely separated networks were linked together over the public telephone system (itself the world's largest electronic network). The huge mainframe and smaller sibling, the minicomputer, were hastily and incorrectly declared dinosaurs on the edge of extinction. The network was hailed as the future of computing.

Here was the crux of the problem in the 1980s: in the 1960s and '70s, a company committed to a certain type of computer—say, an IBM mainframe. IBM then took complete responsibility for the computer installation, including the cabling that connected both peripherals and hundreds of terminals to the mainframe. But with the rise of the PC and networks, no single vendor had responsibility. IBM and a few other companies had the vision to offer a single-source approach to the system, but buyers were wary of being tied to a single vendor. Companies bought PCs from IBM, Compaq, Apple, and a dozen other companies; printers from Hewlett-Packard, Apple, and QMS; network software from Novell and Banyan; network hardware from Synoptics, 3Com, and other startups in this new field; and application software from Microsoft, Lotus, and WordPerfect. (And this namedropping only hints at the number of vendors available!)

Suddenly, there was no single point of responsibility for either the network or the cables that interconnected everything. As networks became standardized, with well-defined rules, cabling did not quite follow suit so readily. In the old days, if a problem occurred with your cabling, you called the electrician, the phone company, or the computer company. With the network, who do you call? "Not my job," said the computer vendors, the network hardware suppliers, and just about everybody else involved. "I don't know about networks," said the architect. In such a situation, everyone began pointing fingers and claiming it was the other guy's equipment causing the problem.

Large companies developed in-house expertise. Those involved with creating and running the network became intimately involved in understanding the cabling. (A sign of the new sophistication: the *wiring* became the *cabling*.) Independent computer/network consultants were also available. Most were quite good and competent, while some presented a dubious gamble at best.

The Physical Plant Is Critical

Most problems in a network can be traced to the so-called physical plant, principally the cables and connectors that tie everything together. By various estimates, anywhere from 50% to 75% of all network problems involve the physical plant. Regardless of whose estimates you use, the importance of reliable cabling systems is obvious. If a failure occurs, first check the cabling system. Disruptions occur from many sources:

- Shoddy workmanship during installation. Incorrectly applied connectors, poor cable routing, cable kinks, and excessive bends are among the most common problems. This is a particularly frequent problem with newer high-speed networks.
- Violations of interconnection rules involving allowable cabling distances and types of cables used.

- Use of substandard components in relation to the demands of the application.
- Failure to maintain the plant as circuits are added, moved, or rearranged.
- Human error in installing equipment and maintaining the plant.
- Having a constrained, rigid system whose growth and evolution can only be accommodated by Rube Goldberg renovations.
- Users who attempt to make the cabling outperform its capabilities or who use the wrong cable. For example, using a low-speed telephone cable to make a high-speed network connection sometimes works, but usually doesn't.
- Electrical noise from surrounding equipment, including power cables, fluorescent lights, motors, and so forth.
- Rodents chewing through cable. Buildings have mice, and they will nibble on cable.

Standards Are Us

A related trend was the powerful move toward standards-based open systems. Companies no longer wanted to be tied to single-company, proprietary schemes. They wanted choice, and safety in the knowledge that the equipment they bought from one company would work with the equipment purchased from another company. The PC architecture became a *de facto* standard. Networks, too, were standardized by bodies like the Institute of Electronic and Electrical Engineers (IEEE), American National Standards Institute (ANSI), International Electrotechnical Committee (IEC), International Standards Organization (ISO), Telecommunication Industry Association (TIA), and Electronic Industries Association (EIA).

So why not standardize cabling systems? AT&T and IBM both offered structured cabling systems. These offered two significant benefits: flexibility and coherency. By coherency, we mean that the system was, in fact, a system tied together in an orderly and rational fashion.

Coherency, in turn, provided flexibility for moves, adds, and changes. One thing in building cabling is certain: it isn't static. People change offices, workgroups change personnel, and companies grow (or contract) as they succeed (or don't succeed). A structured system or open system allows moves, adds, and changes in wiring to be done easily. Rather than ripping out the wiring (a practice more common than you might think) and installing new cabling, simple changes at different points in the system could be easily made by simply rearranging components or plugging in new ones. Equally important is the idea of "futureproofing" a building against premature obsolescence of the cable plant. The cabling should meet tomorrow's needs as well as today's.

The IBM cabling system had limited success, but it largely failed because it did not adequately meet the evolving needs of users. It still seemed too focused on IBM equipment and systems: its greatest popularity is with the IBM-backed Token Ring network. And the shielded twisted-pair cable favored is undeniably a good performer. It is, however, being edged out by lower cost alternatives that still meet the performance requirements of different applications. In other areas, IBM guessed wrong; the type of optical fiber they specified is different from the type adopted by the rest of the industry. To its credit, IBM has moved away from its own cabling system and has adopted open cabling systems.

The cabling industry has been through a lot of consolidation and reorganization since the first edition of this book was published. AT&T's SYSTIMAX system, which was built on twisted-pair and fiber-optic cables and components that grew out of its telecommunications and data networking expertise, was the pioneer in structured cabling. In the last few years, SYSTIMAX Solutions has moved from AT&T to Lucent Technologies, then to Avaya, and is now a Commscope company. It continues to be a leader in the structured cabling field.

Amp was another early adopter of the structured cabling paradigm. The AMP NETCONNECT system is now a part of Tyco Electronics.

By the beginning of the '90s, the industry was ready for standards defining how buildings should be cabled for both telephone and data. Both TIA/EIA-568, in the United States, and ISO/IEC 11801, internationally, address cabling of commercial buildings. The importance of these standards is that they are *performance* driven rather than *application* driven. They don't recommend cabling specifically for Ethernet or Token Ring. Rather the guidelines recommend methods of installing cable for, say, 20-MHz operation. This cabling, if properly installed, will permit either (or both) Ethernet or Token Ring to run on it. The hot area in networking today is high-speed networks operating at gigabit rates. By installing a high-performance cabling system, you don't have to know what network you are going to use. The cabling should accommodate FDDI, ATM, Fast Ethernet, Gigabit Ethernet, or even a network scheme not yet created.

Newer networks are more open in their cable requirements; that is, they run on common twisted-pair or fiber-optic cable, rather than special cables specific only to a given network. Both the TIA/EIA and ISO/IEC standards recognize the need to support "legacy" LANs, those with cabling needs that predate and differ from those recommended by the standards. For example, the earliest versions of Ethernet used coaxial cable. While some users with heavy investments in coaxial Ethernet may prefer to continue with coaxial cable, few new installations will choose to do so. Twisted-pair cable—cheaper, smaller, and easier to work with—has largely replaced coaxial cable. A structured cabling system should be able to meet both the application-independent requirements of TIA/EIA-568B and ISO/IEC 11801 and those of legacy applications.

Because TIA/EIA-568 and ISO/IEC 11801 have become dominant standards for cabling buildings, we will spend a fair portion of this book looking at their requirements. The two standards are very similar, but differ in some important points. It is possible to certify performance to one standard and fail to meet the requirements of the other. TIA/EIA-568B is a subset of ISO/IEC 11801. With a few minor exceptions, following 568 also means meeting the requirements of 11801. The converse is not true. The 11801 standard permits things that the 568 standard does not.

The TIA/EIA and ISO/IEC committees have expended considerable effort to harmonize the two standards. Future editions should have fewer differences and incompatibilities. Some differences do exist. For example, ISO/IEC 11801 includes Category 7 shielded cable while TIA/EIA-568B does not. More important for compatibility, however, is the lessening of differences in component and system specifications that would mean a cabling system that passes one specification will fail the other.

This book concentrates on TIA/EIA-568B. Because the TIA/EIA and ISO/IEC standards are moving closer together, nearly everything in this book also applies to ISO/IEC 11801. The 568B standard actually includes four documents:

- TIA/EIA-568B.1: System-level considerations
- TIA/EIA-568B.2: Unshielded twisted-pair cable and connectors
- TIA/EIA-568B.3: Fiber-optic cable and connectors
- TIA/EIA-568B.4: Shielded twisted-pair cable and connectors

The last document for shielded cable and connectors is somewhat neglected here since shielded cables are not widely used. Most premises cabling standards use unshielded twisted-pair cable or fiber-optic cable.

There have been a couple of important additions to TIA-568B.2:

- TIA-568-B.2-1, which was issued in February, 2003, specifies performance of cabling and components for Category 6 UTP. Cat 6, as we will see in the next chapter, has become the most widely installed type of UTP cable.
- TIA-568-B.2-10, which is currently in draft form as this edition is being written (mid-2005), will define cabling and component specifications for Augmented Category 6 UTP (which will be referred to as Cat 6A).

One important thing to remember about 568B and 11801 is that they represent the consensus of experts from both the telecommunications and network industries. Many of the same people who worked to create the cabling standards for commercial buildings also serve on committees defining network and telecommunications standards. Thus the standards represent an important effort to accommodate the real needs of users. Building cabling standards, on one hand, attempt to create a flexible structure to satisfy the needs of emerging applications. These same emerging applications are defined with the cabling standards in mind. For example, a significant design feature of Gigabit Ethernet was to allow high-speed network communications over cabling already installed in the building. Users don't want a network that requires special cabling, or one that isn't open and flexible enough to meet growing and evolving needs.

But before we begin looking at the elements of building cable in detail, this chapter will offer a quick overview of building cabling and of networks. Regardless of the application independence of building cabling standards, a practical approach must consider the needs of the network. Because we will look at networks in some detail later and will refer to them throughout, we'll end this chapter with a brief overview of some popular networks.

The Bandwidth Crunch

A recurring consideration with networks is to provide enough bandwidth to accommodate all users. As the needs of more and more users became more sophisticated, low-speed networks became a bottleneck in passing information from station to station. If users are only trading files and messages of modest size, a low-speed network suffices. But applications quickly grew until they started consuming a significant portion of the bandwidth. For example, the word processing file of the text for this chapter is only a modest 28,000 bytes. A magazine ad with a large color photograph can run to several million bytes. Computer-aided design files can also consume millions of bytes. Multimedia, and especially video, also produce large files running to billions of bytes.

Applications like video and multimedia also place demands on networks. Not only do they require a high bandwidth, they also are real-time applications. They cannot be delayed by other files being transferred or by bottlenecks in the network, at least not if you want something more than jerky, flaky images.

The quest for bandwidth leads naturally to higher speed networks. Early networks operated at 16 Mbps or less, while emerging networks run at speeds up to 10 Gb/s. Other techniques, which we will discuss in Chapter 5, also make more efficient use of bandwidth.

The Internet has also increased the thirst for bandwidth. The Internet has become a necessary business tool for research and communication. Related to the Internet are Intranets, which are inside-the-company versions of the Internet used for internal communications.

Although the exact impact of the Internet varies from company to company, the importance of the Internet on bandwidth demands can be seen in one simple fact: *today Internet traffic on the public telecommunications network exceeds voice traffic.* That is, there are more bits of Internet communications than there are bits of telephone calls being carried on long-distance telephone lines. Figure 1–1 shows the worldwide network demand in Gb/s for voice and data signals. Note that the data traffic exceeded the voice traffic for the first time in 1999 and continues to grow at a much higher rate[1]. In the future, as more voice traffic is packetized and carried over the Internet, the disparity will grow even larger. This increase in data traffic means that local neworks will be constantly increasing in speed.

The point is this: as networks operate faster, the demands they place on the cabling system also increase. A high-speed network is a trickier animal to tame than a low-speed network. As a result, the cable plant and its proper installation become increasingly critical to the reliable operation of the network.

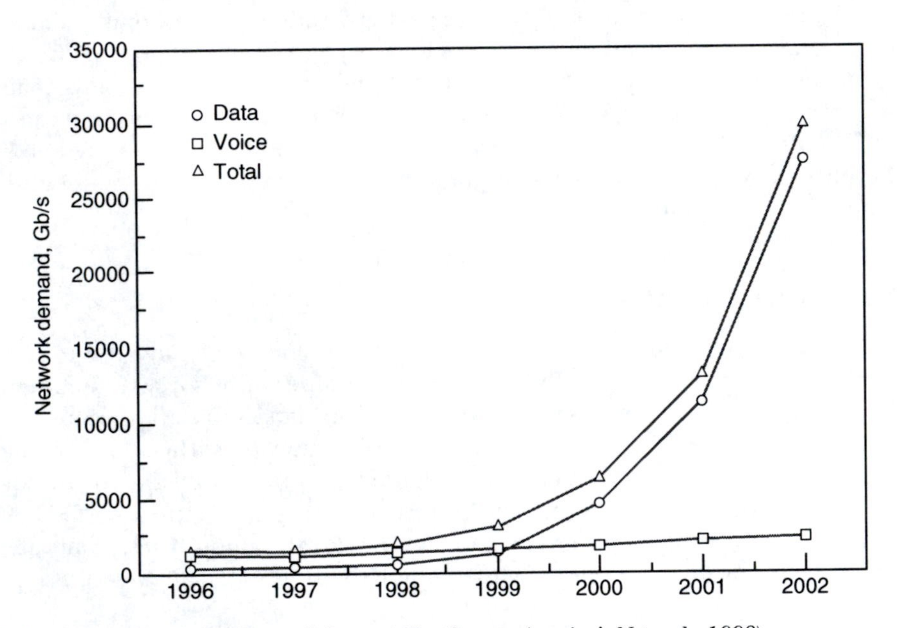

Figure 1–1. Comparison of voice and data traffic *(Source: America's Network: 1998)*

[1] *In this figure, the compound annual growth rates (CAGR) are roughly 17% for voice and a little over 100% for data. There has been a wide range of growth rates published over the years. These are fairly typical values.*

BUILDING CABLING

The cabling in a building can be divided into two sections (Figure 1–2): the backbone cabling running between floors and between buildings and the horizontal cabling running to different offices on a floor. Cables are usually consolidated at different points and then distributed through the buildings. These areas where cabling is consolidated are known as *telecommunications rooms* (TR). They were historically known as *wiring closets* (WC—not to be confused with water closets). Although the term "wiring closet" is obsolete, it is still frequently encountered.

There are three levels of distribution: the main cross connect, intermediate cross connect, and horizontal cross connect. A cross connect is a point where cables are connected; it forms a demarcation between different parts of the cabling system. Depending on the size of the building, not all three levels are necessary. Figure 1–3 shows the hierarchy of wiring closets.

The *main cross connect* is the first level of backbone cable. It is the area where outside cables enter the building for distribution through the building. For example, the outside telephone lines will be terminated in the main cross connect. There is usually one main cross connect per building or group of buildings.

The *intermediate cross connect* is a point dividing the first and second levels of backbone cabling. In large installations, it is more convenient for the main cross connect to feed to an intermediate cross connect, which in turn feeds several telecommunications closets.

The *horizontal cross connect* is the point at which the backbone and horizontal cables meet.

Backbone cables also can connect cross connects in different buildings. Some companies, schools, and universities have several buildings close together—within several hundred yards—in what is called a campus. It is often desirable to have all these buildings essentially wired

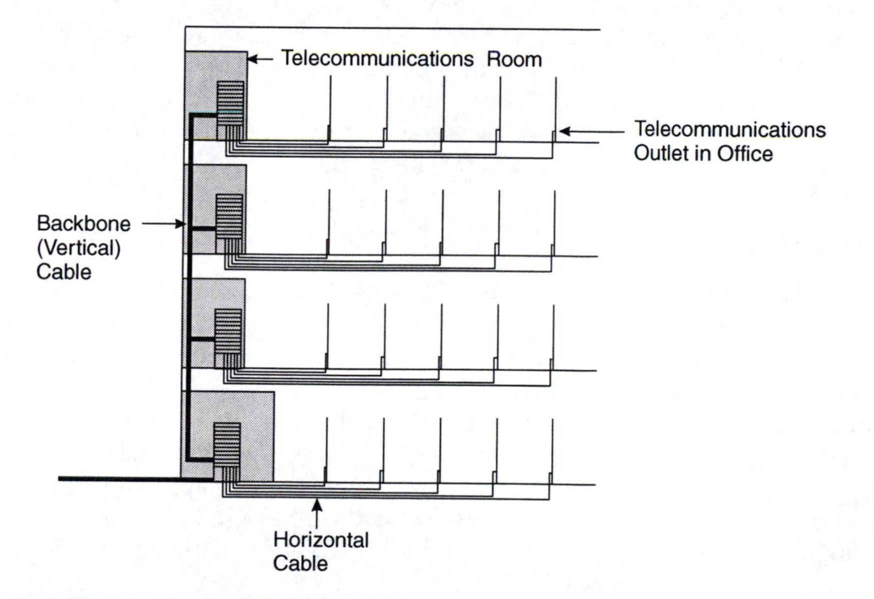

Figure 1–2. Backbone and horizontal cabling

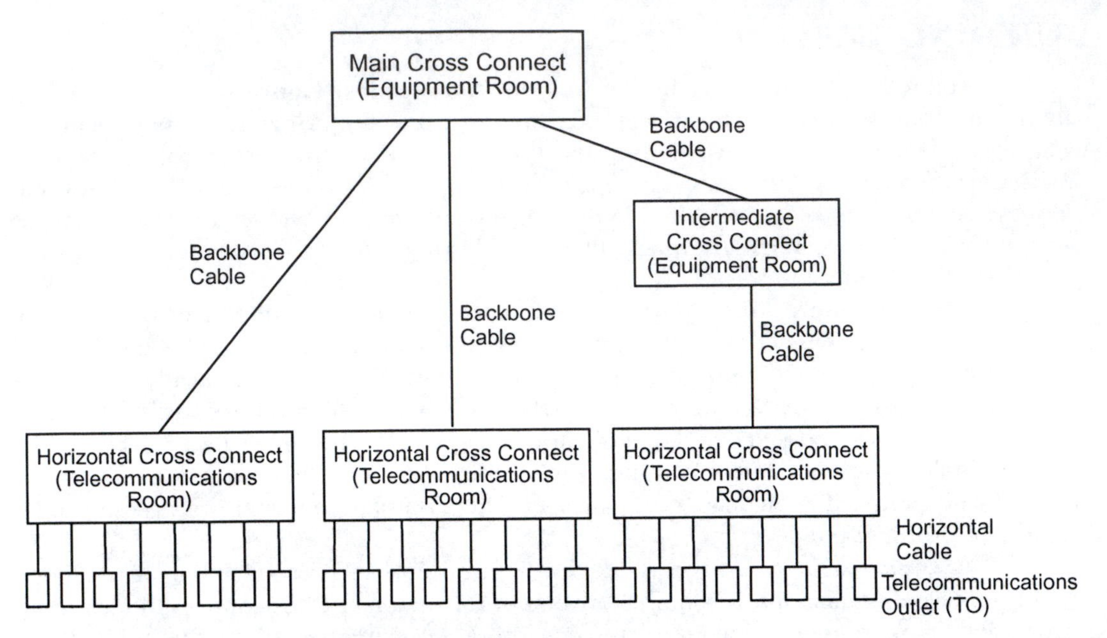

Figure 1–3. Hierarchy of cabling and wiring closets

as a single entity. When buildings are separated by significant distances—as far apart as corporate offices in different cities or countries—they can still be connected, but usually through the public telephone system or through leased lines. Such distances do not allow direct connection of main cross connects through a backbone cable.

We should mention that TIA/EIA and ISO/IEC standards are recommendations; they do not have the force of law. You can cable your building any way that works. You can use cabling rules from a specific vendor, you can design your own system, or you can run your cables and cross your fingers. But by following widely recognized standards, you simplify planning and ensure a cabling plant that will perform to specification.

Networks

A local area network is simply a means of interconnecting computers and related equipment like printers. That's simple in concept; a lot of ingenuity went into making networks practical. Over the years a number of different networks have evolved. From the perspective of premises cabling, the two most important things about networks are the types of cables they require and the speed at which they operate. Indeed, the trend in networking has been to simplify the cabling while increasing the operating speed. Figure 1–4 gives a short summary of the networks, their data rate, the types of cables they require, and the approximate year they were introduced. Notice that all network types use twisted-pair cable and optical fibers. FDDI and ATM are notable in that they are networks specifically designed to use optical fibers; other networks have been adapted to allow use of fiber.

(The dates indicate the year a substantially complete specification was published for the network. Products for the type of networks were marketed and networks built before that date.

Network	Cable Types	Data Rate	Standard	Year
Ethernet	Unshielded twisted pairs Coaxial Optical fiber	10 Mbps	IEEE 802.3	1980
Fast Ethernet	Unshielded twisted pairs Shielded twisted pairs Optical fiber	100 Mbps	IEEE 802.3	1995
Gigabit Ethernet	Unshielded twisted pairs Optical fiber	1000 Mbps	IEEE 802.3	1998
10-Gigabit Ethernet	Optical fiber Twinax	10 Gbps	IEEE 802.3	2002
10GBASE-T	Unshielded twisted pairs	10 Gbps	IEEE 802.3an	~2006
Token Ring	Shielded twisted pairs Unshielded twisted pairs Optical fiber	4 and 16 Mbps 100 Mbps	IEEE 802.5	1985
FDDI	Optical fiber Unshielded twisted pairs Shielded twisted pairs	100 Mbps	ANSI X3T9.5	1990
Fibre Channel	Optical fiber Coax Unshielded twisted pairs	1, 2, 4, 10 Gbps	ANSI X3.230	1994
ATM	Optical fiber Twisted pairs	155 and 622 Mbps	ATM Forum	1995

Figure 1–4. Popular types of networks

However, these did not necessarily meet the final published specifications. Additional details may be published later. For example, ATM devices were built in 1993 and 1994, but the final specification for local-area ATM networks was not approved until 1995. The earliest adopters faced the possibility of ending up with equipment that did not meet specifications. Remember, too, that networks continue to evolve.)

Ethernet was the earliest network to gain widespread popularity; today it is available in more flavors than any other. The original, and still popular, version of Ethernet runs at 10 Mbps.

Fast Ethernet is a significant upgrade to bring 100-Mbps speed to Ethernet networks.

Gigabit Ethernet, at 1000 Mbps, increases the speed of Ethernet ten times over Fast Ethernet and one hundred times over original Ethernet. Today Ethernet, in its different variations, is by far the most popular network type, commanding over 90% of the market. During the early 1990s, it emerged as the clear winner over competing network schemes.

10-Gigabit Ethernet (10-GbE) was originally defined to operate on fiber-optic cabling in 2002. A UTP-based version, 10GBASE-T, is under development and should be standardized by about 2006.

In the early 1990s, Ethernet emerged as the clear winner among LAN technologies and today commands over 90% of the LAN market.

Token Ring was popularized by IBM. Originally operating at 4 Mbps and later at 16 Mbps, Token Ring networks follow Ethernet in popularity, principally because of the huge installed base of IBM equipment. Although Token Ring was upgraded to 100 Mbps, it continued declining in popularity. It is encountered today only in legacy systems.

Fibre Channel (FC) is optimized for storage area networks (SAN). The original version operated at 1 Gbps. Subsequent versions increased the bit rate to 2, 4, and 10 Gbps.

FDDI (Fiber Distributed Data Interface) was the first network specifically designed for fiber optics. It was also the first network to operate at 100 Mbps. The network can also use copper cables in certain segments. FDDI never became as popular as early proponents had hoped and has been eclipsed by newer competing approaches.

ATM (Asynchronous Transfer Mode) is a network scheme whose future is not clear. It operates at 155 Mbps and 622 Mbps in LAN environments. (Standards for slower speeds of 25 Mbps and 51 Mbps have been created for connections to the desktop, but these have not caught on.) ATM was once seen as the hands-down favorite as the next-generation networking system, but slow standards development, relatively high costs, and the appearance of Gigabit Ethernet have made ATM less attractive. Still ATM is finding use in telecommunications applications and should not be ignored.

The progression of Ethernet shows an important point about the growth of networks and the demand for bandwidth. Every time the speed of Ethernet increased, it was by a factor of 10. Almost 15 years had passed between original 10 Mbps Ethernet and 100 Mbps Fast Ethernet, in part because the demand for bandwidth had not yet grown and in part because the technology was not ready to build Fast Ethernet. The time between the release of the standard for Fast Ethernet and Gigabit Ethernet was only about 3 years, and 10-Gb/s Ethernet followed 4 years later.

Building Cabling: The System

There are several considerations involved in planning and implementing a premises cabling system:

1. Choose the level of performance required, which will determine the type of cabling that is needed. In general, choose the highest-performance cabling system that is affordable. For today's installations, that would generally mean Category 6 UTP.

2. Decide what installation rules to follow: TIA/EIA-568B, for example.

3. Decide whether any legacy applications will need to be supported. This will determine if any additional cabling considerations apply.

4. Install the cable, being careful to follow applicable applications guidelines and practices.

5. Certify the cabling plant. This means testing the installation to ensure that the performance meets requirements.

6. Document the cabling. When moves, adds, changes, or repairs are required, you'll need to know what's what and what's where.

7. Use the system (enjoyably).

Summary

- A poorly designed or improperly installed cabling system can be a major source of network problems.
- A structured cabling system aims to offer application independence.
- The two main structured cabling standards are TIA/EIA-568B in the United States and ISO/IEC 11801 in Europe.
- The two main types of cable used in premises systems are unshielded twisted pairs and fiber optic.
- A structured cabling system has a hierarchical structure that includes the main cross connect, intermediate cross connect, and horizontal cross connect.
- The backbone cable typically provides connections between cross connects (between network equipment).
- The horizontal cable connects between the horizontal cross connect (or telecommunications closet) and the work area.
- Ethernet is the most popular network type, offering speeds of 10, 100, 1000 Mbps, and 10 Gbps.

Review Questions

1. Name the two main standards for structured cabling systems.
2. Define structured cabling system.
3. What is another term for the vertical cable running between floors in a building?
4. What is the term for the cable running from a closet to a work location?
5. Name the three levels of cross connects.
6. Names five causes of problems in cabling systems.
7. Distinguish between an application-driven cabling system and a performance-driven system.
8. Why is a performance-driven structured system preferred?
9. Are you connected to a network at work, school, or home? What type? What speed?
10. What is the most popular type of LAN, and at what speeds does it operate?

CHAPTER 2

Copper Cables

Objectives

After reading this chapter, you should be able to:

- Distinguish among the three main types of copper cables used in premises/network applications: unshielded twisted pairs, shielded twisted pairs, and coaxial.
- Define the main categories of unshielded twisted-pair cable and tell how the cables differ from one another, including Categories 3, 5, 5e, and 6A.
- Understand the specifications for Augmented Category 6 UTP.
- Tell how Category 7 cable differs from other categories in construction and performance.
- Describe the main performance characteristics of unshielded twisted-pair cable: NEXT, ELFEXT, characteristic impedance, attenuation, return loss, structural return loss, propagation delay, and delay skew.
- Understand alien crosstalk and its effect on the Augmented Cat 6 specifications.
- Explain why frequency and bit rate are not the same.

Because cables form the heart of the premises cabling system, we will look closely at the types and properties of cables in detail. The great division among cables is between copper cables carrying electrical signals and optical-fiber cables carrying light signals. This chapter covers copper cables; the next chapter looks at optical cables.

Many cables have been developed or adapted for network and premises wiring applications. The two main categories of cable are *coaxial cable* and *twisted-pair* cable. A cable has a simple purpose: to carry a signal a given distance at a given data rate. In doing so, the cable must protect the signal from distortion due to noise. Sufficient noise can turn a meaningful signal into gibberish, cause circuits to trigger falsely, ruin an important file, or crash a computer.

In this chapter, we will be discussing the properties of copper cable that affect signal transmission and the ability to transmit a signal without undue impairment. In terms of importance, the most fundamental—the Big Three, as it were—are:

- Attenuation
- Characteristic impedance
- Near-end crosstalk (NEXT)

These three are important in every application, from the modest-performing network to the latest and greatest. In addition, a number of secondary properties are important, although their importance tends to vary depending on the application. For 10-Mbps Ethernet, they are of negligible significance; for Gigabit Ethernet, they are essential to proper operation. These secondary properties include:

- Return loss
- Structural return loss
- Far-end crosstalk (FEXT)
- Propagation delay
- Delay skew
- Balance

At the higher frequencies used by 10-Gigabit Ethernet networks, yet another property becomes important.

- Alien crosstalk

We will discuss each property, beginning with noise and crosstalk.

Noise

When the first personal computers were introduced in the late 1970s, they radiated enough energy to interfere with televisions several hundred feet away—not just in the house where it was used, but in neighboring houses as well. This interference is known as electromagnetic interference (EMI). EMI is a form of environmental pollution with consequences ranging from the merely irksome to the deadly serious. EMI making television reception snowy is one thing; its interfering with a heart pacemaker is quite another story.

A copper wire can be a main source of EMI. Any wire can act as an antenna, radiating energy into the air where it can be received by another antenna. A simple demonstration of EMI can be done by placing an AM radio near a computer. You can tune into the computer at some point on the dial. Try running different programs to hear the differences each makes. What you are hearing is EMI.

Just as a wire can be a transmitting antenna, it can be a receiving antenna, picking up noise out of the air. Obviously, this received noise can interfere with the signal being carried. Of special significance is noise that couples from one wire to another in the same cable.

Crosstalk

A special type of noise is crosstalk. Crosstalk is energy coupled from one conductor to another in the same cable or between cables. Crosstalk was once fairly prevalent in telephone systems, where you could faintly hear another conversation.

When current flows through a wire, it creates an electromagnetic field around the wire. This field can induce a current in an adjacent wire within the field. This coupling is crosstalk.

Figure 2–1 shows the idea of crosstalk. A signal is transmitted down a wire (called the active or driven line). A certain portion of the energy couples on a second wire (called the quiet line). This energy on the quiet line is the crosstalk. Crosstalk is measured either in percentages or decibels. In percentages, the magnitude of the signal on the quiet line is given as a percentage of the signal on the driven line. Thus, the crosstalk might be 3% or 10%. If the driven line signal was 5 volts, a 10% crosstalk means that 0.5 volt appeared on the quiet line. In communication cabling, crosstalk is more often expressed in decibels, a term we'll turn to in detail shortly.

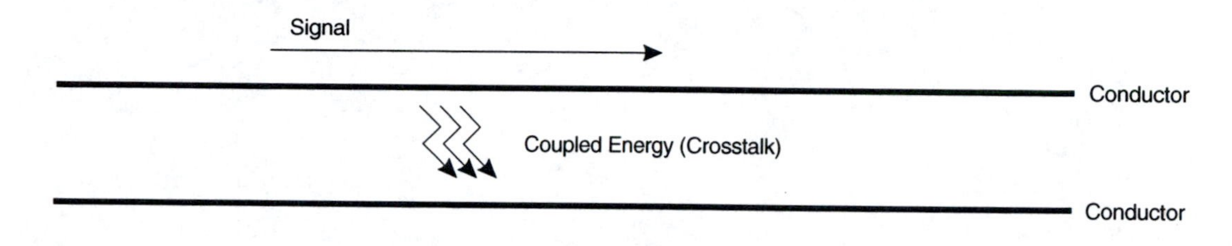

Figure 2–1. Crosstalk

Crosstalk can be measured at the near end or far end of the line. In near-end crosstalk (NEXT), the energy on the quiet line is measured at the same end as the source of the signal. In far-end crosstalk (FEXT), the energy is measured at the opposite end, away from the signal source.

NEXT

NEXT is of interest in virtually every premises cabling application. Consider a signal on the driven line. As it travels down the wire, it becomes attenuated, losing energy. At the opposite (far) end, it is weakest. So the greatest amount of energy is available for coupling at the near end. Similarly, crosstalk on the quiet line is usually strongest at the near end for two reasons. First, it can receive the greatest energy here because energy on the active line has not yet become attenuated and, second, it will not be attenuated as it travels to the far end.

NEXT has an additional practical consequence. In a typical network, one pair of wires is used to transmit, while another pair is used to receive, as shown in Figure 2–2. The signal on the transmit line is strongest at the transmission end. A signal on the receive line is also the weakest at this point. There is a greater likelihood of the coupled noise interfering with this weakened signal.

The formula for calculating NEXT in a cable is:

$$NEXT\ (dB) = 10 \log \left(\frac{P_q}{P_a}\right)$$

where P_q is the noise power measured on the quiet line and P_a is the driving power on the active line. In other words, crosstalk is the ratio of crosstalk power to signal power, measured in decibels.

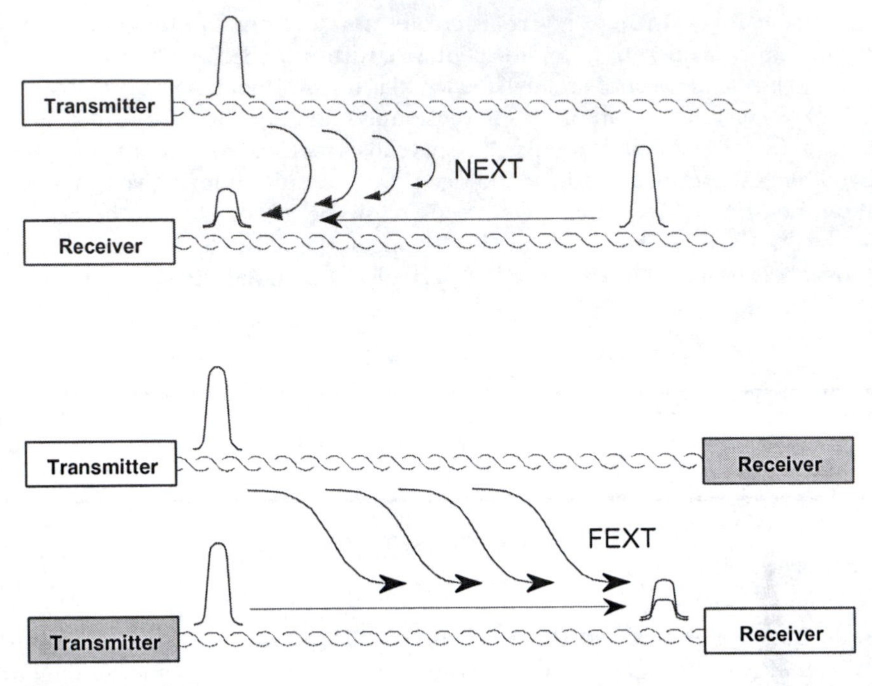

Figure 2–2. NEXT and FEXT

FEXT

The original premises cabling specifications for twisted-pair cable only specified NEXT. Most applications use only two pairs, one for transmitting and one for receiving. So you are only concerned with the effect of crosstalk from the transmitting line onto the receiving line at the near end. With some high-speed networks, notably Gigabit Ethernet, all four pairs of a cable are used to transmit and receive. Now FEXT becomes important. Consider again our active and quiet lines. We will now measure the crosstalk at the far end. The crosstalk will be the total of any crosstalk that occurs at any point along the line.

Although FEXT can be directly measured, cabling specifications do not use this value. Rather, they use a value for the *equal-level* FEXT or ELFEXT. ELFEXT is the ratio of the desired received signal to the undesired crosstalk at the far end:

$$ELFEXT = \frac{Signal}{FEXT}$$

Because the signal will be attenuated from the point it is launched at the near end and its arrival for measurement at the far end, ELFEXT can be calculated as the difference between FEXT loss and attenuation (both in dB):

$$ELFEXT = FEXT - Signal\ Attenuation$$

Because both the FEXT and the signal are attenuated as they travel, FEXT measurements are very low, lower than NEXT. But even the smaller power in the FEXT is important, because the signal is also much smaller. In NEXT loss, you compare how much smaller the NEXT is compared to a relatively strong signal. In ELFEXT, you are comparing the loss to a weakened signal.

Another consequence of ELFEXT is that crosstalk can occur at any point along the path from transmitter to receiver. NEXT occurs near the transmitter, but FEXT anywhere. FEXT is the sum of all crosstalk that occurs along the entire path. Thus, it can be harder to isolate ELFEXT problems because a bad component or other problem can be anywhere in the path. With NEXT, you can usually assume that problems are occurring near the transmitting source.

There are two basic generalizations that we can make about noise and crosstalk:

> ***Noise and crosstalk become more severe and harder to deal with at higher frequencies than at lower frequencies.***

> ***Noise and crosstalk exist in every copper-cable system. The trick is to minimize them to such an extent that they can be ignored.***

Power-Sum NEXT and ELFEXT

Standard NEXT and ELFEXT measurements simply compare the crosstalk from one pair to another pair. Only two pairs are involved. For many applications, this makes sense. An application like 10-Mbps Ethernet only uses two pairs, one for transmitting and one for receiving. Thus, pair-to-pair crosstalk is of interest.

In *power-sum* NEXT and ELFEXT, we assume that energy is coupled from all other lines in the cable onto a single quiet line. Gigabit Ethernet uses all four pairs of a cable to carry a signal. A power-sum NEXT or ELFEXT measurement energizes three pairs and measures crosstalk on the fourth pair. We are comparing multipair-to-pair crosstalk. Another instance where power-sum measurements are important are in backbone or zone cables, which have more than four pairs. One popular type of cable uses 25 pairs in a single jacket. The power-sum test will energize 24 pairs and measure the crosstalk into the remaining quiet pair. Ideally, you should individually test all 25 pairs as the quiet line, so power-sum measurements can be quite involved.

In any crosstalk test, the area of greatest concern is the worst-case measurement. If the worst case passes, your cable passes. If the worst case fails, the cable fails.

Alien Crosstalk

Both NEXT and FEXT are concerned with interference between pairs in the same cable. As transmission speeds move into the multi-gigabit range, another type of crosstalk has recently become of interest. Alien crosstalk (AXT) involves interference between different cables that are located close together, as illustrated in Figure 2–3.

Although this figure only shows the two cables, there could be many cables potentially generating alien crosstalk at the same time. Measurement of alien crosstalk typically uses a configuration known as "six-around-one," which consists of 6 disturbing cables around a "victim" cable, as illustrated in Figure 2–4.

Depending on where the source of the alien crosstalk is, it can be classified as either ANEXT or AFEXT. As indicated in Figure 2–5, if the interfering transmitter is at the same end of the channel as the receiver, it is called ANEXT. Similarly, it is called AFEXT if the interfering transmitter is at the far end.

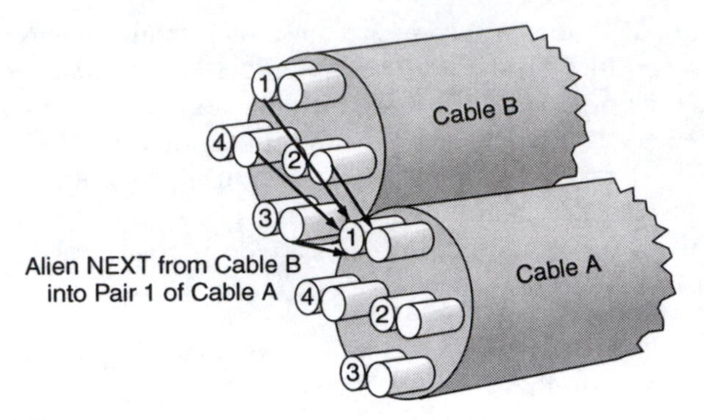

Figure 2–3. Alien crosstalk

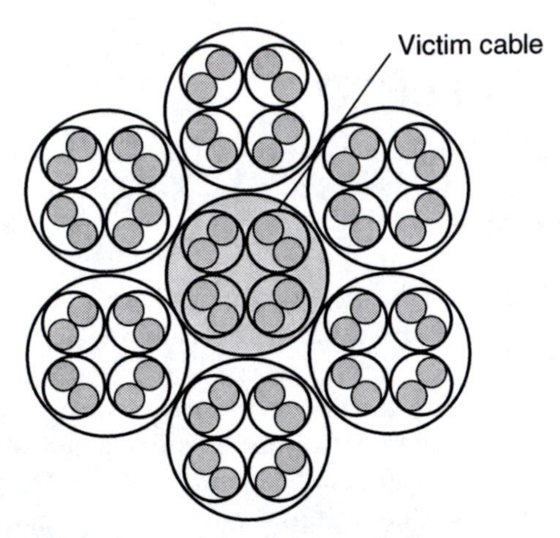

Figure 2–4. "Six-around-one" configuration for alien crosstalk measurement. The outer 6 cables are energized and noise measurements are made on the victim cable (shaded)

An additional complication with AFEXT is that the channels may be of different lengths. Figure 2–6 shows a 10m channel generating AFEXT into a 100m channel. This AFEXT from Transmitter A2 encounters much less attenuation than the desired signal from Transmitter B2. Unless the Transmitter A2 on the 10m channel reduces its power level, this can generate much more interference than if it was on a long channel. In other words, the AFEXT path is much shorter (and therefore has much less attenuation) than the signal path. Note that this cannot happen with regular FEXT, because the FEXT path length is always identical to the signal path length.

Alien crosstalk is typically smaller in magnitude than NEXT and FEXT. However, when NEXT cancellation is used, it can become one of the dominant transmission impairments. There has been some theoretical investigation of the possibility of canceling alien crosstalk, but there are currently no practical techniques for doing so.

Unlike the other cabling parameters, the environment in which the cable is installed plays a key role in AXT. Clearly, if there are no other cables around, there is no alien crosstalk. On the other hand, AXT is much more of an issue in densely cabled areas like data centers.

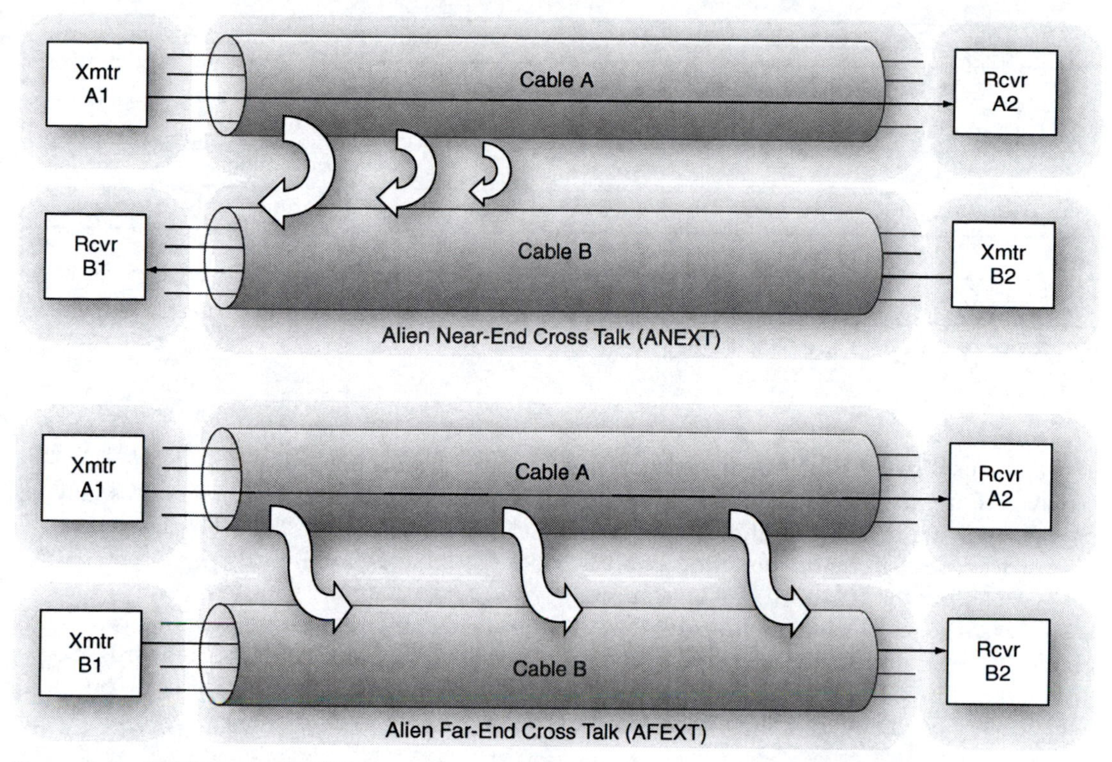

Figure 2–5. ANEXT and AFEXT. Transmitter A1 on cable A is generating ANEXT into Receiver B1 and AFEXT into Receiver B2 on cable B

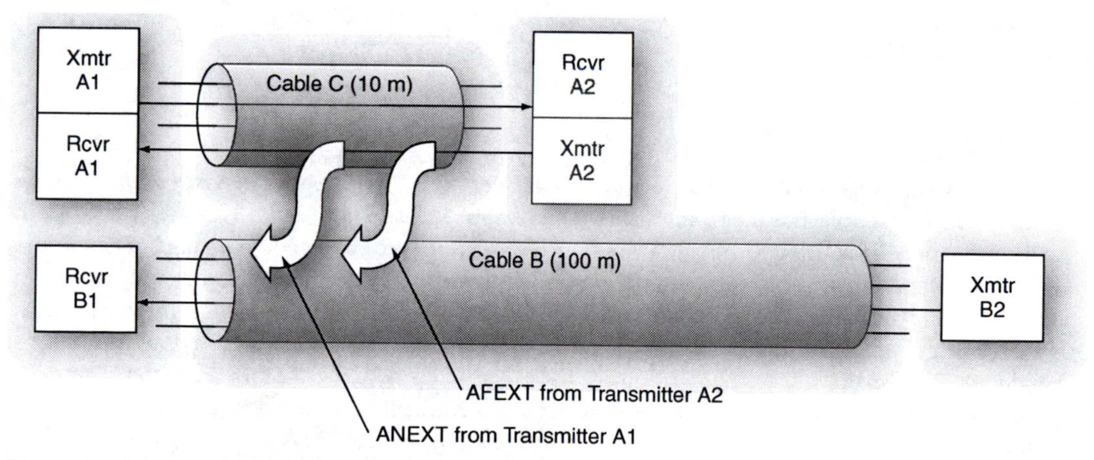

Figure 2–6. AFEXT from a short channel

Characteristic	Better Performance in dB
Crosstalk (NEXT/ELFEXT)	Higher number
Alien Crosstalk (ANEXT, AFEXT)	Higher number
ACR	Higher number
Structural Return Loss	Higher number
Cable Attenuation	Lower number
Connector Insertion Loss	Lower number

Figure 2–7. Putting the decibel in perspective

The specifications for Augmented Category 6 cabling will include requirements on alien crosstalk. For existing installations of Category 6 cabling, AXT mitigation procedures may be needed in some cases when the cabling is used at frequencies above 250 MHz. These mitigation procedures will be documented in TSB-155, "Additional Guidelines for 4-pair 100 Ω Category 6 cabling for 10GBASE-T Applications."

Current strategies for minimizing ANEXT and AFEXT involve making sure that cables are physically separated from each other, or using special ANEXT-mitigating patchcords. Some installation practices, such as tightly tie-wrapping cables and patchcords into large bundles, may need to be changed if alien crosstalk is a problem.

The Decibel

All types of crosstalk (NEXT, FEXT, and alien), as well as many other parameters of cabling systems, are measured in decibels (dB.) Appendix F gives a brief review of how decibels are used in calculations. Figure 2–7 shows how decibels are used with respect to some common cabling system parameters.

Impedance

There is one other characteristic that can affect noise and signal transmission quality: impedance. Every cable has a property known as *characteristic impedance.* Characteristic impedance is determined by the geometry of the conductors and the dielectric constant of the materials separating them. For the coaxial cables we will be discussing next, determining characteristic impedance is easy because the cables have a uniform and regular geometry. In other cases, such as twisted-pair cables discussed later in this chapter, characteristic impedance is harder to determine because the geometry is more irregular. Figure 2–8 shows an example of how characteristic impedance can be determined in simple geometries.

The most efficient transfer of electrical energy occurs in a system where all parts have the same characteristic impedance. If a signal traveling down a transmission path meets a change in impedance, a portion of the energy will be reflected back toward the source. This reflected energy is also noise; that is, it is unwanted energy in the system that can distort signals.

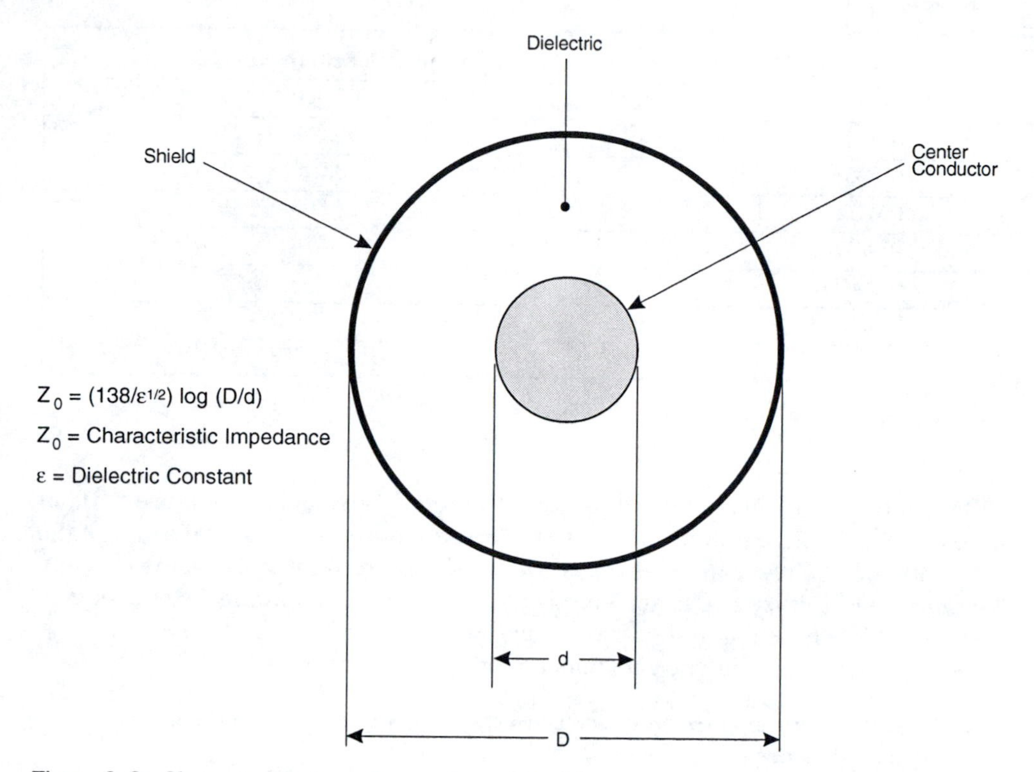

Figure 2–8. Characteristic impedance of coaxial cable

How much energy is reflected depends on the severity of the impedance mismatch. If the mismatch is large in terms of the difference in ohms, a greater amount of energy will be reflected.

A component's electrical dimensions are an important concept in understanding signal transmission. Electrical length is compared to the wavelength of the signal traveling through the components. Figure 2–9 shows the relationship between a signal's frequency and its wavelength. A 10-MHz signal, for example, has a wavelength of 30 meters, while a 100-MHz signal has a wavelength an order of magnitude shorter—3 meters. If a component is much smaller than the wavelength, it is electrically insignificant. The signal doesn't "see" the component.

It's important to remember that this relationship between the wavelength of the signal and the physical size of components applies only to characteristic impedance and reflections. Crosstalk and attenuation, on the other hand, are not similarly affected by a component's electrical dimensions.

The significance of a component's size can only be seen in comparison to the wavelength of the signal passing through it. As the signal speed increases, the wavelength decreases and the component becomes electrically larger. Therefore, at very high speeds, even a small component becomes electrically significant. For example, at 1 MHz, a 10-foot patch cable would have an insignificant electrical length. At 100 MHz, the same cable would be highly significant.

Frequency	Wavelength (meters)
1 kHz	300,000
10 kHz	30,000
100 kHz	3,000
1 MHz	300
10 MHz	30
100 MHz	3
1 GHz	0.3
$f = c/\lambda$	$\lambda = c/f$

Figure 2–9. Frequency versus wavelength

Notice that energy can reflect in both directions. Consider the situation in Figure 2–10, in which the line has three different impedances. At the first discontinuity, a small portion of the energy is reflected and the rest continues on. At the second, some energy is again reflected. But what happens when this back-reflected energy meets the first discontinuity? Some energy is re-reflected so that it is traveling in the forward direction again.

Again, the most important thing to remember about impedance is this:

> ***The most efficient transfer of energy occurs when all parts of a circuit have the same characteristic impedance.***
>
> ***Impedance mismatches are of greater concern at higher frequencies than at lower frequencies. The higher the frequency, the more important impedance matching becomes.***

The reason is that at high frequencies, components tend to be electrically shorter and therefore more electrically significant than at lower frequencies.

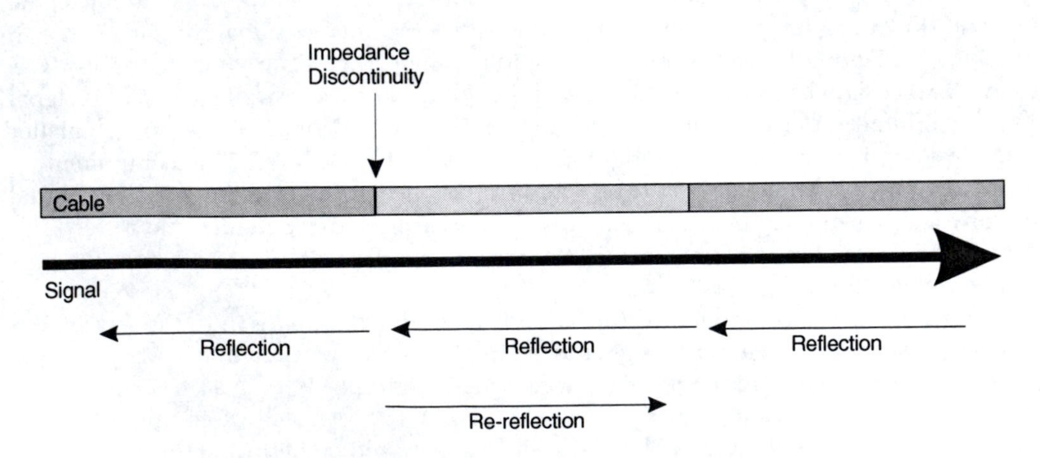

Figure 2–10. Impedance mismatches and reflections

The task of cable design is to devise cables that transmit signals in as noise-free a manner as practical. Once signal transmission quality is achieved, other issues such as cost and ease of use can be addressed.

Attenuation

Attenuation is the loss of signal power along the length of the cable. A significant feature of attenuation in copper cable is that it increases with frequency. Attenuation per unit length is greater at 10 MHz than at 1 MHz. Figure 2–11 shows typical attenuation for coaxial and twisted-pair cables.

Attenuation must not become severe enough that the receiver can no longer distinguish the signal from the noise. Attenuation significantly limits practical cable lengths, especially at high frequencies where attenuation is more severe.

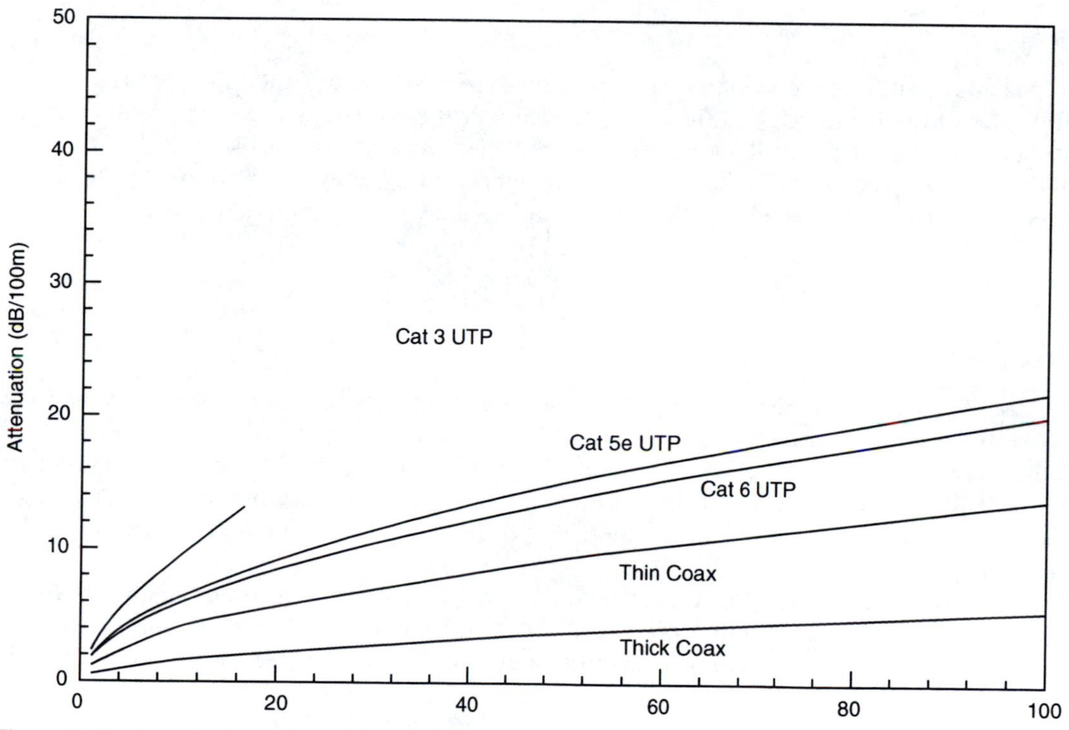

Figure 2–11. Attenuation of coaxial and UTP cables

Return Loss

Impedance mismatches cause energy to reflect back toward the source. Return loss is a measure of how much energy is reflected. It measures the reflected energy in relationship to the incident or original energy. Like NEXT loss, return loss expresses how much lower the reflected energy is than the original signal. Return loss is somewhat similar to NEXT. With NEXT, you launch a signal and measure the energy on another pair. With return loss, you launch a signal and measure the energy on the same pair.

Propagation Delay

Propagation delay is the time it takes a signal to travel the length of the cable. To calculate signal propagation delay times in a network requires that you know the delays of all components in the system, including hubs, patch cables, and so forth. A network switch, for example, represents a propagation delay. Depending on the switch, its design, and the amount of processing of the signal that must occur, this delay can vary from small to relatively large. Most vendors can supply propagation delay values for their equipment if you need to calculate network delays.

Skew

Skew is the difference in time it takes for signals launched at the same time to arrive at the other end of the cable. Slight differences in each pair means that not all signals will arrive at the other end simultaneously. Skew becomes important in applications like Gigabit Ethernet, which uses four pairs to transmit the signal. The signals, even if they do not arrive exactly simultaneously, must arrive nearly at the same time so that the receiver can keep everything synchronized.

Coaxial Cable

Coaxial cable offers the best high-frequency performance of copper cables. Figure 2–12 shows the basic structure of a coaxial cable: center conductor, dielectric, shield, and jacket. These components are coaxial: they share the same center axis. As was shown earlier in Figure 2–5, the characteristic impedance of a coaxial cable is determined by three things: diameter of the center conductor, diameter of the shield, and the dielectric constant of the material separating them.

A coaxial cable is a closed transmission line. The shield serves to confine the energy within the cable. While we often think of a wire as carrying electrons through the conductor, it is equally important to think about the electromagnetic energy propagating down the wire. This

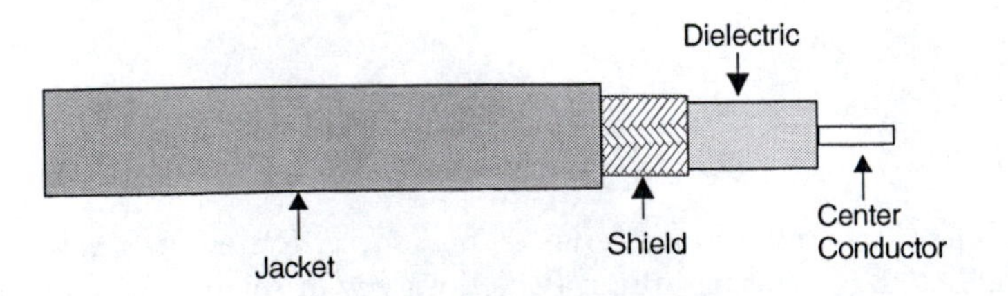

Figure 2–12. Basic construction of a coaxial cable

electromagnetic energy is the magnetic and electric fields generated by the movement of the electrons. These fields travel on the outside of the center conductor, in the dielectric, and on the inside of the outer conductor. This system is called "closed" because the shield serves to keep the energy trapped within the cable.

In a coaxial cable, crosstalk is of very little concern. The shield works well at keeping the energy inside from radiating out or coupling to adjacent cables. Conversely, it protects the signal by keeping outside noise from getting in. Still, the problem remains of noise and signal distortion that results from impedance mismatches. This reflected energy, to a great degree, is also trapped within the cable.

But the shield isn't perfect. Some shields are woven from very thin wires. Such braided shields provide coverage on the order of 85% to 95%. There are still gaps from energy to enter or exit. Other shields are foils, similar to aluminum foils. While these tend to provide 100% coverage, they are thin enough for some of the energy to penetrate. In practice, a combination of foil and braid shields are used to achieve better shielding effectiveness.

Two variations of coaxial cable are twinaxial cable and triaxial cable, shown in Figure 2–13. Twinax has two center conductors running parallel within the dielectric. IBM midrange computers use a 93-ohm twinax cable to connect to terminals. Twinax is somewhat similar to the twisted-pair cable we will discuss shortly in that it uses differential transmission.

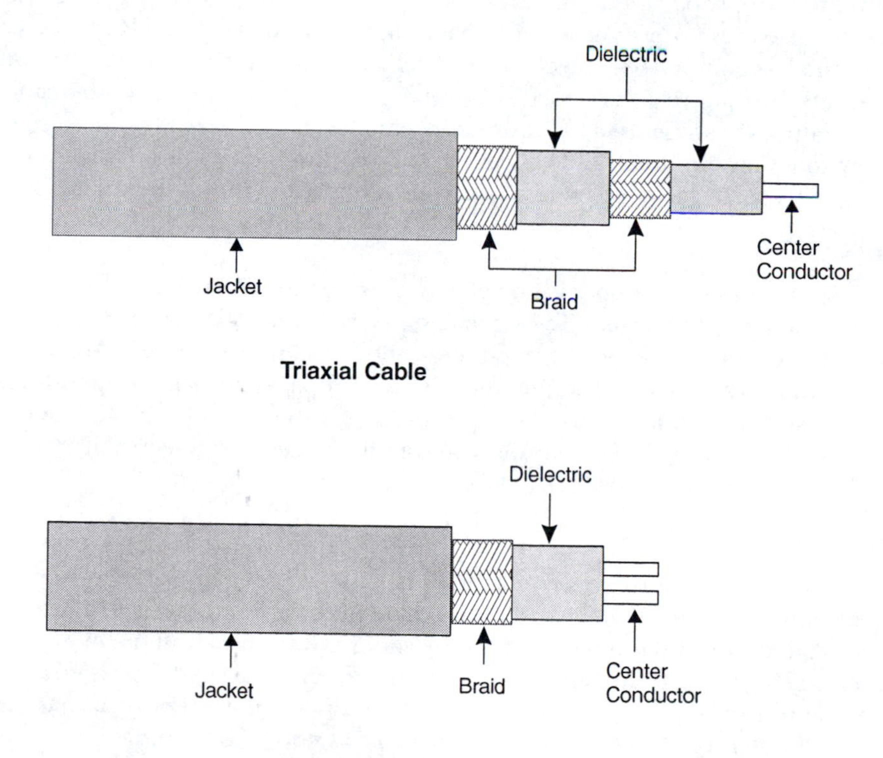

Figure 2–13. Twinax and triax cable

Coaxial Cable Characteristics

Characteristic Impedance

Most coaxial cables used in LAN applications have a characteristic impedance of 50 ohms. CATV and other video systems typically use 75-ohm cable.

Velocity of Propagation

The speed at which a signal propagates through the cable is determined by the properties of the dielectric material. While electromagnetic energy travels at the speed of light, what we commonly call the speed of light—300,000 kilometers per second or 186,000 miles per second—is the velocity of light in free space. Electromagnetic energy travels slower in other mediums.

A cable's velocity of propagation is usually expressed as a percentage of the energy free-space speed: 70% or .7*c*—equivalent to 210,000 kilometers per second or 130,200 miles per second. In terms more practical to premises wiring, this speed translates to 210 meters or 688 feet in 1 microsecond.

Velocity of propagation is handy in several ways. Networks have certain rules about how long a signal can travel along the network. For a given distance, velocity of propagation can determine how long it takes the signal to travel the path. Some network designers will calculate transit times for signals.

Velocity of propagation can also allow you to measure the length of a cable from one end if the other end is unterminated. A signal is injected into the cable. It travels to the other end, where the large impedance mismatch will reflect it back to the source. The round-trip time is measured. If you know the velocity of propagation, you can calculate the cable's length. The same technique can be used to locate cable faults such as breaks: you measure the time it takes energy to reach the break and reflect back. From the velocity of propagation, you can determine the distance (or, even better, have your test instrument do the calculations for you).

Propagation Delay

Closely related to propagation velocity, propagation delay is the time it takes a signal to travel the length of the cable. To calculate signal propagation delay times in a network requires that you know the delays of all components in the system, including hubs, patch cables, and so forth. A network hub, for example, represents a propagation delay. Depending on the hub, its design, and the amount of processing of the signal that must occur, this delay can vary from small to relatively large. Most vendors can supply propagation delay values for their equipment if you need to calculate network delays.

Termination

A coaxial cable must be terminated in its characteristic impedance. Some networks use a bus structure in which neither end of the cable necessarily connects to a device. A special connector called a terminator is placed on the cable end to absorb all the energy reaching it. This prevents reflections. If the cable is not terminated, the cable end represents an infinite impedance that will reflect nearly all the energy. If the shield and center conductor are shorted, the signal will also reverse and travel down the cable as noise. So termination is essential to proper network operation. Always be sure that the terminator has the correct resistance value. Terminating a 50-ohm cable with a 75-ohm terminator still causes reflections.

Coaxial Cables for Networks

In 1973, the Xerox Corporation began developing a network they named Ethernet. The first prototype was running successfully by 1976 at the Xerox Palo Alto Research Center. It used a large, sturdy, almost unwieldy coaxial cable with a diameter of either 0.405″ for plenum applications or 0.375″ for nonplenum applications. The cable uses a four-layer shield: braid, foil, braid, foil.

The cable has a 50-ohm (±2 ohms) characteristic impedance and uses N-type connectors. Its large diameter makes the cable stiff and, therefore, hard to work with.

By 1988, a new version of Ethernet had been developed that used a thinner, more flexible, and less expensive coaxial cable. This 50-ohm cable, RG-58A/U, uses a stranded center conductor and a single-layer braided or foil shield. This version of Ethernet became popularly known as thinnet or cheapernet. The earlier version, in turn, was called thicknet. Thinnet cables use BNC connectors.

Broadband cable for carrying video signals uses 75-ohm cables and the same F-connectors that are used in the CATV industry. These cables find little use today in networks except in specialized applications. Digital Equipment Corporation used 75-ohm cable in its DECnet system to carry video signals. Residential cabling uses 75-ohm cable.

Still, as Ethernet grew in popularity, network designers searched for additional ways to reduce the cost of cabling. At the same time, IBM introduced its network system called Token Ring. Token Ring networks originally used shielded twisted-pair cable. Soon designers also developed unshielded twisted-pair cable—like standard telephone cable but better. Today, twisted-pair cable is the favored copper cabling system for premises.

Coaxial cable is also used in residential structured cabling systems to carry video signals such as cable or satellite TV. The recommended cable is Series 6 cable, a quad shielded cable with a 1.5-GHz or 2.2-GHz bandwidth.

One note about coaxial cable naming. The original method was to use military terminology with the RG nomenclature: thus, RG-59 cable. More recently, the cables have used the term *series* rather than *RG*. You may see Series 6 cable sometimes called RG-6 cable or RG-59 cable called Series 59 cable.

Figure 2–14 shows thicknet, thinnet, and Series 6 cables.

Twisted-Pair Cables

Twisted-pair cables are so named because each pair of wires is twisted around one another as shown in Figure 2–15. Twisted-pair cables help reduce crosstalk and noise susceptibility in two ways.

First, the twists reduce inductive coupling between pairs. Inductive coupling is caused by the expanding and contracting magnetic fields caused by a signal through the wire. The twists create coupling among the two wires in the pair such that opposing electromagnetic fields are canceled. A greater twist ratio—the number of twists per unit length—brings greater coupling and less chance of crosstalk to nearby pairs.

The optimum twist rate involves a tradeoff between crosstalk and other cable characteristics. More twists lower crosstalk, but increase attenuation, propagation delay, and costs. High-quality Category 5e cables have about one to three twists per inch. For best results, the twist length should vary significantly from pair to pair.

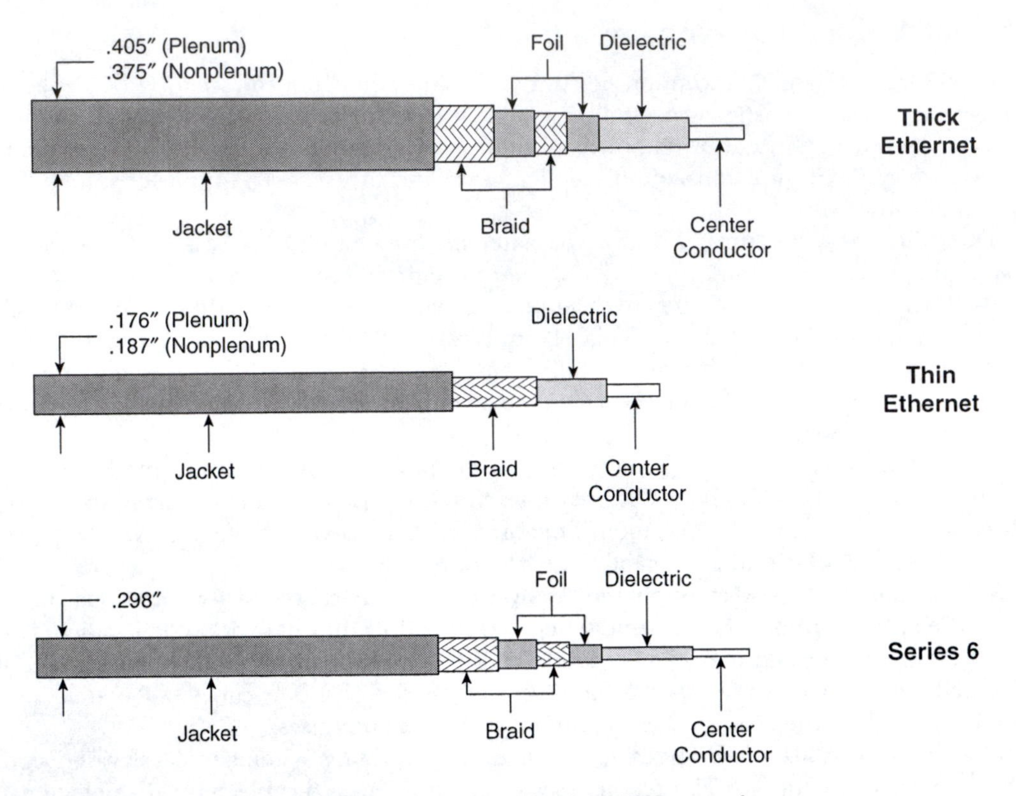

Figure 2–14. Coaxial cable used in Ethernet and residential CATV networks

Figure 2–15. Twisted-pair cable

Second, networks using twisted-pair wiring use balanced transmission. The balanced circuit uses two wires to carry the signal. The conductors and all circuits connected to them have the same impedance with respect to ground and other conductors. Ground serves as a reference potential, rather than as a signal return.

Each conductor in a balanced line carries a signal (and noise) that is of equal potential with the other conductor, but of opposite polarity. If one conductor carries a 2.5-V signal, the other conductor carries a –2.5-V signal. The receiver detects only the voltage *difference* between the two conductors—in this case, 5 V. If a +1-V noise is introduced onto each conductor, the resulting signal is 3.5 V and –1.5 V. The receiver still sees a 5-V difference. A perfectly balanced system is difficult to achieve, so some difference in noise levels may appear between the two conductors. Still, the effect is to effectively cancel the noise.

Categories of Twisted-Pair Cable

Twisted-pair cables are as old as the telephone industry. Alexander Graham Bell, the inventor of the telephone, patented twisted-pair cable in 1881 (Figure 2–16). The patent makes clear that Bell understood quite well the advantages of twisting pairs together in order to achieve better transmission. It's common to say that networks run over telephone cable. This is wrong. Not all telephone cable is twisted pairs, especially in homes. Quad wiring in houses has four parallel wires, not two twisted pairs. Likewise, silver satin cable, which is flat with a silver jacket, is unsuited for high-speed data communications. Silver satin cable is often used to connect computer modems to wall jacks: it's fine for the slow speeds of modems. Even a fast 56-kbps modem requires a cable with a bandwidth of only 3 kHz.

Twisted-pair cables designed for low-speed analog telephony are unsuitable for high-speed network applications. Telephone cables were designed to carry voices with frequencies of up to only 4 kHz. Because of this low frequency, the twists in the wire can be long and "sloppy." By sloppy, we mean the precision and uniformity of the twists are not of great importance. What is important is that the wires *are* twisted. A telephone twisted-pair cable has about one twist per foot.

By putting more twists per unit length and making them more uniform, cable makers developed cables capable of supporting data rates well above the 4 kHz of telephone cables. Soon, different manufacturers were offering a wide range of better cables under terms like *data grade, high frequency, extended frequency,* and so forth. There was some confusion since these

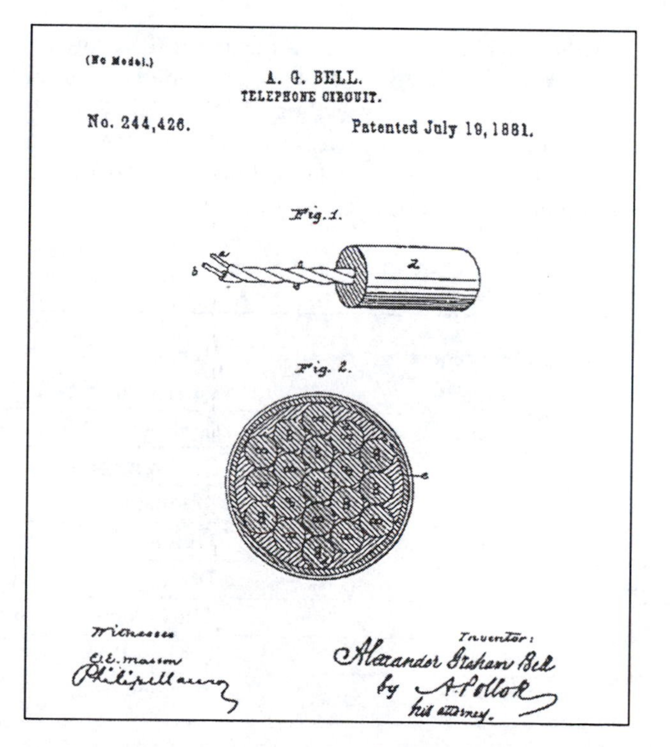

Figure 2–16. Bell's patent for twisted-pair cable

improved cables were offered in various grades, with slightly different specifications from different manufacturers, and each with different terms.

A major distributor, in an attempt to bring order to its offerings, offered twisted-pair cable in four levels of performance. Later a fifth level was added. These cables were termed Level 1, Level 2, Level 3, Level 4, and Level 5. Particularly, Levels 3, 4, and 5 had a rigorous set of specifications governing their performance. Standard-setting bodies, such as UL, IEEE, and TIA/EIA, began building specifications around these levels.

Just as the idea of levels became popular, the distributor decided to trademark the term *Level* in reference to twisted-pair cable. In other words, they wanted to own the term *Level.* The industry quickly adopted the term *Category* to replace *Level.* So today we have Categories 1 through 6 cable. (The trademark was rejected by the U.S. Trademark and Patent Office, incidentally. So the whole problem was much ado about nothing.)

The basic idea of categories of cables is to provide different grades to meet different data rates. But to meet these data rates, other performance characteristics like NEXT must also be specified. The categories of cable are shown in the Figure 2–17.

Categories 1 and 2 are only suited for analog telephony. Neither Category 1 nor 2 cables are recognized by either TIA-568 or ISO/IEC-11801.

Category 3 cable was the only type of UTP specified in the first edition of TIA-568. Today it is only recommended for telephony applications and is not often used.

Category 4 UTP was aimed specifically at 16 Mb/s Token Ring applications. It was never widely used. It has been removed from the cabling standards and is no longer being manufactured.

Category 5 UTP was the workhorse cable for a few years. Its properties were specified to 100 Mb/s to support Fast Ethernet. Cat 5e UTP was developed in the late 1990s to support Gigabit Ethernet networks. Like Cat 5, it is specified to 100 MHz. However, Cat 5e also specifies power-sum crosstalk and delay skew parameters which are needed to support networks that transmit on multiple pairs (like 1000BASE-T). Cat 5 is no longer compliant with TIA-568 and has been totally supplanted in the marketplace by Cat 5e.

Category	Maximum Frequency	Remarks
1	Not rated	Not specified by TIA/EIA
2	64 kHz	Not specified by TIA/EIA
3	16 MHz	Declining in popularity
4	20 MHz	Obsolete
5	100 MHz	Obsolete
5e	100 MHz	Declining in popularity
6	250 MHz	The emerging favorite
6A	500 MHz	Highest-Performance UTP
7	600 MHz	Shielded cable

Figure 2–17. Categories of cable

Category	Max. Freq. (MHz)	Parameters	Typical Application
3	16	Attenuation, NEXT, Characteristic Impedance	Voice
5e	100	Cat 3 plus: RL, SRL, PSNEXT, PSELFEXT, Prop. Delay, Delay Skew	Fast Ethernet (100BASE-T)
6	250	Cat 5e plus: LCL	Gigabit Ethernet (1000BASE-T)
6A	500	Cat 6 plus: alien crosstalk	10-gigabit Ethernet (10GBASE-T)

Figure 2–18. Comparison of UTP categories

Category 6 UTP was developed to provide additional headroom for Gigabit Ethernet networks as well as other high-speed networks to be developed in the future. The properties of Cat 6 are specified to 250 MHz and also include, for the first time, specifications on the balance of the cabling system to reduce radiated emissions. The Cat 6 specifications are in TIA-568-B.2.1, "Transmission Performance Specifications for 4-Pair 100 Ω Category 6 Cabling." Although no additional applications have yet been developed specifically for Cat 6 UTP, it has become the most popular cable for new installations in commercial buildings.

In 2005, specifications for Augmented Category 6 UTP (which we will usually refer to as Cat 6A) are being developed by the TIA to support the 10GBASE-T standard which is concurrently being developed in IEEE 802.3. The properties of Cat 6a will be specified to 500 MHz, and will also include alien crosstalk.

In Europe, Category 7 cable was defined in ISO/IEC-11801 with an operating range up to 600 MHz. Although it uses the UTP numbering scheme, Cat 7 is a shielded system. Cat 7 cable is not included in the TIA cabling standards and has not caught on in the United States.

Figure 2–18 summarizes the 4 currently defined categories of UTP cable.

UTP Market Share

Figure 2–19 illustrates the cumulative worldwide installed base of horizontal cabling in commercial buildings. Note the rapid decline of Cat 5 cabling after Cat 5e came on the market in 1997. The installed base of Cat 5e peaked in 2004 at 51% and began to decline in 2005. After a rather slow ramp up, Cat 6 cabling has steadily gained market share and in 2005 will account for about 75% of new installations. By about 2007 (give or take a year), the installed base of Cat 6 cabling should surpass Cat 5e. At about that time, Cat 6A cable will start to establish a presence in the market.

Megahertz Versus Megabits

At this point, we should take a short detour to discuss the relationship between the frequency rating of a cable and the data rate of the digital signals carried by the cable. For

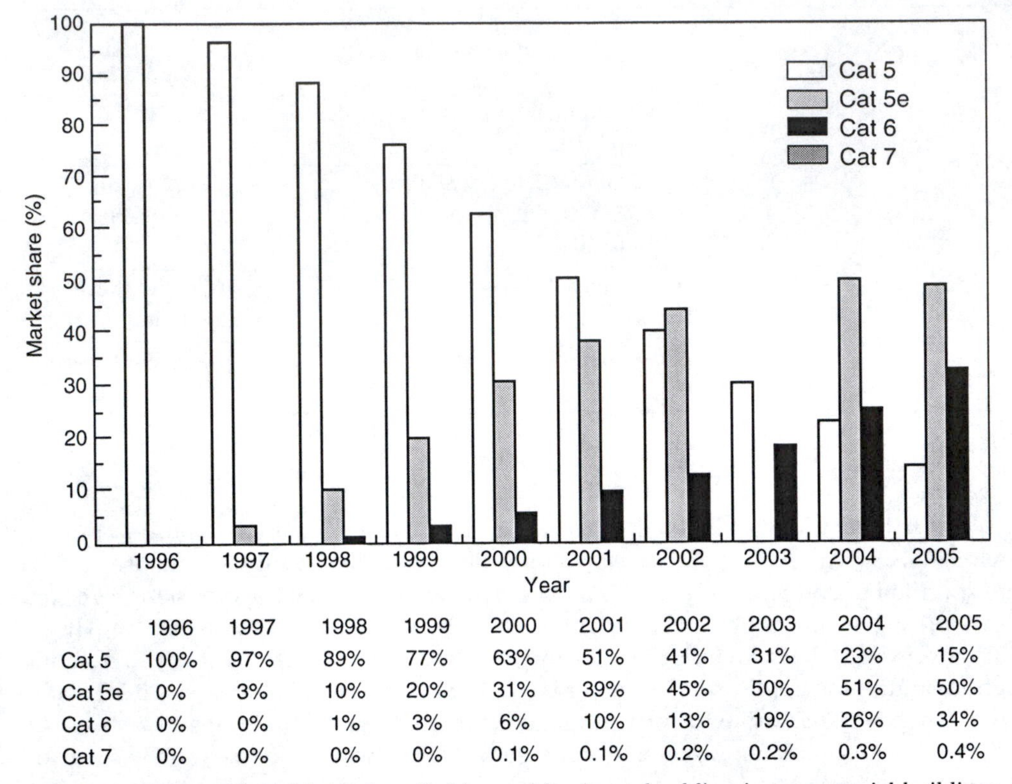

	1996	1997	1998	1999	2000	2001	2002	2003	2004	2005
Cat 5	100%	97%	89%	77%	63%	51%	41%	31%	23%	15%
Cat 5e	0%	3%	10%	20%	31%	39%	45%	50%	51%	50%
Cat 6	0%	0%	1%	3%	6%	10%	13%	19%	26%	34%
Cat 7	0%	0%	0%	0%	0.1%	0.1%	0.2%	0.2%	0.3%	0.4%

Figure 2–19. Estimated worldwide installed base of horizontal cabling in commercial buildings *(Source: BISRIA and LAN Technologies)*

example, Category 5e cable is rated at 100 MHz, while FDDI has a data rate of 100 Mbps. Are the frequency rating and data rate equivalent? Sometimes yes, sometimes no. The reason is the encoding method used.

A string of digital highs and lows is often unsuitable for transmission over any appreciable distance. Where a digital system uses simple high and low pulses to represent 1s and 0s of binary data, a more complex format is often needed to transmit digital signals over a network. A modulation code is a method for encoding digital data for transmission.

Figure 2–20 shows several popular modulation codes. Each bit of data must occur within its bit period, which is defined by the clock. The clock is a steady string of pulses that provide basic system timing. Some codes are self-clocking and others are not. A self-clocking code means that the clock or timing information is contained within the code. In a non-self-clocking system, this timing information is not present.

Clock information is important to a receiver. One purpose of a receiver is to rebuild signals to their original state. In order to do so, the receiver must know the timing information. There are three alternatives:

1. The transmitted information can also contain the clock information; in other words, the modulation code is self-clocking.

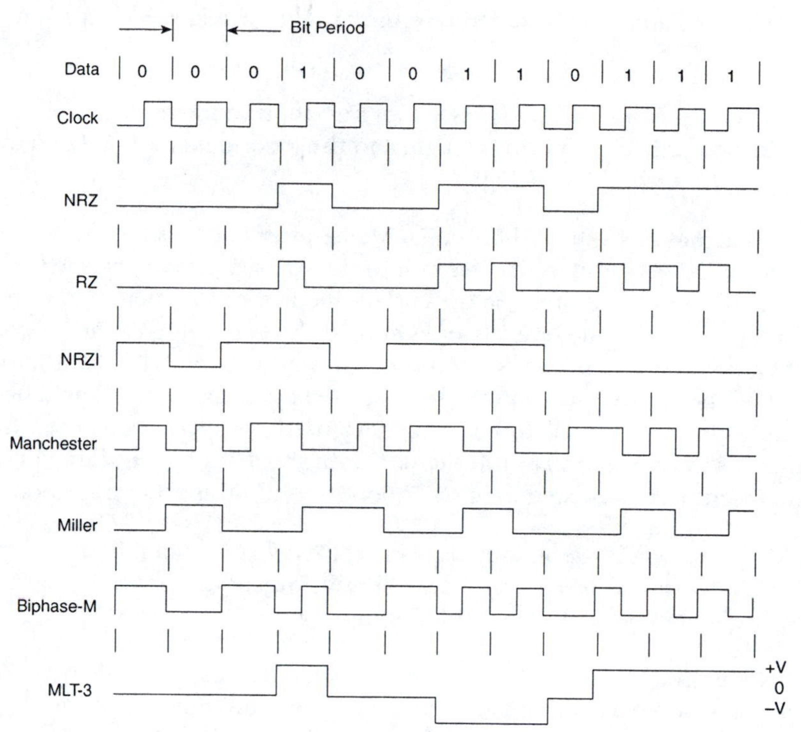

Figure 2–20. Modulation codes

2. The clock or timing information must be transmitted on another, separate line. This, of course, adds to system complexity by increasing the number of lines from transmitter to receiver.

3. The receiver provides its own timing and does not rely on clock signals from the transmitter.

Modulation Codes

NRZ Code. The *NRZ (nonreturn-to-zero) code* is similar to "normal" digital data. The signal is high for a 1 and low for a 0. For a string of 1s, the signal remains high. For a string of 0s, it remains low. The level changes only when the data level changes.

RZ Code. The *RZ (return to zero) code* remains low for 0s. For a binary 1, the level goes high for one half of a bit period and returns low for the remainder. For a string of three 1s, for example, the level goes high-low, high-low, high-low in three bit periods.

NRZI Code. In an *NRZI (nonreturn-to-zero inverted) code,* a 0 is represented by a change in level; a 1 is represented by no change in level. Thus, the level will go from high to low or from low to high for each 0. It will remain at its present level for each 1. An important thing to notice here is that there is no firm relationship between the 1s and 0s of data and the highs and lows of the code. A binary 1 can be represented by either a high or a low. So can a binary 0. NRZI is sometimes called NRZ-S (NRZ-space).

Manchester Code. A *Manchester code* uses a level transition in the middle of each bit period. For a binary 1, the first half of the period is high, and the second half is low. For a binary 0, the first half is low, and the second half is high.

Differential Manchester Code. A differential Manchester code also uses a transition in the middle of the bit period. The difference lies in how the transition is interpreted. To represent a 1, the level remains the same as it was at the end of the previous bit period and then changes at midpoint. To represent a 0, the level changes both at the beginning of the bit period and at the end. In other words, there is always a transition in midperiod. A 1 has a transition at the beginning of the bit period; a 0 has no transition at the beginning. The Manchester code also defines two code violations called J and K, which are used for special purposes. A J violation occurs when there is no transition at both the beginning and midpoint of the bit period. A K violation occurs when there is a beginning transition, but no midperiod transition.

Miller Code. In the *Miller code* or *delay modulation,* each 1 is encoded by a level transition in the middle of the bit period. A 0 is represented either by no change in level following a 1 or by a change at the beginning of the bit period following a 0.

Biphase-M Code. In the *biphase-M code,* each bit period begins with a change of level. For a 1, an additional transition occurs in midperiod. For a 0, no additional change occurs. Thus, a 1 is both high and low during a bit period. A 0 is either high or low (but not both) during the entire bit period.

MLT-3 Code. Unlike other codes, MLT-3 code uses three levels of transmission, rather than two levels. The three levels are –1 V, 0 V, and +1 V. Any change in a bit means a change in level. The level stays the same for consecutive 1s and 0s. The movement is upward, then downward as bit changes occur: 0, +1, 0, –1, 0, +1, 0, –1. MLT-3 coding is widely used in high-speed networks using UTP as the cabling medium because the code keeps the frequency content of the signal low to minimize crosstalk and attenuation.

PAM-5 Code. PAM-5 is the code used with Gigabit Ethernet over UTP. It uses five levels. Four of the levels are used for the signal; the fifth level is used for error correction. Unlike most applications, Gigabit Ethernet uses all four pairs. The basic data rate is 1000 Gbps. But because the signal is divided into four pairs, each pair only carries one quarter of the rate or 250 Mbps. The four-level coding of PAM-5 reduces the frequency to 125 MHz, although most of the energy is under 100 MHz, allowing Gigabit Ethernet to run on Category 5 cable.

nB/nB Encoding

Most low-speed networks use Manchester or differential Manchester encoding. At higher speeds, Manchester encoding falls from favor because it requires a clock rate twice that of the data rate. A 100-Mbps network requires a 200-MHz clock. NRZI data transmissions have no transitions when all zeros are present, which eliminates self-clocking.

Application	Coding Scheme	Data Rate (Mbps)	Transmission Rate (Mbps)	Bandwidth Efficiency (Bits/Hz)
Ethernet	Manchester	10	10	1
Token Ring	Differential Manchester	16	16	1
TP-PMD (Fast Ethernet)	MLT-3	100	125	4
Gigabit Ethernet*	PAM-5	1000 (250/pair)	1250 (265.5/pair)	2
ATM	16-CAP	51.84	51.84	2
ATM	NRZ	155.52	155.52	2

**Gigabit Ethernet has a data rate of 1000 Mbps, but breaks the signal into four separate 250-Mbps channels for transmission.*

Figure 2–21. Relationship between data rate and frequency for different modulation codes and network types

High-speed networks use a group encoding scheme in which data bits are encoded into a data word of longer bit length. For example, the 4B/5B method encodes four data bits into a 5-bit code word. The receiver decodes the 5-bit word into 4 bits. This scheme guarantees that the data never have more than three consecutive 0s. Such encoding requires much less bandwidth than Manchester encoding. The overhead is only 20%. Thus high-speed networks using 4B/5B, such as FDDI or Fast Ethernet, transmit at a 125-Mbps rate, but the data rate is 100 Mbps. The 25-Mbps difference is due to the fifth bit in the 4B/5B encoding. Even with 4B/5B, the data is still transmitted with one of the encoding schemes in Figure 2–20. Other encoding schemes include the 8B/10B method used by IBM in its fiber-optic interconnection system for large computers.

Different encoding schemes are more efficient in their use of bandwidth than others. By bandwidth, we mean the comparable *frequency* of the signal. There is not necessarily a one-to-one relationship between data rate (in Mbps) and frequency (in MHz). Figure 2–21 summarizes the relationship between common network encoding schemes, their data rates, and their frequency. Notice that TP-PMD uses MLT-3 encoding with a 125-Mbps transmission rate and a 100-Mbps data rate, but has a frequency of only 31.25 MHz. In other words, TP-PMD is quite efficient, requiring only one third of the bandwidth available on Category 5 cable. *Bandwidth efficiency* is found by dividing the transmission rate by the frequency requirement (bandwidth used).

Even though Fast Ethernet has a transmission rate of 125 Mbps, it requires a cable capable of carrying only 31.25 MHz. But you don't know the relationship between transmission rate and frequency unless you know the encoding scheme used.

The important thing to remember is that the transmission rate and frequency bandwidth of the cable are not always the same.

Carrierless Amplitude and Phase Modulation

The encoding techniques above are digital; that is, they directly use pulses. CAP modulation is related to the techniques used in modems. A frequency is modulated by changing its amplitude

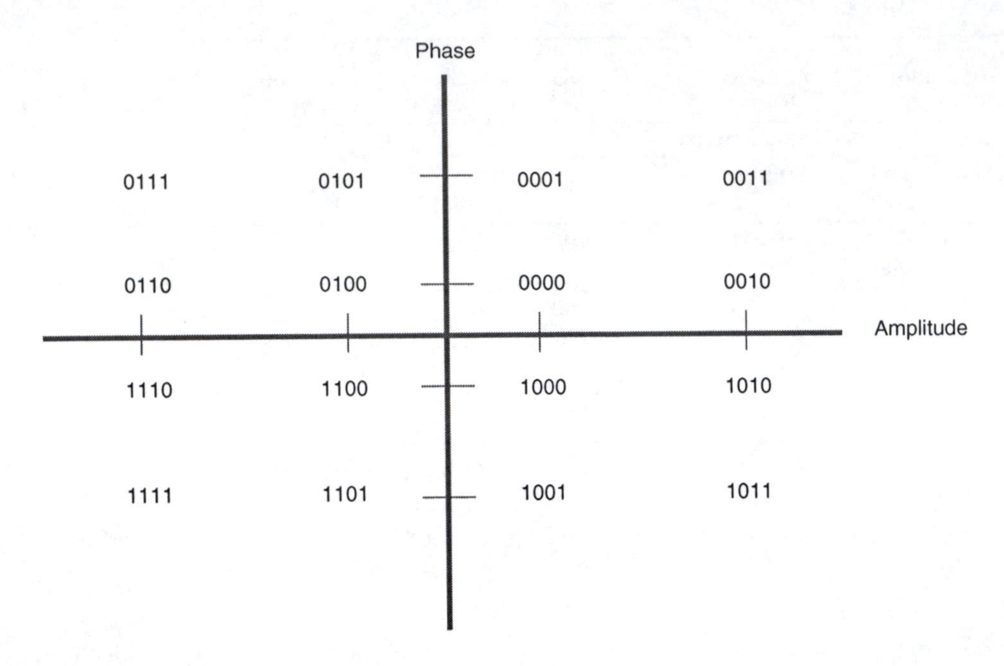

Figure 2–22. CAP-16 encoding

and phase. For example, CAP-16 uses 16 different combinations of phase and amplitude to represent bits. Each specific combination represents four bits, as shown in Figure 2–22. This allows high bit rates at low frequencies. For example, 51.8 Mbps ATM, which is used to carry ATM to the desktop, uses CAP-16 encoding. While the CAP-16 systems theoretically require only a 12.96 MHz frequency, the ATM system uses double that frequency or 25.92. This keeps the frequency spectrum under 30 MHz. 30 MHz is a "magic" number because FCC regulations regarding EMI emissions from equipment begin at 30 MHz. Thus keeping the frequency under 30 MHz not only eliminates considerations involved in complying with government regulations, but it also minimizes EMI problems in the first place.

Another CAP scheme is CAP-64, which can represent 6 bits for every discrete combination of phase and amplitude. The number of bits that can be represented is equal to 2 raised to the power of the number of bits. For CAP-16, $2^4 = 16$. In other words, if you have 16 amplitude/phase combinations, you can represent 4 bits per combination. Similarly CAP-64 is $2^6 = 64$, so you can have 6 bits per combination if you have 64 possibilities.

Unshielded Twisted Pairs

The properties of UTP cables have been standardized in a number of documents in the TIA-568 series, including:

- TIA-568-B.2 gives component specifications for Category 3 and 5e cables and connecting hardware.
- TIA-568-B.1 gives end-to-end specifications for Category 3 and 5e channels and permanent links.

- TIA-568-B.2-1 gives both component and end-to-end specifications for Category 6 cabling.
- TIA-568-B.2-10 (which is in draft form as this edition is being prepared) gives both component and end-to-end specifications for Augmented Category 6 cabling.

In the future, all of the permanent link and channel specifications will probably be pulled into a single 568-C.1 document and all of the component specifications will be consolidated into a 568-C.2 document.

UTP is the favored copper cable for premises wiring for several reasons. It is simple, flexible, and easy to work with; connectors and related hardware are the familiar modular plugs and jacks long used in the telephone industry. UTP is considered to be inexpensive, well known, and widely available.

The Special Case of Categories 5 and 6

Even with these advantages, Category 5e and higher cables require special attention. They can be quite sensitive to incorrect installation. For example, their performance rests on maintaining twisting of pairs as close to the connector as possible—½ inch or less. Likewise, clinching a cable tie too tightly when securing cables can also degrade performance. To the installer used to working with Category 3 cable, which is insensitive to generous untwisting and rugged to abuse, Categories 5e and 6 cables can be a new experience.

With a premises cabling plant meant to operate at 100 MHz (or 250 MHz with Category 6), other interconnection hardware—connectors, patch panels, wall outlets—must also be rated for the same performance. You can't simply go down to the local Builder's Square or Radio Shack and load up on the same components you'd use to wire telephones.

Better grades of cable, of course, mean higher costs. Category 6 components cost more than Category 5e components, which in turn are more expensive than Category 3 parts. However, Cat 5e cabling has dropped in price to the point that it is usually considered the standard option for price-sensitive installations. Cat 5e cables and connecting hardware are routinely available at electrical distributors, home centers, and electronics stores.

Twisted-Pair Construction

The basic configuration of UTP cable is a four-pair cable: four individually jacketed twisted-pair cables are held within a single outer jacket as shown in Figure 2–23. The wire size is 24-gauge stranded or solid conductors (although larger 22-gauge wire is sometimes used). Solid conductors are recommended because they offer lower attenuation. Each conductor's insulation is color coded for easy identification (see Appendix D for color coding information). Four-pair cables are used mainly in the horizontal wiring.

For vertical backbone wiring, a 25-pair cable is often used. A two-pair cable is sometimes used for telephone cables. However, it is recommended that four-pair cable be used in the horizontal subsystem of new buildings to ensure maximum flexibility. For example, a four-pair Category 5 cable is equally suited to analog telephone, digital telephone, and network applications. If only a two-pair wire were installed, the cable would not be capable of being used in some applications. Figure 2–24 shows a cable having several 25-pair subcables. Each 25-pair unit is grouped together in a binder group.

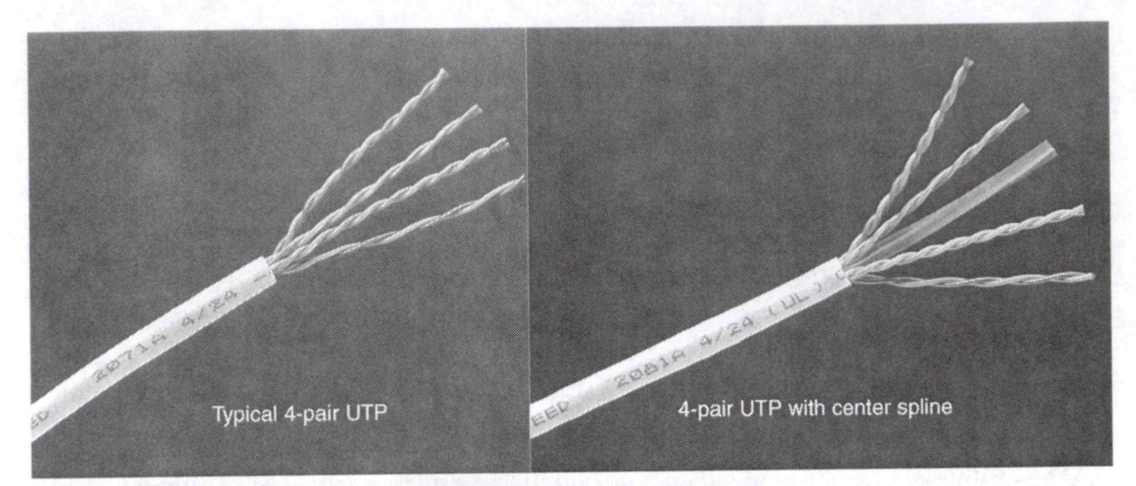

Figure 2–23. Four-pair Category 6 UTP cable *(SYSTIMAX Solutions graphics are courtesy of SYSTIMAX Solutions™, a CommScope company)*

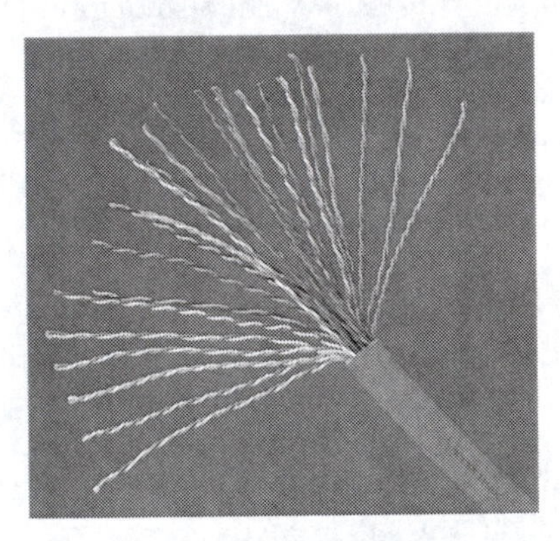

Figure 2–24. 25-pair UTP cable *(SYSTIMAX Solutions graphics are courtesy of SYSTIMAX Solutions™, a CommScope company)*

For high-performance cables (especially Categories 5e and 6), the number of twists per inch, the direction of the twist, and the relationship between the twists of different pairs are all important to achieving and maintaining performance. For example, Category 6 cable often has three or four twists per inch. Some pairs are twisted in one direction, while other pairs are twisted in the other. Some manufacturers bond the cables to retain the twist and the lay of one twisted pair in relation to another. The twisting of pairs is also staggered so that the twisting is not identical along the length of the cable. Some Category 6 cables have a plastic spline running down the cable. The spline serves to keep the various pairs physically separated to reduce crosstalk.

With Augmented Cat 6 cable, meeting the alien crosstalk specs is the most challenging task. Cat 6A cables often use constructions different from those employed in lower-category cables, such as an elliptical-shaped jacket with the cables offset from the center. To accommodate these

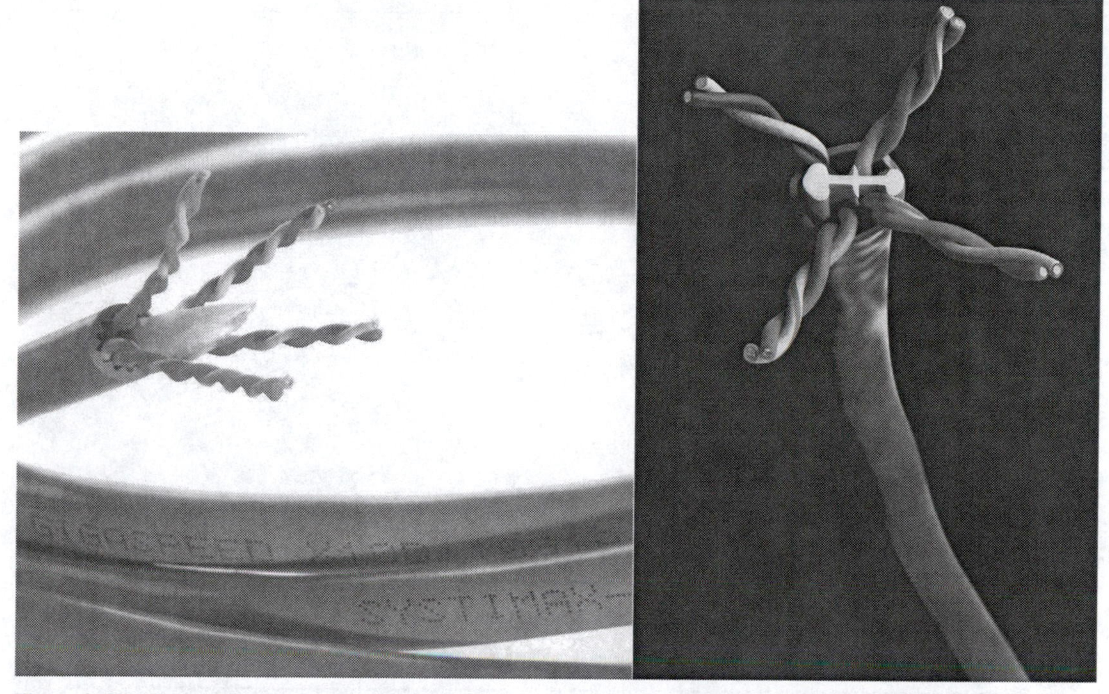

Figure 2–25. Examples of an Augmented Category 6 UTP cable. On the left, note the flat center member and the ribbed inner wall of the jacket. *(SYSTIMAX Solutions graphics are courtesy of SYSTIMAX Solutions™, a CommScope company)* On the right, note the asymmetric center member. *(Courtesy of ADC)*

new designs, the maximum diameter of UTP cable has been increased to 9mm. Figure 2–25 shows two examples of Category 6A cable construction.

UTP Characteristics

A listing of pertinent selected UTP characteristics for Categories 3, 5e, 6, and 6A cables, as defined in TIA/EIA-standards, is given in Appendix A. Those listed are the properties most often tested during installation and certification of premises cabling.

Four items are of interest here: NEXT and ELFEXT, structural return loss, attenuation, and attenuation-to-crosstalk ratio. These deserve some discussion.

NEXT and ELFEXT Losses

Figure 2–26 shows graphically the NEXT and ELFEXT loss requirements for horizontal cable. Appendix A supplies further information on how NEXT and ELFEXT are calculated.

The NEXT loss requirements for Category 5e cable are 3 dB tighter than for Category 5 at 100 MHz. NEXT levels must be 3 dB lower. For Category 6 cable, NEXT is 9.3 dB lower than Category 5e cable and 12.3 dB lower than Category 3 cable. Notice that we use the term *lower* even though the dB value is a larger number. Remember that a high NEXT loss value means that the crosstalk is lower compared to the signal. A higher NEXT loss value means lower crosstalk.

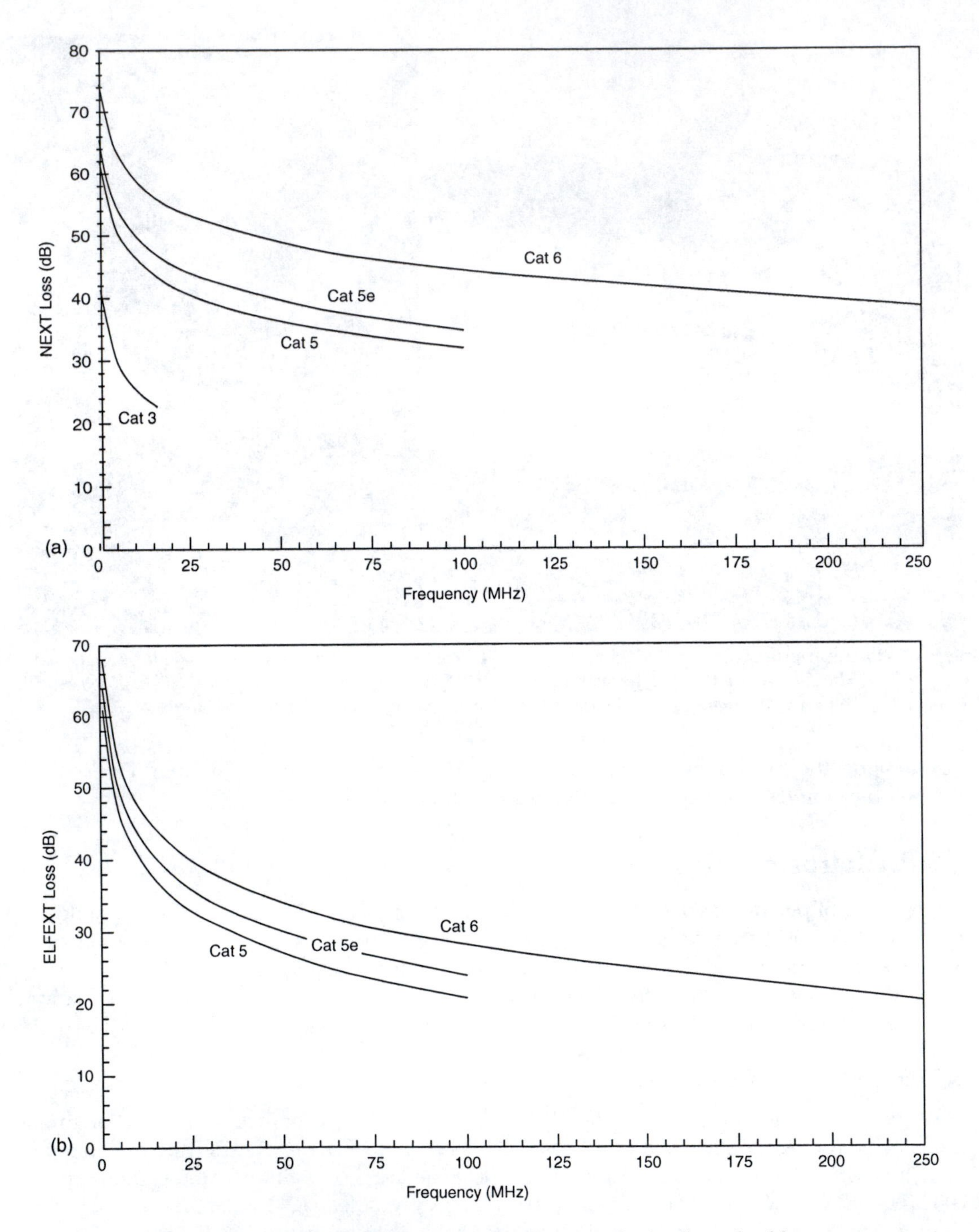

Figure 2–26. UTP NEXT (a) and ELFEXT (b) loss requirements for horizontal cable

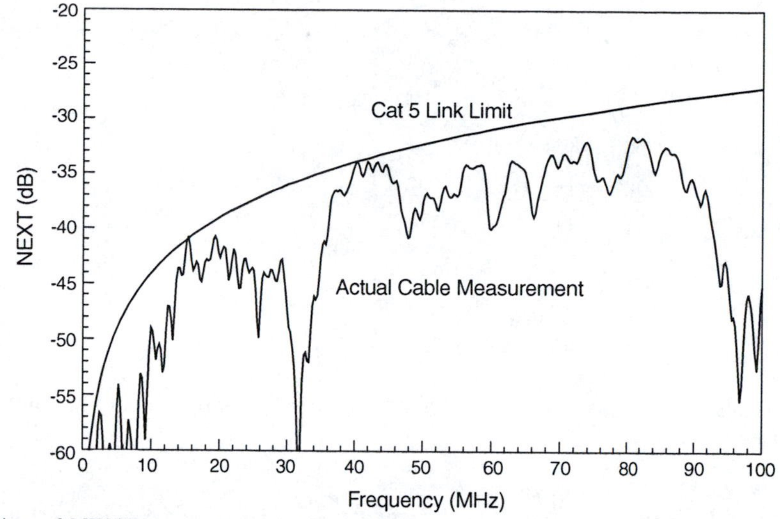

Figure 2–27. Actual NEXT measurement *(Courtesy of Fluke Networks)*

If you actually measure NEXT across the entire frequency, you'll see that crosstalk jumps around with changes in frequency. Figure 2–27 shows measured NEXT loss for a Category 5 cable. Notice that it does not increase or decrease linearly with frequency. It shows sharp peaks and valleys across the spectrum. The figure also shows the smooth curve that relates to 568B requirements. Similar variations in loss across the frequency range will be seen for ELFEXT and power-sum measurements.

The reason for the peaks and valleys is that the crosstalk is occurring at different points along the path. You can think of the cable as being divided into numerous small sections, with each section contributing crosstalk. Depending on the frequency and the length along the cable, the crosstalk can either add to or subtract from other crosstalk. When the crosstalk adds together, you get the peaks of high crosstalk. When the crosstalk subtracts, you get the valleys of low crosstalk.

If you were to perform a measurement and get a smooth curve rather than peaks and valleys, you would know that nearly all the crosstalk is coming from a single point, such as a bad connector or serious fault in the cable. This single point of crosstalk swamps the other sources so that there is no significant adding or subtracting along the path, just one large point of crosstalk flowing back to the measurement instrument.

Similar differences in loss values between types of cable are also found in ELFEXT, power-sum NEXT, and power-sum ELFEXT. Field testers can typically measure these values for four-pair cable. With 25-pair cable, as shown in Figure 2–28, measurements become complex since the crosstalk from all pairs onto each single pair must be calculated.

In making measurements for power-sum crosstalk, you don't have to energize all 24 pairs at the same time. Rather, each pair is energized individually and the crosstalk on the single quiet line is measured. For example, to find the power-sum crosstalk on pair 1, you would drive pair 2 with a signal and measure the NEXT on pair 1. Then you would drive pair 3 and measure

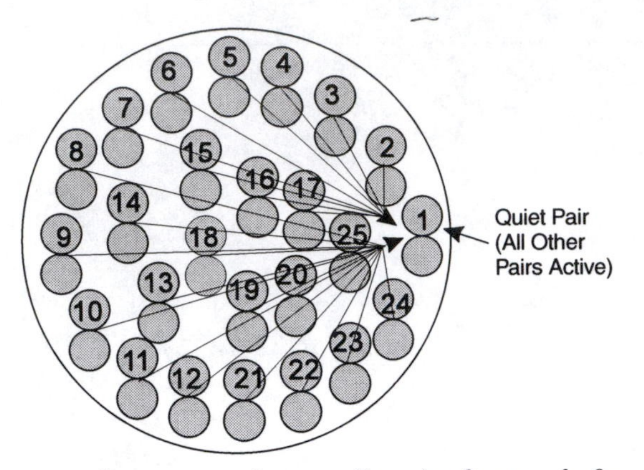

Figure 2–28. Power-sum crosstalk accounts for coupling simultaneously from all other pairs onto a single pair

the NEXT on pair 1. Then drive pair 4 and measure NEXT on pair 1. This would continue to pair 25. The power-sum crosstalk is then calculated:

$$NEXT_{PS} = (NEXT_{PR2-1}{}^2 + NEXT_{PR3-1}{}^2 \ldots + NEXT_{PR25-1}{}^2)^{1/2}$$

Next, perform the same procedure on the other 24 pairs. Measure NEXT coupled onto pair 2, then pair 3, and so forth. The power-sum calculations are typically 3 to 6 dB worse than a straight pair-to-pair calculation. The 6-dB difference is important. A cable that barely meets Category 5 pair-to-pair NEXT will fail the power-sum Category 5 test.

Unfortunately, there exists no practical method of field testing 25-pair UTP in the field. The cable manufacturer tests the cable in the laboratory using a network analyzer. This equipment is expensive, requires careful setup and calibration, and is unsuited to general field use.

Structural Return Loss

Structural return loss (SRL) is a measure of the uniformity of a cable's characteristic impedance. SRL is affected by the design and manufacturing of the cable, because these result in variations in the structure of the cable: variations in twisting, differences in the dielectric, and variations in separation distance between conductors that make the cable less than a perfect cable. SRL is measured in decibels, with a higher number meaning a better cable. Again, the important SRL is the worst one in the cable. Notice that the higher the cable category, the better the SRL is (i.e., the more uniform the impedance). At the same time, for each type of cable, SRL decreases at higher frequencies.

SRL can contribute noise, just as crosstalk does. NEXT is usually the main source of noise. What the SRL specification does is ensure that the SRL-induced noise is never more than 20% of the *total* noise at the receiver.

Attenuation-to-Crosstalk Ratio

Think for a moment about signals and crosstalk. If a given level of crosstalk is coupled onto the cable, its relationship to the signal depends on the signal's power. If the signal is strong, then the crosstalk is only a small percentage of the signal. If the signal is weak, or highly attenuated after traveling down the cable, then the crosstalk is a greater percentage of the signal. If the coupled crosstalk is .5 V and the signal is 5 V, the noise is 10% of the signal. If the signal is attenuated to 2.5 V, the noise is now 20% of the signal.

The attenuation-to-crosstalk ratio is a figure of merit that accounts for the relative strength of the signal. Figure 2–29 shows the relationship between NEXT and attenuation for Categories 5, 5e, and 6 cables. The distance that separates NEXT from attenuation defines the ACR. ACR generally decreases at higher frequencies. In fact, if you extend the lines for attenuation and NEXT, you will see they will eventually cross. The cable then has a negative ACR. At 10 MHz for Category 5 cable, NEXT loss is 47 dB and attenuation is 6.5 dB. The ACR is 31.5 dB. At 100 MHz, where NEXT is lowered to 32 dB and attenuation increases to 22 dB, the ACR is reduced to 11 dB.

You can also see that the better cables offer a better ACR. At 100 MHz, the ACRs are thus:

Category 5:	11 dB
Category 5e:	13 dB
Category 6:	24.1 dB

Category 6 cable offers a 13.1-dB improvement in ACR over Category 5 cable. In fact, Category 6 cable offers a 10-dB ACR at 200 MHz, only slightly lower than Category 5 cable at 100 MHz.

A good rule of thumb is that you need a minimum of 10 dB of ACR for reliable operation. Even more is better, but 10 dB is a cautious minimum. You might get by with less, but you may experience degraded performance in the application.

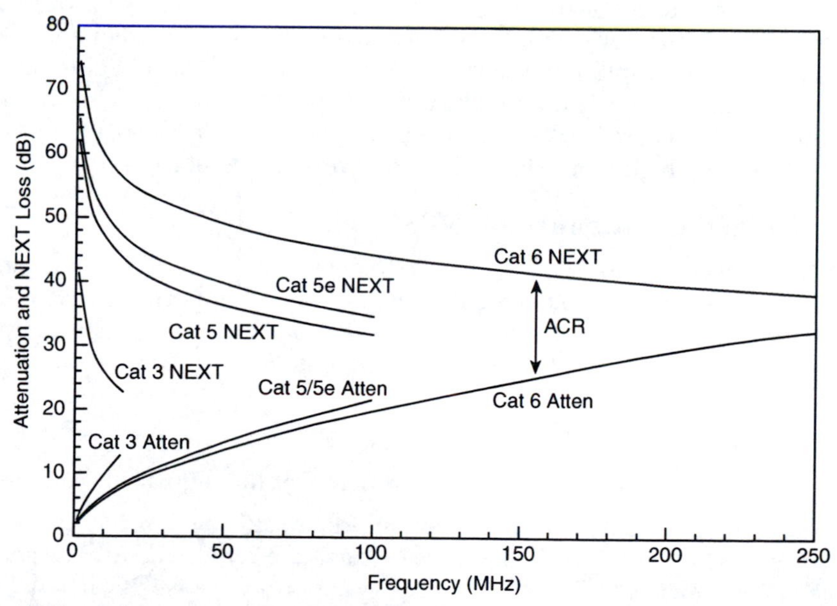

Figure 2–29. ACR

In modern systems (starting with Gigabit Ethernet), NEXT cancellation is often used to improve the effective performance of the cabling system. Using modern digital signal processing (DSP) techniques, the receiver can examine the incoming data stream and look for correlations with the signal recently sent by the local transmitter. These correlations can be subtracted out of the incoming data and the NEXT interference greatly reduced.

Keep in mind that ACR was developed before NEXT cancellation was common. If NEXT cancellation is being used, then the dB value of cancellation should be added to the ACR to get a more accurate view of system performance. For example, if the ACR is -5 dB and 20 dB of NEXT is being cancelled, then the effective system ACR is 15 dB.

This additional ACR margin in Categories 5e and 6 represents the headroom we mentioned earlier. Category 6, by virtue of its better ACR, means that it will perform better. If you push Category 5e cable toward its limits (as Gigabit Ethernet does), Category 6 cable still has lots of reserve performance.

While Category 6 cable is rated for operation to 250 MHz, it is specified to have a positive ACR at 200 MHz—about 10 dB's worth. Category 5 cables reach a zero ACR around 150 MHz.

To improve ACR, you can either reduce the crosstalk or lower the attenuation—or preferably do both. You can see from the ACR graph that Categories 5 and 5e cables have identical attenuation requirements, while Category 6 cables have lower attenuation. Category 6 cables typically use a conductor larger than 24 gauge to lower attenuation.

One other note about ACR: ELFEXT is the counterpart to ACR because it automatically accounts for both attenuation and FEXT.

It's important to realize that ACR is only important when signals will travel in both directions, as is common with telephone or networks. In an application like video distribution, where the signal only has to travel from source to receiver in one direction, ACR is not a factor. Indeed, such applications can operate at a negative ACR. For example, some video distribution systems can carry video signals at frequencies of several hundred megahertz over Category 5 cable. Systems are available, for example, that can distribute video at frequencies up to 550 MHz (77 channels) over a cable supposedly rated for 100 MHz. That's because the demands for such one-way transmission are not as demanding as for two-way transmission—there is no crosstalk consideration. (There is crosstalk, of course; it just doesn't matter too much.) But for applications that require high-speed signals in both directions, ACR can't be ignored.

Putting It All Together: Attenuation, NEXT, and ELFEXT

All the parameters we have discussed are working between all the pairs at the same time in a cable. The interactions can be quite complex. Figure 2–30 summarizes the effects of attenuation, NEXT, and ELFEXT in a cable. You can see that both types of crosstalk are coupling between all the pairs. In addition, the signals are being attenuated. Besides the effects shown in the figure, don't forget two other effects:

- Return losses caused by impedance mismatches
- Skew delay caused by cable imperfections or differences in lengths between the pairs

An important thing to remember about specifications setting limits on crosstalk, attenuation, return losses, and other effects is that they are worst-case requirements. The assumption is that most pairs will perform better, but that no pairs will perform worse. If you consider pair-to-pair crosstalk in a cable, if one combination fails, the cable fails—it doesn't meet specifications.

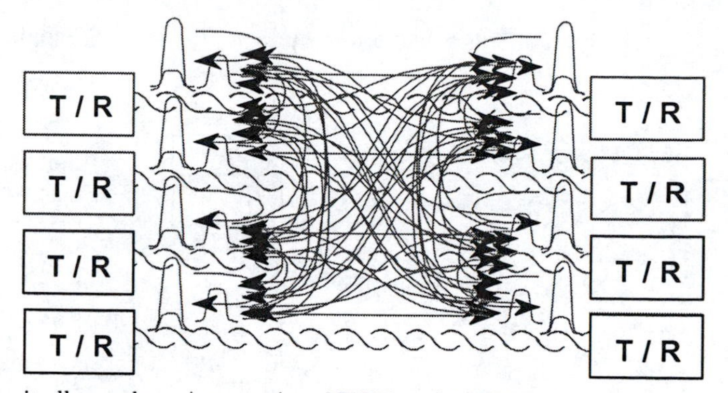

Figure 2–30. Putting it all together: Attenuation, NEXT, and ELFEXT in a power-sum application

It's the old cliché about the weakest link breaking the chain. The worst single test result is how you evaluate a cable. You don't average tests. You don't say that the important pairs passed so the cable passes. One failure means total failure in terms of meeting the specifications and saying your cable meets Category 5e specifications or whatever spec you're testing to.

Fortunately, most test equipment used in the field performs all the required tests automatically and interprets the results.

Figure 2–31 summarizes the important parameters for different types of transmission on a twisted-pair cable. A generic Category 5 application, such as 10 Mbps or Fast Ethernet, has the simplest requirements. It only uses one pair to transmit in each direction. Only attenuation and pair-to-pair NEXT are typically important. As the application becomes more complex, additional parameters must be considered. Parallel transmission adds power-sum NEXT, ELFEXT, and delay skew. Finally, bidirectional transmission on several pairs places the greatest demands on the cable and hence the most parameters that are important. Return loss becomes a factor.

Balance

There are two ways in which a signal can propagate along a twisted-pair cable. The desired mode of propagation consists of having equal and opposite signals on each of the conductors in the pair. In the example shown in Figure 2–32, a 2-volt signal is transmitted by placing +1 volt on one conductor and –1 volt on the other. The desired signal is the difference between the signals on the two conductors, in this case 2 volts. For this reason, this is called the differential signal.

An additional, undesirable mode of propagation is for the same signal to be carried on both conductors. This is known as a common-mode (or longitudinal) signal, and is illustrated by the interference voltage that is added to both conductors in Figure 2–26. When the signals on the two conductors are subtracted at the receiver, the common-mode noise cancels out. If for example, the noise added is 0.25 volt, the conductors now carry 1.25 volts and –0.75 volt. The difference is still 2 volts.

In an ideal cable, signals that start out as differential (or common-mode) stay that way. In practice, there is always some mode conversion—that is, a small amount of the differential

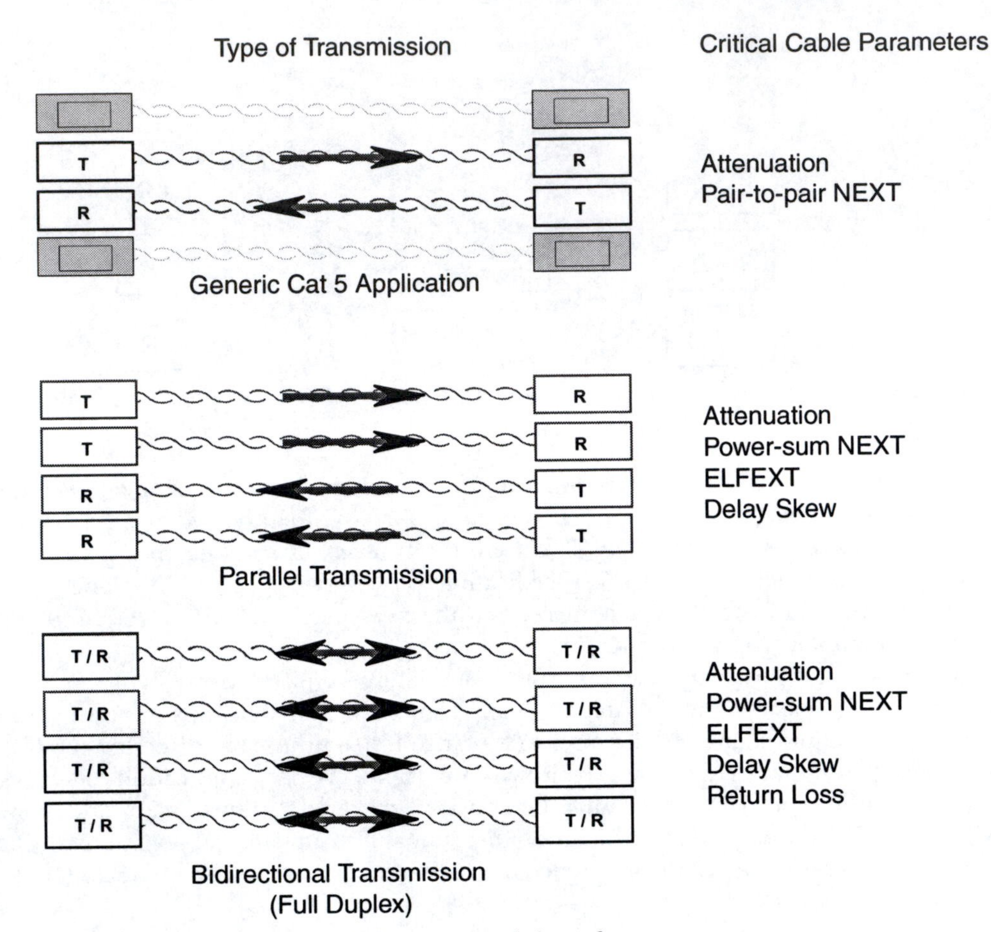

Figure 2–31. Critical parameters for different transmission schemes

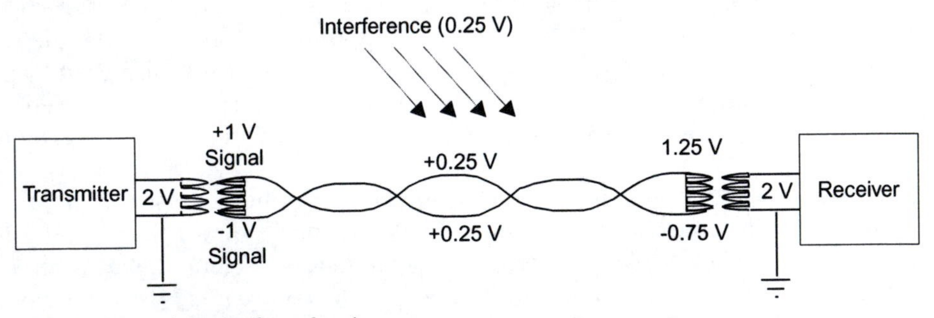

Figure 2–32. Differential signals and noise

signal gets converted to common-mode as it propagates along the cable. Similarly, some of the common-mode signal may be converted to differential mode.

There are several parameters that are used to specify how well balanced a UTP system is. They are:

- LCL (Longitudinal Conversion Loss)—ratio of differential voltage to source longitudinal voltage (measured at transmitter)
- TCL (Transverse Conversion Loss)—ratio of longitudinal voltage to source differential voltage (measured at transmitter)
- LCTL (Longitudinal Conversion Transfer Loss)—ratio of differential voltage to source longitudinal voltage (measured at receiver input)
- TCTL (Transverse Conversion Transfer Loss)—ratio of longitudinal voltage to source differential voltage (measured at receiver input)

A couple of observations may help to make sense out of this array of parameters. First, the *longitudinal* parameters are a ratio of differential to longitudinal voltages, while the *transverse* parameters are the opposite. Second, the *conversion loss* parameters are a one-port measurement made at the transmitter, while the *conversion transfer loss* parameters are two-port measurements with the denominator measured at the transmitter and the numerator measured at the receiver.

As an example, Figure 2–33 shows how an LCTL measurement is made. In this case, the longitudinal voltage at the receiver is measured and compared with the differential voltage that was applied by the transmitter.

Balance is very important because it has a direct effect on the EMC performance of the cabling system (both emission and susceptibility). But a perfectly balanced cable is a very good transmission medium, because differential signals are relatively immune to noise pickup—or, more correctly, at ignoring the picked up noise. If the cable isn't well balanced, it will pick up different amounts of voltage on each conductor, which will cause noise in the receiver. Or, the cable will radiate energy in the form of EMI.

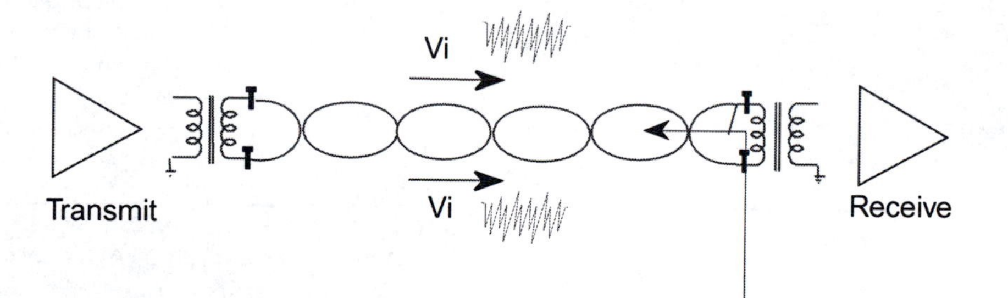

Figure 2–33. Measuring LCTL

Shielded Twisted-Pair Cable

Shielded twisted-pair (STP) cable was popularized by IBM in its Token Ring network. The most commonly used version of STP is called Type 1 cable. IBM created a cabling system, summarized in Figure 2–34, for use with their systems in premises wiring applications. While the system has not caught on as a widely accepted standard, Type 1 cable remains popular as the predominant type of STP.

Name	Type	Impedance (ohms)	Description
Type 1/1A 	STP	150	Two individually shielded solid-conductor, 22 AWG twisted pairs surrounded by an outer braid shield. Both pairs are suited for data transmission.
Type 2/2A 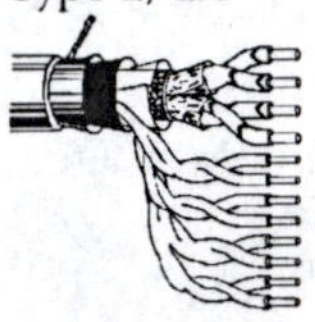	STP/UTP	150 (STP)	Two solid-conductor, 22 AWG twisted pairs surrounded by an outer braid shield. Four solid-conductor, 22 AWG twisted pairs outside the braid for telephone use.
Type 3 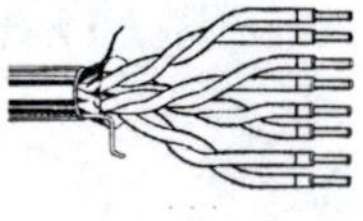	UTP	100	Four solid-conductor 24 AWG twisted pairs, unshielded. Similar to Category 2; not recommended for network applications.
Type 6/6A 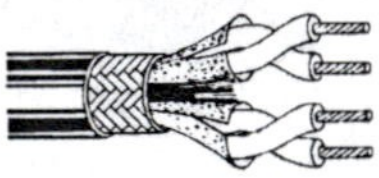	STP	150	Two stranded-conductor, 26 AWG twisted pairs surrounded by an outer braid. Similar to Type 1 except for wire gauge. Used for short patch cords.
Type 9/9A 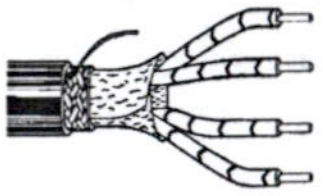	STP	150	A plenum-rated cable with two 26 AWG twisted pairs surrounded by an outer shield.

Figure 2–34. IBM cabling system *(Courtesy of Belden Wire and Cable Company)*

Type 1 cable contains two shielded cable pairs within an outer shield. The original Type 1 cable was rated to 20 MHz; an improved version—Type 1A—is rated to 300 MHz.

STP cable is quite attractive from a performance viewpoint. It allows longer distances and higher speeds than UTP. But UTP is more popular because STP is more expensive and more difficult to work with.

For maximum shielding effectiveness, both ends of a shielded cable should be grounded through a shielded interconnection. One purpose of the shield is to conduct noise to ground. If only one end is grounded, the shield becomes a long antenna, effective only at frequencies for which its length is less than ⅙ of the wavelength. If the shield is not grounded at either end, then it has little effect on noise.

Category 7 STP

Category 7 cable occupies a strange place in the scheme of cabling. While it fits into the Category taxonomy for UTP cable, it is a shielded rather than unshielded cable. Yet it differs from other STP cable in that its characteristic impedance is 100 ohms.

Structurally, Category 7 cable consists of four individually shielded pairs and an overall shield. Again, unlike the IBM cable Types 1A and 6A, which have two pairs, Category 7 is a four-pair cable. The cable is rated to 600 MHz, twice the frequency of IBM-style STP cables. The actual recommended operating frequency of Category 7 cable depends on how many interconnections are in the path. For two connectors in the path, the operating frequency is up to 600 MHz. For four connectors in the path, the upper frequency is 300 MHz.

Category 7 cabling is included in ISO/IEC 11801, but not in the TIA-568 series of standards. Even in Europe, which is much more accepting of shielded cable than most of the world, Category 7 cabling has achieved very limited deployment. At the end of 2005, it is estimated that Category 7 cabling comprises about 0.4% of the worldwide installed base of horizontal cable in commercial buildings.

One of the factors that has contributed to the low acceptance of Category 7 is that, like Category 6 UTP, there have not been any applications which required it. Unlike Category 6 UTP, however, there is no installed base of cabling which Category 7 is backward-compatible with. This situation may change with the advent of 10-gigabit Ethernet (10GBASE-T), which will operate over Category 7 as well as 6 and 6A. Whether this will result in increased popularity for Category 7 remains to be seen.

Unlike other categories of cable, which use the same style of modular plug and jack connectors, Category 7 allows two types of connectors. One is a modular plug and jack that is backward compatible with existing hardware. The other is a connector specifically designed for Category 7 cable.

Screened UTP

Screened UTP is really shielded UTP. But because shielded unshielded twisted pairs is a contradiction of terms, it's called screened UTP (sUTP), screened twisted pair (ScTP), or foil twisted pair (FTP). Standard Category 3, 5e, and 6 cables are available in shielded versions. As shown in Figure 2–35, the shield is typically a foil shield, which is not as effective as the

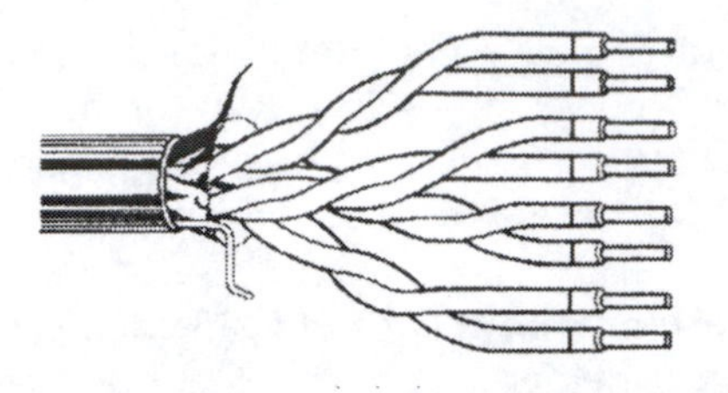

Figure 2–35. Screened UTP *(Courtesy of Belden Wire and Cable Company)*

braided shield used on STP and coaxial cable. Screened UTP cables share the same electrical characteristics as regular UTP, but offer greater immunity to outside noise. Unlike STP, which has each pair individually shielded from each other to reduce NEXT, ScTP has only a single shield around all pairs. The shield does not reduce NEXT; it only protects against outside noise.

Cable Lengths

The rule of thumb for total horizontal cabling is 100 meters. Of this TIA/EIA-568B recommends 90 meters from the telecommunications closet to the work area outlet. In addition, it allows 6 meters total for jumper cables, patch cords, and cross connects inside the telecommunications closet and 3 meters for cable in the work area (from the outlet to the computer). This amounts to 99 meters total. The standard then waffles a bit and allows 10 meters for cable in the telecommunications closet and work area. But the 1-meter difference amounts to a quibble.

More latitude is given for backbone cables. For Categories 3, 5e, and 6 UTP and Type 1 STP, the distance is 90 meters, depending on the bandwidth of the signal. Figure 2–36 summarizes distances for the backbone. Notice that Categories 3, 5e, and 6 distances depend on the frequency they will be used at. If the frequency falls below the range given, the cable can be treated as a voice-grade application. Below 5 MHz, all UTP cables can be run 800 meters.

Cable	Backbone Cables: Main Cross Connect to Telecommunications Room	
	Frequency	**Distance**
UTP	Voice grade	800 m
Cat 3 UTP	5–16 MHz	90 m
Cat 5e UTP	5–100 MHz	90 m
Cat 6 UTP	5–250 MHz	90 m
STP	5–300 MHz	90 m
62.5/125 or 50/125 Fiber	—	2000 m
Single-Mode Fiber	—	3000 m

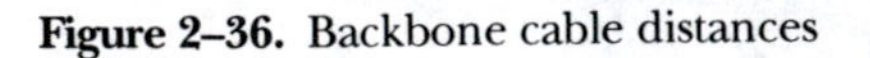

Figure 2–36. Backbone cable distances

At 10 MHz, Category 5e cable can still be run 800 meters, but Category 3 is restricted to 90 meters.

The distance a cable can be run in the backbone depends on the signal frequency it carries.

Flammability Ratings

Cables are protected by an outer jacket. The jacket protects against abrasion, oil, solvents, and so forth. The jacket usually defines the cable's duty and flammability rating. Heavy-duty cables have thicker, tougher jackets than light-duty cables. Equally important in a building is the cable's flammability rating. The *National Electrical Code (NEC®)* establishes flame ratings for cables, while Underwriter's Laboratories has developed procedures for testing cables. The *NEC®* requires that cables that run through plenums (the air-handling space between walls, under floors, and above drop ceilings) must either be run in fireproof conduits or be constructed of low-smoke and fire-retardant materials. For building use, there are three categories of cables depending on the cable type:

Plenum cables can be installed in plenums without the use of conduit.

Riser cables can be used in vertical passages connecting one floor to another.

General-use cables can be used for general installations. They cannot be used in riser or plenum applications without fireproof conduits. These cables can be used in offices spaces—to connect from a wall jack to a computer, for example.

The NEC lists cables by a short acronym describing the cable classification and a final single-letter suffix giving the flammability rating. Figure 2–37 summarizes the classifications for cables discussed in this book. Most cables have their classification printed on the jacket.

Cable Type	Description	Permitted Uses	Substitutions
CMP	Communications plenum	Anywhere, including air ducts and plenums	None
CMR	Communications riser	In vertical shafts, and general use	CMP
CMG CM	Communications general-purpose	General use except for risers and plenums	CMP or CMR
CMX	Communications limited use	Only in raceways and dwellings	CMP, CMR, CMG, or CM
CMUC	Communications under-carpet	For undercarpet use only	None

Figure 2–37. Copper cable classifications and flammability ratings

SUMMARY

- Crosstalk is coupling of energy from one conductor to another.
- NEXT is the ratio of crosstalk power to signal power, measured at the near end; FEXT is crosstalk measured at the far end.
- Pair-to-pair NEXT and FEXT measures crosstalk between individual pairs.
- Power-sum NEXT and FEXT measures crosstalk from multiple pairs onto a single pair.
- Alien crosstalk refers to crosstalk between cables rather than between pairs in a single cable.
- ELFEXT accounts also for attenuation.
- Attenuation is the loss of signal power.
- ACR is the difference between NEXT and attenuation at a given frequency.
- Unshielded twisted-pair (UTP) cable is the most common copper cable used in premises cabling.
- UTP is available in Categories 3, 5e, 6, and 6A.
- The categories of cable are mainly distinguished by their frequency range and by characteristics such as NEXT, ELFEXT, attenuation, and ACR.
- 4-pair and 25-pair (or multiples thereof) are common UTP configurations.
- Category 7 cable is a shielded twisted-pair cable with a 100-ohm characteristic impedance.
- Shielded twisted-pair cable based on the IBM system have a 150-ohm impedance.
- Coaxial cables use 50-ohm characteristic impedance for network applications and 75 ohms for video applications. Cables used in premises applications typically have four layers of shield.
- The rate at which information is being sent is measured in megabits per second (Mbps). The rate at which the signal on a transmission line changes is measured in megahertz (MHz). The encoding algorithm used by a particular protocol determines the relationship between MHz and Mbps. Typically, MHz is less than or equal to Mbps.

REVIEW QUESTIONS

1. Name the three types of cables used in business and residential cabling systems.
2. Distinguish in terms of operating frequency range between Categories 3, 5, 5e, 6, and 6A UTP.
3. Define headroom. What is the headroom offered by Category 6 horizontal cable over Category 5e cable at 100 MHz?
4. How does Category 7 cable differ from Category 6 cable?
5. Which UTP categories (3, 5, 5e, 6, 6A, and 7) are recognized media for horizontal cabling in TIA-568B? For categories that are not recognized, briefly explain why.
6. Define ACR.

7. What does it mean if the ACR is negative?
8. What are the two main differences between NEXT loss and ELFEXT loss?
9. Where is Series 6 coaxial cable typically used?
10. What is the difference between NEXT and power-sum NEXT?
11. What is the difference among UTP, ScTP, and STP?
12. What are the three most fundamental properties of copper cable?
13. When is delay skew important?
14. How does the transmission performance of copper cable vary with frequency?
15. When is return loss a critical parameter?
16. A 100-mW signal is passed through a 20-dB attenuator. What is the output power?
17. A signal is amplified by 6 dB. How much is the signal power increased?
18. A signal's power is attenuated by a factor of 40. How many dB is this?
19. Category 6 UTP cable has a maximum attenuation of 19.8 dB per 100 m at 100 MHz. If a 10-mW signal at 100 MHz is launched into a 50-m length of this cable, how much power can be received at the far end?
20. The same Category 6 UTP cable as in problem 19 has an attenuation of 6.0 dB per 100 m at 10 MHz. If a 10-mW signal at 10 MHz is launched into a 50-m length of this cable, how much power can be received at the far end?
21. What is the wavelength of a 300-MHz signal in free space?
22. What frequency does a wavelength in free space of 1 cm correspond to?
23. What frequency has a wavelength in free space the same as the diameter of the earth (8000 miles)?
24. In what way does alien crosstalk (AXT) differ from all previously defined cabling parameters?
25. If a Cat 6A cabling installation has a problem with alien crosstalk, what steps can be taken to mitigate the problem?
26. In terms of flammability ratings, what type of cables can be substituted for riser cable?
27. What technique can be used in a LAN interface to improve the effective ACR of the cabling?
28. As of the end of 2005, what percentage of the worldwide installed base of horizontal cable is Category 6? Category 5e? What is the trend?
29. Balance has a direct effect on what properties of the cabling system?
30. What determines the amount of bandwidth used by a signal of a particular bit rate? Or, in other words, what determines the mapping between bits per second and Hz?

CHAPTER 3 Fiber Optics

Objectives

After reading this chapter, you should be able to:

- List the main types of optical fibers.
- Describe how light travels through a fiber.
- List seven advantages of an optical fiber.
- Describe a basic fiber-optic link.
- Identify the differences between LED and laser sources.
- Discuss the differences between multimode and single-mode fibers as they apply to premises cabling.
- Sketch a simple link power budget.
- Define dense wavelength-division multiplexing.

Fiber-optic technology is fundamentally different from copper cabling. Fiber-optic systems transmit photons through hair-thin optical fibers, while copper systems transmit electrons through copper wire. Fiber-optic transmission provides higher bandwidth with lower loss than copper-based transmission which allows higher data rates to be transmitted over longer distances.

The Advantages of Fiber

Fiber optics offer distinct advantages over traditional copper media.

Information-Carrying Capacity

Fiber offers bandwidth well in excess of that required by today's network applications. The 62.5/125-μm fiber commonly used in buildings has a minimum bandwidth of 160 MHz-km. Because a fiber's bandwidth scales with distance, the bandwidth at 100 meters is over 1.5 Gbps or 5 Gbps (depending on the wavelength of light transmitted). In comparison, Category 5e cable is specified only to 100 MHz over the same 100 meters. A new generation of multimode fibers and light sources supports 10 Gbps over 300 meters.

When high-performance single-mode cable used by the telephone industry for long-distance telecommunications is deployed in local networks, the bandwidth is essentially infinite. That is, the information-carrying capacity of the fiber far exceeds the ability of today's electronics to exploit it.

Low Loss

An optical fiber offers low power loss. Low loss permits longer transmission distances. Again, the comparison with copper is important: in a network, the longest recommended copper distance is 100 meters; with fiber it is 2000 meters or more.

A drawback of copper cable is that loss increases with the *signal* frequency. Attenuation in twisted-pair cable is higher at 100 MHz than at 10 MHz. This means high data rates tend to increase power loss and decrease practical transmission distances. With fiber, loss does not change with the *signal* frequency. Attenuation does change with the frequency of the light transmitted through it, but not for the data rate. Thus, a 10-MHz and 100-MHz signal are attenuated alike in the fiber. Figure 3–1 shows the relationship between signal frequency and attenuation for both copper and fiber cables.

Electromagnetic Immunity

By some estimates, 50 to 75% of all copper-based network outages are caused by cabling and cabling-related hardware. Crosstalk, impedance mismatches, and EMI susceptibility are

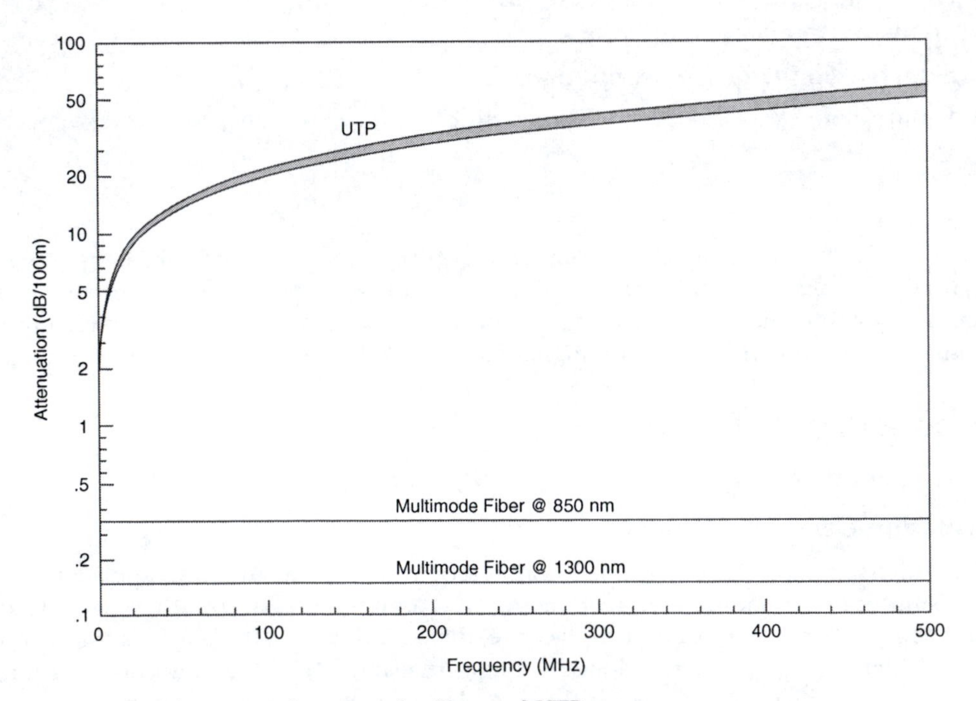

Figure 3–1. Attenuation versus frequency for fiber and UTP

major factors in noise and errors in copper system. What's more, such problems can increase with incorrectly installed Category 5 cable, which is sensitive to poor installation.

Because a fiber is a dielectric, it is immune to electromagnetic interference. It does not cause crosstalk, which is a main limitation of twisted-pair cable. What's more, it can be run in electrically noisy environments, such as a factory floor, without concern, because noise will not affect the fiber. There's no concern with proximity to noise sources like power lines or fluorescent lights. In short, fiber is inherently more reliable than copper.

Light Weight

Fiber-optic cable weighs less than comparable copper cable. A two-fiber cable is 20% to over 50% lighter than a comparable four-pair Category 5 cable. Lighter weight makes fiber easier to install. Typical weights for 1000 feet of different cables are:

2- fiber cable:	11 lb
12-fiber cable:	33 lb
4-pair UTP:	25 lb
25-pair backbone UTP:	93 lb

Smaller Size

Fiber-optic cable is typically smaller than the copper cables it replaces. Again, relative to 4-pair UTP, an optical fiber takes up about 15% less space.

Safety

Because the fiber is a dielectric, it does not present a spark hazard, nor does it attract lightning. What's more, cables are available in the same flammability ratings as copper counterparts to meet code requirements in buildings.

Security

Optical fibers are quite difficult to tap. Because they do not radiate electromagnetic energy, emissions cannot be intercepted. And physically tapping the fiber takes great skill to do undetected. Thus the fiber is the most secure medium available for carrying sensitive data.

The Disadvantages of Fiber

Fiber also has a few disadvantages.

- The cost of transceivers is higher. Because LAN equipment (PCs, hubs, etc.) is electronic, electronic-to-optical and optical-to-electronic (E-O and O-E) conversions must take place on every interface to the fiber. This makes optical transceivers inherently more expensive than the interface to copper cables.
- Although pulling fiber is often easier than pulling copper cable (due to fibers' small diameter and lighter weight), the total installation process for fiber is more difficult and requires more expertise and special equipment. Advances have been made in easy-to-install fiber connectors, but they remain more exacting and labor-intensive than copper connectors.

- Fiber systems require more routine maintenance. For example, fiber connectors must be cleaned before inserting because a little dust or dirt can seriously affect the performance of the connector.

As a result, fiber systems[1] tend to have a higher total cost than copper systems when considering all factors: cable; connectors; installation; and transceivers.

Fibers and Cables

Fiber optics is a technology in which signals are converted from electrical into optical signals, transmitted through a thin glass fiber, and reconverted into electrical signals. As shown in Figure 3–2, the basic optical fiber consists of three concentric layers differing in optical properties:

Core: the inner light-carrying member.

Cladding: the middle layer, which serves to confine the light to the core.

Buffer: the outer layer, which serves as a "shock absorber" to protect the core and cladding from damage.

Total Internal Reflection

A basic understanding of how light travels through the optical fiber is essential to understanding many of the issues involving fiber properties. Light travels by a principle known as *total internal reflection.*

Light injected into the core and striking the core-to-cladding interface at an angle greater than the critical angle will be reflected back into the core. Because angles of incidence and reflection are equal, the light ray continues to zig-zag down the length of the fiber. The light is trapped within the core. Light striking the interface at less than the critical angle passes into the cladding and is lost. Figure 3–3 shows light traveling through the fiber by total internal reflection.

The rays of light do not travel randomly. They are channeled into modes, which are possible paths for a light ray traveling down the fiber. A fiber can support as few as one mode and as many

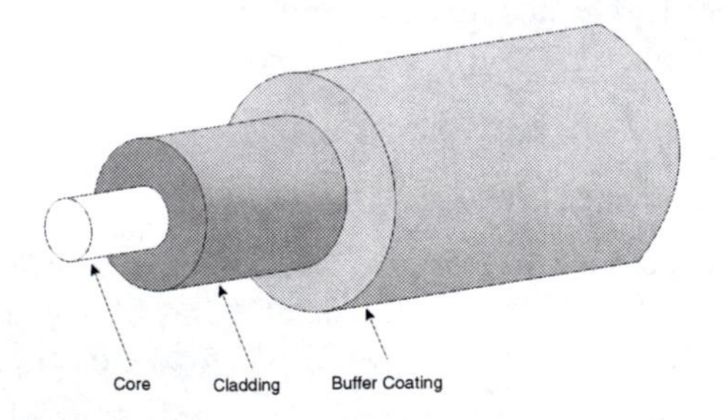

Figure 3–2. Parts of an optical fiber

[1] *Plastic optical fiber (POF) systems are somewhat easier to install and maintain than traditional glass fiber systems, as we will discuss later in this chapter.*

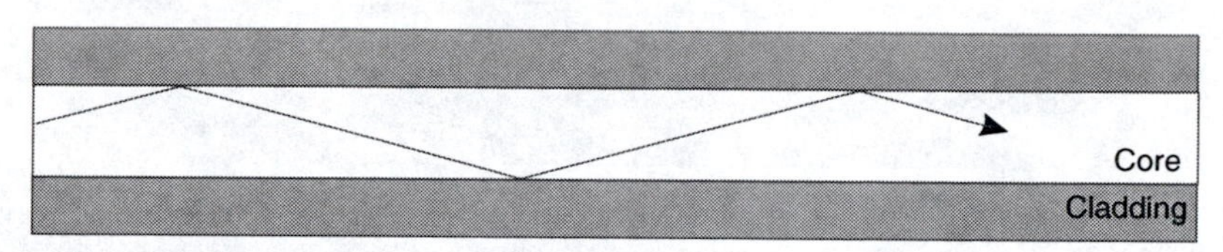

Figure 3–3. Total internal reflection

as tens of thousands of modes. While we are not normally interested in modes *per se,* the number of modes in a fiber is significant because it helps determine the fiber's bandwidth. More modes typically mean lower bandwidth. The reason is dispersion.

As a pulse of light travels through the fiber, it spreads out in time. While there are several reasons for such dispersion, two are of principal concern. The first is modal dispersion, which is caused by different path lengths followed by light rays as they bounce down the fiber. Some rays follow a more direct route than others. The second type of dispersion is called chromatic or material dispersion: different wavelengths of light travel at different speeds. By limiting the number of wavelengths of light, you limit the material dispersion.

Dispersion limits the bandwidth of the fiber. At high data rates, dispersion will allow pulses to overlap so that the receiver can no longer distinguish where one pulse begins and another one ends.

Types of Fibers: Single Mode or Multimode?

The basic structure of an optical fiber can be modified in several ways to achieve different signal transmission characteristics. Modifications affect bandwidth, attenuation, and the practicalities of coupling light into and out of the fiber. Figure 3–4 shows the basic types of fiber: step-index multimode, graded-index multimode, and step-index single mode.

In the simplest optical fiber, the relatively large core has uniform optical properties. Termed a *step-index multimode fiber;* this fiber supports thousands of modes and offers the highest dispersion—and hence the lowest bandwidth.

By varying the optical properties of the core, the *graded-index multimode fiber* reduces dispersion and increases bandwidth. Grading makes light following longer paths travel slightly faster than light following shorter paths. The net result is that the light does not spread out nearly as much. Nearly all multimode fibers used in networking and data communications have a graded-index core.

But the ultimate in high-bandwidth, low-loss performance is a *single-mode fiber.* Here the core is so small that only a single mode of light is supported. The bandwidth of a single-mode fiber far surpasses the capabilities of today's electronics. Not only can the fiber support speeds tens of gigabits per second, it can carry many gigabit channels simultaneously. This is done by having each channel be a different wavelength of light. The wavelengths do not interfere with one another. Single-mode fiber is the preferred medium for long-distance telecommunications. It finds use in networks for interbuilding runs and will eventually become popular for high-speed backbones.

The most popular fibers for networking are the 62.5/125-μm and 50/125-μm multimode fibers. The numbers mean that the core diameter is 62.5 μm and the cladding is 125 μm.

The relative diameters of the core and cladding of the three most common sizes of fiber are illustrated in Figure 3–5.

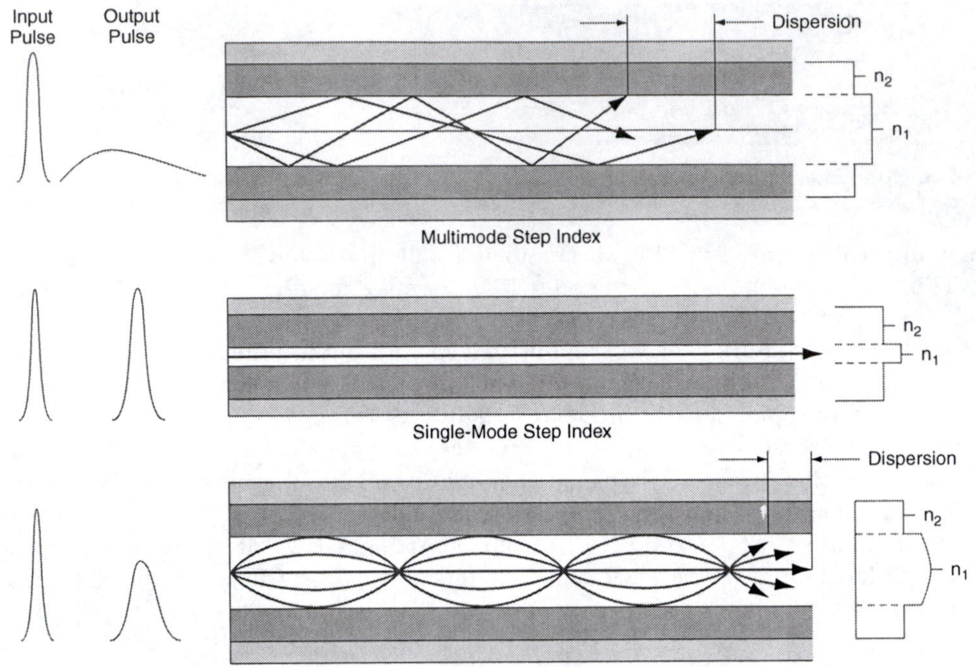

Figure 3–4. Type of fibers *(Courtesy of AMP Incorporated)*

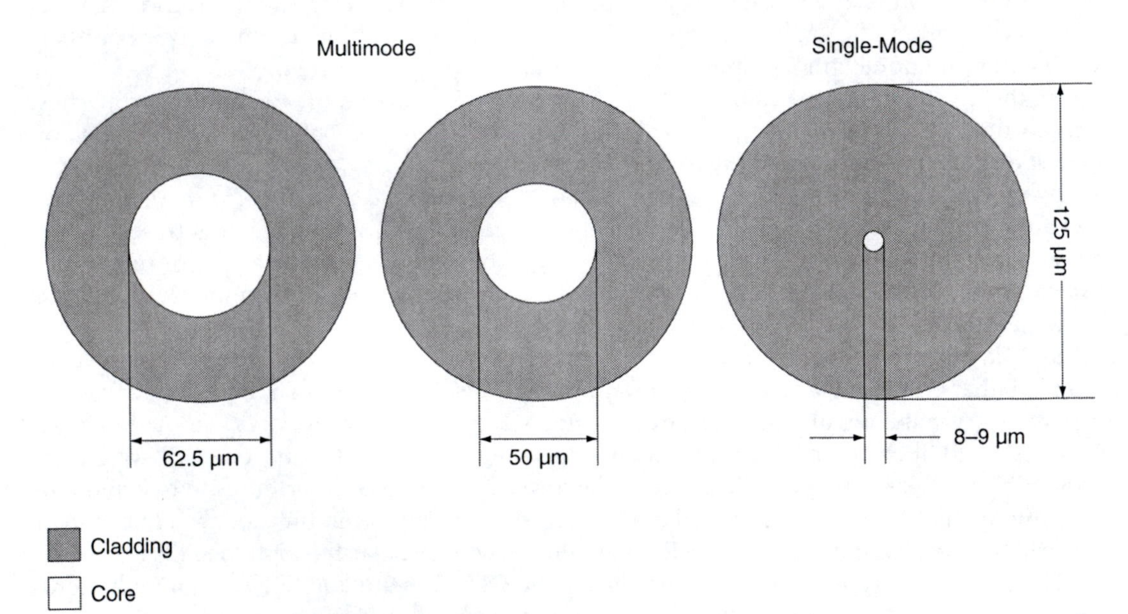

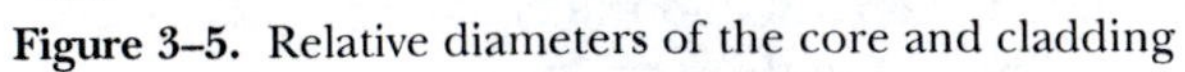

Figure 3–5. Relative diameters of the core and cladding

The quick summary:

> ***Graded-index multimode fibers are the preferred fibers for horizontal cable and some backbone applications.***
>
> ***Single-mode fibers, by virtue of their immense bandwidth and long transmission capabilities, are for applications requiring their additional performance, including backbone applications and interbuilding runs.***

Numerical Aperture (NA)

The NA of the fiber defines which light will be propagated and which will not. As shown in Figure 3–6, NA defines the light-gather ability of the fiber. Imagine a cone coming from the core. Light entering the core from within this cone will be propagated by total internal reflection. Light entering from outside the cone will not be propagated.

NA has an important consequence. A large NA makes it easier to inject more light into a fiber, while a small NA tends to give the fiber a higher bandwidth. A large NA allows greater modal dispersion by allowing more modes for the light to travel in. A smaller NA reduces dispersion by limiting the number of modes.

> ***The higher the NA and core diameter, the lower the bandwidth of the fiber. Conversely, the lower the NA and core diameter, the higher the bandwidth—and the more difficult it is to inject light into the fiber.***

Attenuation

Attenuation is loss of power. During transit, light pulses lose some of their energy. Attenuation for a fiber is specified in decibels per kilometer (dB/km). For commercially available fibers, attenuation ranges from under 1 dB/km for single-mode fibers to 1000 dB/km for large-core plastic fibers.

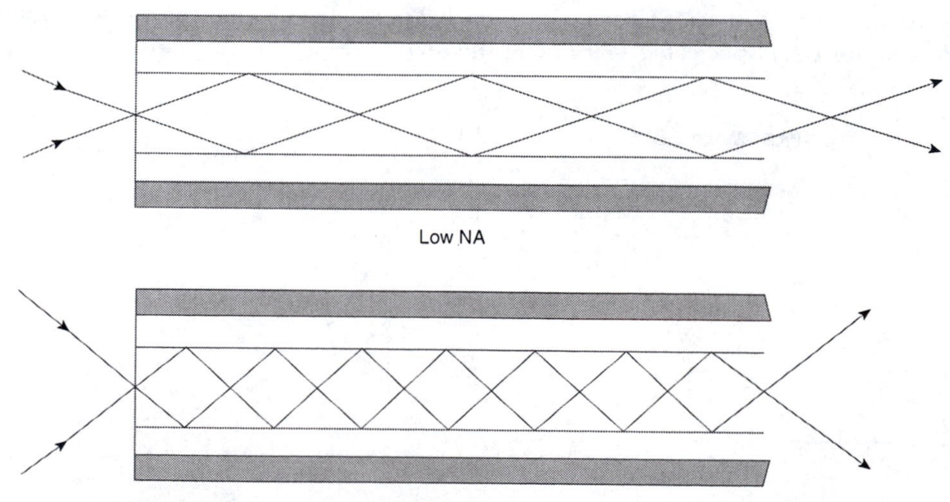

Figure 3–6. Numerical aperture *(Courtesy of AMP Incorporated)*

Attenuation varies with the wavelength of light. There are several "windows" of low loss, where fiber is suited. The three traditional windows are 850 nm, 1300 nm, and 1550 nm. These windows date back to the earliest days of fiber and represent wavelengths where the fiber offers low attenuation. Other windows have been introduced in the late 1990s. While the bands are being formalized as this book is being written, the regions break out like this:

- 850 nm (the first window; traditionally used for low-cost LAN interfaces)
- 1260 to 1360 nm (the second window: the 1300-nm region)
- 1525 to 1562 nm (the third window, also known as the C (for conventional) band, because this was the original longer wavelength region; the 1550-nm region; this is the band most commonly used for DWDM systems)
- 1575 to 1610 nm (also known as the L (for long) band, because it uses wavelengths right above the C band; sometimes called the fourth window)
- 1440 to 1500 nm (also known as the S (for short) band, because it uses wavelength below the C band)
- 1350 to 1450 nm (made available by improved manufacturing that eliminates high loss traditionally found around 1400 nm; the fifth window)

Figure 3–7 shows attenuation of the function of wavelength for single-mode and multimode fibers. The bump at 1400 nm is not present in fifth-window fibers.

Bandwidth

Fiber bandwidth is given in MHz-km. Bandwidth scales with distance: if you halve the distance, you double the bandwidth. If you double the distance, you halve the bandwidth. What does this mean in premises cabling? For a 100-meter run (as allowed for twisted-pair cable), the bandwidth for 62.5/125-μm fiber is 1600 MHz at 850 nm and 5000 MHz at 1300 nm. For the 2-km spans allowed for some fiber networks, bandwidth is 80 MHz at 850 nm and 250 MHz at 1300 nm.

The discussion of bandwidth efficiency discussed in the last chapter for copper cable does not hold true for fiber-optic cable. With fiber, the relationship between frequency and bit rate

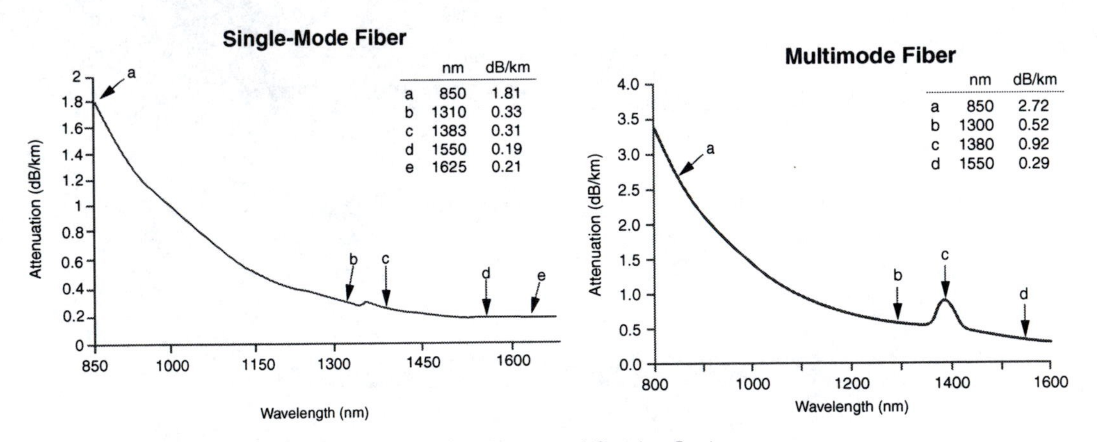

Figure 3–7. Attenuation versus wavelength *(Courtesy of Corning Inc.)*

is much closer and can be considered 1:1 in most cases. The fiber has so much available bandwidth that sophisticated and complex coding schemes like MLT-3 aren't needed.

Figure 3–8 summarizes the main characteristics of several types of common (legacy) fibers. Note that 100/140 μm fiber is seldom encountered anymore. Modern laser-optimized fibers are covered in Figure 3–9.

Fiber Type	Attenuation (Max., dB/km)		Bandwidth (Min., MHz-km)		NA
	850 nm	1300 nm	850 nm	1300 nm	
Single Mode	—	1.0	—	See note	.1
50/125	3.5	2.0	400	400	.20
62.5/125	3.5	1.5	160	500	.275
100/140	5.0	4.0	100	200	.29

Notes: 1) The bandwidth of a single-mode fiber is essentially infinite in that it surpasses the abilities of today's electronics to exploit its capabilities.

2) 62.5μ multimode fiber with 200 MHz-km bandwidth at 850 mm is available, but is not specified in TIA-568B.

Figure 3–8. Common fiber characteristics

Return Loss

Just as a signal in copper cable can reflect backward from changes in impedance, discontinuities in an optical fiber can also cause reflections. The greatest source of reflections is at fiber-to-fiber interconnections, especially if the fibers are separated by an air gap. This reflected energy can be propagated back to the transmitter, where it can interfere with operation of a laser source. LEDs, on the other hand, are not adversely affected by this backscattered light. Even with LEDs, return reflections are still measured because they can be used to evaluate the mating of two connectors. An air gap between fibers will increase reflections, so by measuring return energy you can tell whether there is a gap between fibers.

The energy reflected must be minimized. While there are several approaches to reducing the reflected energy, the most common are the physical contact (PC) and angled physical contact (APC) end finishes. In the PC end finish, the fiber and connector end are polished to a radius to ensure that the fibers touch on the high side of the radius. In the APC finish, a slight angle is added to the radius. The fibers touch significantly and this reduces reflection due to a mismatch in the index of refraction between glass and air. The radius and angle of the end finish also tend to reflect light so that it is not propagated by internal reflection.

Return loss refers to how far below the incident light the reflected light is. A 30-dB loss means that the reflected light is 99.9% less than the incident light. A 50-dB loss means that the light is 99.999% less than the incident light. In practical terms, imagine a 50-μW optical signal. The reflected energy would be 0.5 μW for a 30-dB return loss and only 0.005 μW for a 50-dB return loss. Therefore, a higher return loss figure is better.

The TIA/EIA-568B requirements for return loss in connectors are as follows:

Multimode fiber:	20 dB or greater
Single-mode fiber:	26 dB or greater

Other applications, notably high-speed telecommunications, have much higher return loss requirements for single-mode fiber—as high as 67 dB.

More About Cable Distances and Bandwidth

The 62.5/125-µm fiber was the original fiber of choice in premises cabling applications. It offered the best balance of performance, including bandwidth and compatibility with LED-based network electronics. Then came Gigabit Ethernet to throw a small wrench in the works. Just as copper cables have a "magic" distance figure of 100 meters, so has fiber a distance figure of 300 meters. But some 62.5/125-µm installations could not operate properly over this distance at gigabit data rates. Because traditional fibers did not have perfectly graded indices of refraction (they had a dip in the center), they sometimes failed to operate properly. Remember how different modes cause spreading? This is called differential mode delay (DMD). DMD could cause pulses to improperly overlap if the light was launched dead center into the fiber.

Unfortunately, DMD problems were hard to predict because it depends on both the transmitter and the fiber. But the end result was to limit the transmission distance. Indeed, the recommended distance for fiber operating at an 850-nm short wavelength was 220 meters. (Notice that most 62.5/125-µm cable systems would run the full 300 meters, but there was no guarantee. You hooked up your system and crossed your fingers.) In addition, a special "mode-conditioning" patch cable to correct for DMD is required to overcome certain characteristics in some cables. A mode-conditioning cable is simply a cable that causes an offset launch of the light into the main backbone or horizontal cable.

With its smaller diameter and high bandwidth, 50/125-µm fiber can easily handle Gigabit Ethernet over 300 meters. The reason that this fiber wasn't chosen originally was that its small core made it more difficult to couple light into the fiber. However, newer sources—notably VCSELs that will be discussed shortly—are more compatible with 50/125-µm fibers. TIA/EIA-568B allows either fiber.

There is one other thing to remember about fiber. The attenuation and bandwidth specifications in TIA/EIA-568B are for "plain-vanilla" fibers. Many manufacturers offer fiber with better performance. Just as UTP manufacturers offered enhanced Category 5 with superior performance, fiber makers offer enhanced 50/125-µm and 62.5/125-µm fibers with even better performance. In a sense, you can order fiber-optic cables in a good, better, best manner to get standard or increased bandwidth and lower attenuation. For example, Figure 3–9 shows the characteristics of laser-optimized 50 µm multimode fiber which is specified in TIA-568-B.3-1.

Notice, in particular, in Figure 3–9 the dramatic increase in bandwidth for 50/125-µm multimode fibers, from 500 MHz-km to 2000 MHz-km. This increase stems from two sources: improved manufacturing of the fiber and newer lower-cost laser sources called VCSELs.

Fiber type	**Wavelength (nm)**	**Maximum Attenuation (dB/km)**	**Minimum Bandwidth (MHz-km)**	
			Overfilled Launch	**Laser Launch**
Laser-optimized	850	3.5	1500	2000
50/125 µm	1300	1.5	500	N/A

Figure 3–9. Laser-optimized multimode fiber

Notice, also, that the fiber's bandwidth depends greatly on the source: LED or laser. We will discuss VCSELs later in this chapter.

Figure 3–10 shows the recommended maximum operating distances for 50/125-μm and 62.5/125-μm fibers operating at short (850-nm) and long (1300-nm) wavelengths, according to the Gigabit Ethernet standard. You can see that the bandwidth of the cable influences possible distances, as does the wavelength of light transmitted. Enhanced fibers will allow longer distances, but these longer runs do exceed the recommendations of TIA/EIA-568B.

The earlier generation of fibers were designed for use with LED sources operating at more modest data rates—say 100 Mbps or lower. A newer generation of fibers is optimized for operation with low-cost lasers, such as VCSELs, to handle not only Gigabit Ethernet but the emerging need for 10-Gbps Ethernet.

So why does TIA/EIA-568B use the more modest bandwidth standard of 160/200 MHz-km for 62.5/125-μm fibers instead of the newer high-bandwidth cables? One reason is inertia. This has been the standard performance for the cables and remains so. Another reason is installed base: lots of buildings have this cable installed and want to leverage that cable with newer applications. Yet another reason is that the newer gigabit fibers are exactly that—new. And new typically means more expensive and not yet widely deployed.

Again, we're back to Gigabit Ethernet. Because the network can choke on plain-vanilla 62.5/125-μm cable operating at an 850-nm wavelength, most manufacturers offer a better grade of cable at this shorter wavelength. The drive is to guarantee that the cable will handle Gigabit Ethernet over 300 meters. The newer enhanced cables also look to the future and 10-Gbps requirements.

	1000BASE-SX	
Fiber Type	**Modal Bandwidth @ 850 nm (MHz-km)**	**Minimum Range (m)**
62.5 μm MMF	160	220
	200	275
50 μm MMF	400	500
	500	550
10 μm SMF	N/A	Not supported
	1000BASE-LX	
Fiber Type	**Modal Bandwidth @ 850 nm (MHz-km)**	**Minimum Range (m)**
62.5 μm MMF	500	550
50 μm MMF	400	550
	500	550
10 μm SMF	N/A	5000

Figure 3–10. Gigabit Ethernet fiber transmission distances

10-Gigabit Ethernet

The 10-Gigabit Ethernet specification (IEEE 802.3ae), which was issued in 2002, contains many different configurations. They fall into 3 families:

- 10GBASE-S: short-wavelength systems operating over multimode fiber at 850 nm
- 10GBASE-L: long-wavelength systems operating over singlemode fiber in the 1300 nm band
- 10GBASE-E: extra-long-wavelength systems operating over singlemode fiber in the 1550 nm band

A summary of the transmission distance for each type is given in Figure 3–11.

Note that, with the exception of laser-optimized 50 μm MMF, the transmission distances for multimode fiber are very short (26 to 82 meters.) To address this issue, in 2004 the IEEE 802.3 committee approved project 802.31aq to develop an interface that would operate at 10 Gb/s for at least 220 m over the installed base of 500 MHz-km multimode fiber. This specification, which will be called 10GBASE-LRM, had not been completed as this edition was being published.

Wavelength-Division Multiplexing: Increasing a Fiber's Capacity

One way to increase a fiber's capacity is by injecting more signals of different wavelengths into it. For example, you could simultaneously send 850-nm and 1300-nm signals through the fiber. Suppose you send a 1-Gbps signal at 850 nm and a 1-Gbps signal at 1300 nm. In effect, the fiber is now carrying 2 Gbps. This technique is called wavelength-division multiplexing (WDM).

The WDM example just presented operates with signals in different optical windows. In telecommunications, a more sophisticated form of WDM is used. Here the wavelengths are separated by less than nm. Because the wavelengths are so close together, this form of WDM is called dense wavelength-division multiplexing (DWDM). With the 850- and 1300-nm example, you are in effect sending different colors of light. With DWDM, you are transmitting different shades of the same color. Telecommunications companies use DWDM to increase the capacity of fibers. Systems typically multiplex 4, 8, or 16 channels (separate wavelengths) in the 1550-nm region. Systems with up to 64 channels operating at 10 Gbps have been demonstrated over

Type	Fiber	Wavelength (nm)	Modal Bandwidth (MHz-km)	Range (m)
10GBASE-S	62.3 μm MMF	850	160	26
			200	33
	50 μm MMF	850	400	66
			500	82
			2000	300
10GBASE-L	10 μm SMF	1300	N/A	10,000
10GBASE-E	10 μm SMF	1550	N/A	30,000

Figure 3–11. 10-Gigabit Ethernet fiber transmission distances

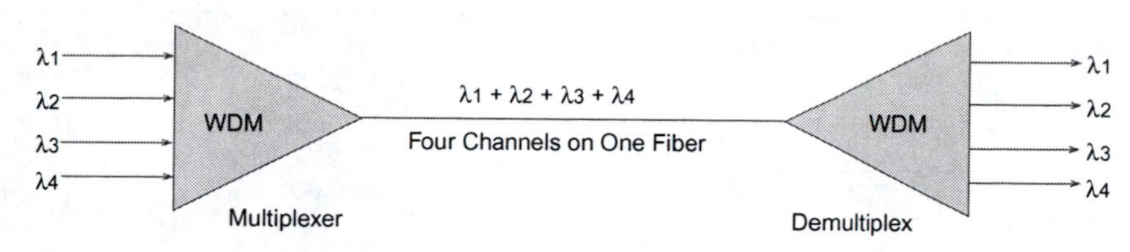

Figure 3–12. Dense wavelength-division multiplexing

single-mode fiber. The 64-channel system yields a total interconnection capacity of 640 Gbps! At each end of the system, you need an optical multiplexer that will either combine or separate the wavelengths.

Long-haul DWDM systems have been demonstrated that transmit hundreds of channels with a total capacity approaching 10 Tb/s (that's *terabits*, or trillions of bits per second). Notice that you increase capacity in two ways: increasing the data rate of each wavelength and increasing the number of wavelengths multiplexed. Data rates for telecommunications are rapidly moving from 2.5 Gbps to 10 Gbps to 40 Gbps. For example, Bell Labs demonstrated a system using 40 channels at 40 Gbps per channel to achieve a total throughput of 1.6 Tbps.

DWDM is one of the hottest areas in applying optical communications to telecommunications. DWDM significantly increases capacity while it dramatically lowers the cost of a system in long-haul communications. Figure 3–12 shows the advantage of a DWDM system. Not only does it eliminate the need for numerous fibers, it also significantly reduces the number of amplifiers that will be needed along the route.

DWDM has not yet found widespread application in premises cabling systems. But you should be aware of it for two reasons. First, DWDM can find use in campus applications over single-mode cable running between buildings. Second, as optical-based DWDM systems become more widely used, telephone companies may actually deliver specific wavelengths to a business. Instead of having three fibers coming to your business, you could have a single fiber and three separate channels (wavelengths) on the fiber. One proposal for 10-Gigabit Ethernet involves dense wavelength-division multiplexing of four 2.5-Gbps signals.

Plastic Optical Fiber (POF)

It is also possible to make fiber out of various plastic materials instead of glass. In general, POF has much higher attenuation and lower bandwidth than glass optical fiber (which is sometimes referred to as GOF to distinguish it from POF). POF tends to have a much larger core diameter than GOF (typically from about 200 to 1000 microns), which facilitates splicing and connectorization because fiber and connector alignment doesn't have to be nearly as precise. Larger diameter fibers are also much more tolerant of dust and dirt, because a small spec of dust blocks a much smaller percentage of the light than a smaller fiber.

Figure 3–13 compares the size of the core and cladding for some common types of plastic and glass fiber.

POF systems have historically had lower bandwidth and higher cost than UTP systems. Therefore, they have not been widely used except in certain specific markets such as industrial networks and automobiles. In Europe and Japan, there seems to be growing interest in using POF for residential networks.

Abbreviation	Name	Diameter, Microns	
		Core	Cladding
SI POF	Step Index POF	980	1000
GI POF	Graded Index POF	500	750
PCS	Polymer Coated Silica	200	230
PF POF	Perfluorinated POF	120	500
MM GOF	Multi Mode GOF	62.5	125
SM GOF	Single Mode GOF	8 to 10	125

Figure 3–13. Common fiber core and cladding diameters

In recent years, there has been considerable improvement in the performance of POF. Modern plastic fiber is usually made from either PMMA (polymethyl methacrylate) or CYTOP™ (trademark of Asahi Glass) and can deliver more than 1 Gb/s over 100 meters. Fibers made from PMMA and CYTOP are sometimes referred to as acrylic and perfluorinated fiber (PF-POF), respectively.

There have also been advances in optical transceivers for use with POF, such as the resonant cavity LED (RCLED), which can provide greater optical output power and higher bandwidth than a conventional LED. POF systems typically operate at 650 nm, which is in the visible part of the spectrum. PF-POF systems can also operate at 850 or 1300 nm and can therefore use transceivers designed for GOF as well.

Despite these technological advances, POF has so far remained a niche player in the commercial cabling market.

Cables

The fiber, of course, must be *cabled* or enclosed within a protective structure. This usually includes strength members and an outer jacket. The most common strength member is Kevlar Aramid yarn, which adds mechanical strength. During and after installation, strength members handle the tensile stresses applied to the cable so that the fiber is not damaged. Steel and fiberglass rods are also used as strength members in multifiber bundles. Figure 3–14 shows typical fiber constructions.

The first layer of protection is the buffer, available in loose and tight styles. The tight buffer construction is a plastic layer applied directly to the fiber coating. This buffer is typically 250 or 900 μm thick. Widely used for indoor applications, tight-buffer cables provide great crush and impact resistance, but lower isolation of the fiber from extreme changes in temperature.

The breakout cable is a special form of tight buffered cable—cables within a cable. It consists of several individual tight-buffered cables (fiber, buffer, strength member, and outer jacket) with an outer jacket and additional strength members. Useful for both riser and horizontal applications, breakout cables can simplify routing and installation of connectors.

In the loose-buffer cable, the fiber lies within a plastic tube with an inside diameter many times that of the fiber. A tube can hold more than one fiber. The tube is usually filled with a gel material that serves to keep water out. Because the fiber floats within the tube, it is isolated

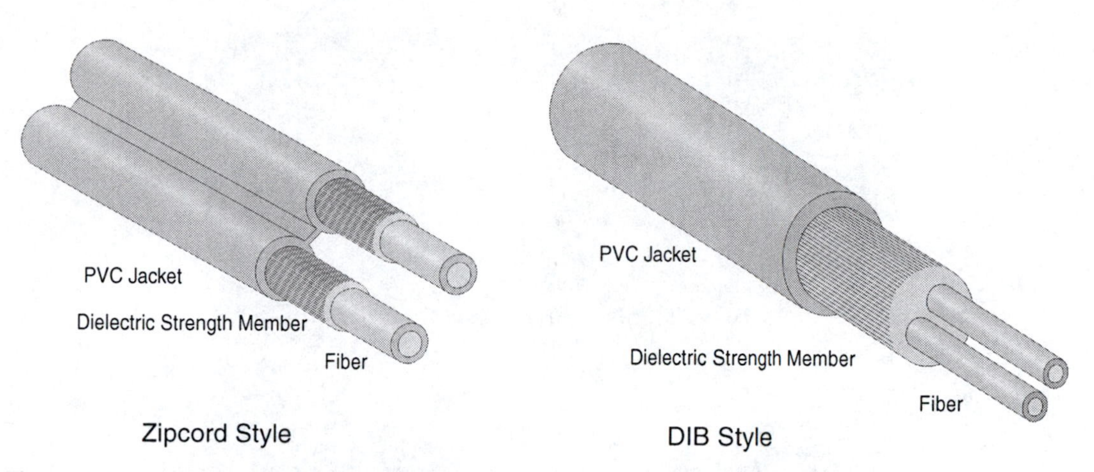

Figure 3–14. Common duplex cable constructions *(Courtesy of Siecor Corporation)*

from external stresses, including expansions and contractions of the cable due to temperature extremes. Widely used in outdoor applications, loose-buffer cables tend to have a larger bend radius, larger diameter, higher tensile strength, and lower crush and impact resistance.

Most multifiber cables have fibers arranged in a circular manner; Figure 3–15 shows a representative multifiber breakout cable. Each ribbon has up to 12 fibers in parallel, with up to 12 ribbons (144 fibers) in a cable. Figure 3–16 shows a ribbon cable.

Duplex cables are used for patch cables and many horizontal cables. For backbone and riser cables, cables with 12 or more fibers are recommended.

Transmitters and Receivers

Getting the Information In and Out

Source and detectors are the fundamental element of the electro-optic interface. The source is either an LED or laser that converts the electrical signal into an optical signal and injects the light into the fiber. The detector is a photodiode that performs the opposite task: it accepts the light from the fiber and converts it back into an electrical signal.

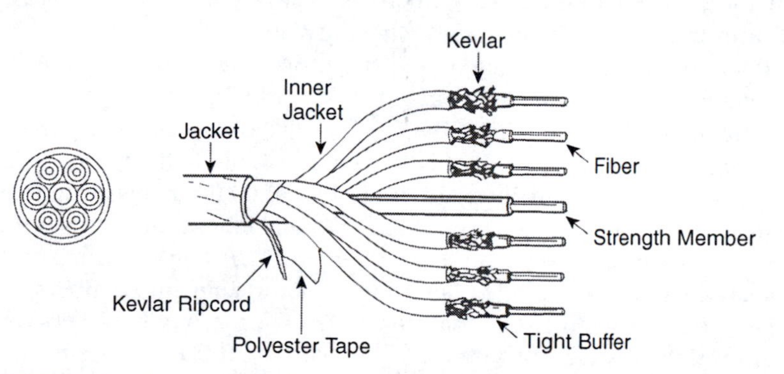

Figure 3–15. Multifiber breakout cables *(Courtesy of Belden Wire and Cable Company)*

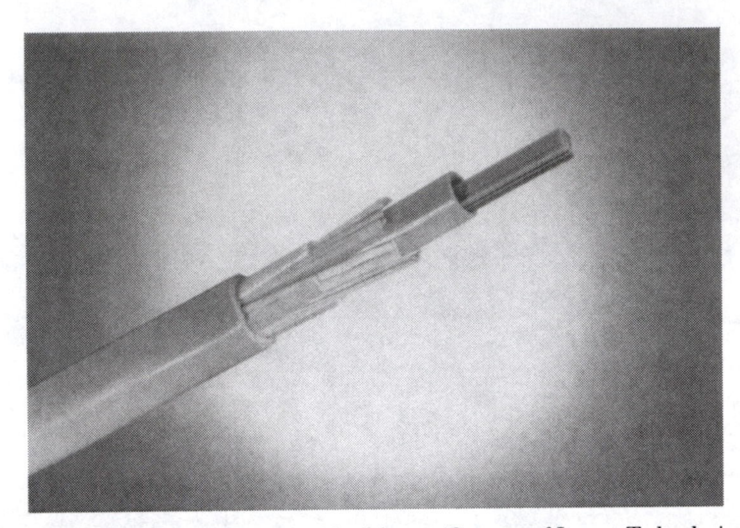

Figure 3–16. Ribbon cable with 144 fibers in 12 ribbons *(Courtesy of Lucent Technologies Bell Labs Innovations)*

Lasers and Leds

The choice of LED or laser for a transmitter depends on the application requirements. Both are small semiconductor chips that emit light when current passes through them. The significant differences between an LED and laser greatly affects the performance of an optical link.

The following are important characteristics of the optical source:

Output Power

How much optical power does the source couple into the fiber? Both LEDs and lasers used in network/premises cabling applications couple tens of microwatts of optical power into a fiber. The amount coupled from a given device depends on the fiber's core diameter and NA. The light from an LED spreads out in a wide pattern so that only a small portion of the total light emitted is actually coupled into the fiber. A fiber with a large core or a higher NA gathers more of this light. For example, an LED might couple 45 μW into a 62.5/125 fiber, but only 35 μW into a 50/125 fiber. The difference is due to the larger diameter and NA of the 62.5/125 cable.

A single-mode fiber requires a laser source. Unlike an LED, a laser produces a very narrow beam of light matched to the small core of the fiber. While lasers used in premises cabling applications couple roughly the same amount of light into a multimode fiber as an LED, they couple much more light into a single-mode fiber. This is because of their narrow beam that more closely matches the small core of the single-mode fiber. The efficiencies of laser sources and single-mode fibers permit longer transmission distances. In an FDDI network, for example, the span between stations is 2 km for multimode fiber and 40 km for single-mode fibers.

An interesting development is the adaption of lasers used in CD players for use in fiber-optic systems. Although CDs use lasers operating at 780 nm, the lasers have been also modified to emit at 850 nm. Both wavelengths are used in fiber optics. While the main attraction of CD lasers is their low cost, their performance characteristics make them well suited to optical communications. They are fast, have a narrow, coherent spectral width, and have good output power. They can outperform a 1300-nm LED at a much lower cost in some applications.

Spectral Width

What is the spread of wavelength in the light? An LED emits a range of light, while a laser emits a single wavelength. Different wavelengths travel at different speeds through a fiber, which contributes to dispersion and thus limits the fiber's bandwidth. LEDs have spectral widths of around 25 to 40 nm. For an LED with a nominal wavelength of 850 nm, a 40-nm spectral width means that the wavelengths actually range from 830 nm to 870 nm. Figure 3–17 compares the typical spectral width and output power of an LED and laser.

Speed

How fast can the device be turned on and off? The speed at which the source can be modulated determines the data rate it can support. A laser can be turned on and off faster than an LED, so lasers are preferred for very high-speed applications.

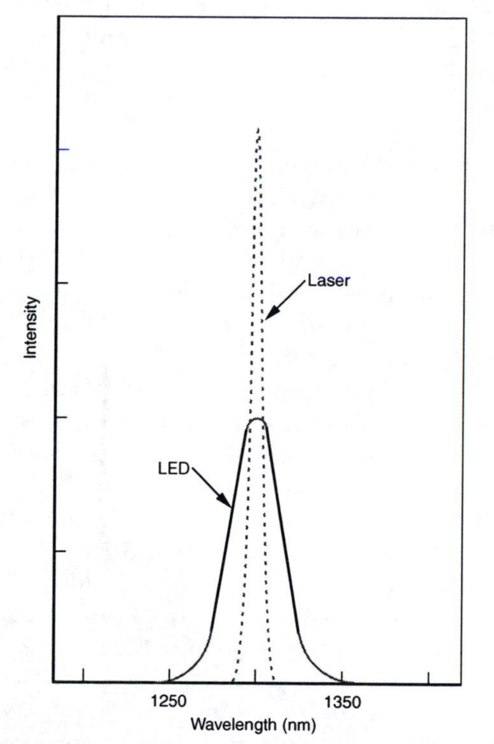

Figure 3–17. Source spectral width *(Courtesy of Siecor Corporation)*

VCSELs

Even more significant is the vertical-cavity, surface-emitting laser, or VCSEL (vik-sell). The VCSEL is one of the newer types of semiconductor lasers, only becoming commercially available in the mid- to late 1990s. But it has several characteristics that make it attractive as a low-cost, high-performance device.

When a semiconductor laser is manufactured, it is "grown" on a circular disk called a wafer. A single wafer can contain hundreds or thousands of individual lasers. Traditional edge-emitting lasers cannot be tested until the wafer is sliced into the individual lasers. Yields—the ratio of good lasers to the total number manufactured—is low. Therefore, a great deal of effort is required to even begin separating the good lasers from the rejects. VCSELs can be tested at the wafer stage so that only good ones need to be separated. This lowers costs at the initial manufacturing stage.

The first crop of VCSELs operate in the 850-nm window and are intended for use with multimode fibers. Future generations will probably operate at 1300 nm and other wavelengths on both single-mode and multimode fibers. One additional benefit of a VCSEL over an edge-emitting laser is that its output is round rather than elliptical. The round output means that most of the output energy is easily coupled into a 50- or 62.5-μm core. While the output beam of the VCSEL is wider than that of a high-performance laser, it still offers exceptional coupling efficiency. Thanks to the VCSEL, 50/125-μm fibers have found renewed interest for higher bandwidth LAN applications. The 62.5/125-μm fiber was preferred because it allowed more light from an LED to be coupled into the fiber. The drawback was a low bandwidth at the shorter 850-nm wavelength. VCSELs offer superior coupling into a 50/125-μm fiber. At 850 nm, a 62.5/125-μm fiber has a bandwidth of 160 MHz-km; the 50/125-μm fiber has a bandwidth of 500 MHz-km. This higher bandwidth means that the 50/125-μm fiber can give better high-speed performance.

One reason the 62.5/125-μm fiber was chosen for premises cabling is that it accepted more optical power from a LED. It was thought that the 50/125-μm fiber was not efficient enough for use with low-cost LEDs. The VCSEL changes all that. The small, round output pattern of the VCSEL means that it works very well with either 62.5/125-μm or 50/125-μm fibers. Figure 3–18 compares the output patterns of LEDs, VCSELs, and edge-emitting lasers.

An edge-emitting laser has a spot size at the fiber of only 8 to 10 μm, making it well matched to the very small core of a single-mode fiber. At the other extreme is the LED, which has a spot size of over 100 μm (and often well over), which is larger than the core of 50-μm or 62.5-μm cores of multimode fibers. This condition, known as an overfilled fiber because the light floods both the core and the cladding, is what causes the lower efficiency of the fibers at high bandwidth and limits the transmission distance.

The VCSEL, with its 30 to 40-μm spot size, is well matched to multimode fiber. Not only is the VCSEL fast to permit speeds of several gigabits, it underfills the core. This condition helps allow longer transmission distances. The narrow spectral width of the VCSEL also increases transmission speeds, because the effects of material dispersion are significantly lowered.

You can think of a VCSEL as bridging the gap between LEDs and lasers. They have the lower costs and compatibility with multimode fibers that you associate with LEDs. They have the speed, narrow spectral width, and narrow output pattern you associate with lasers. As such, they are having significant impact on the economics and performance of systems, especially in the area of high-speed networks.

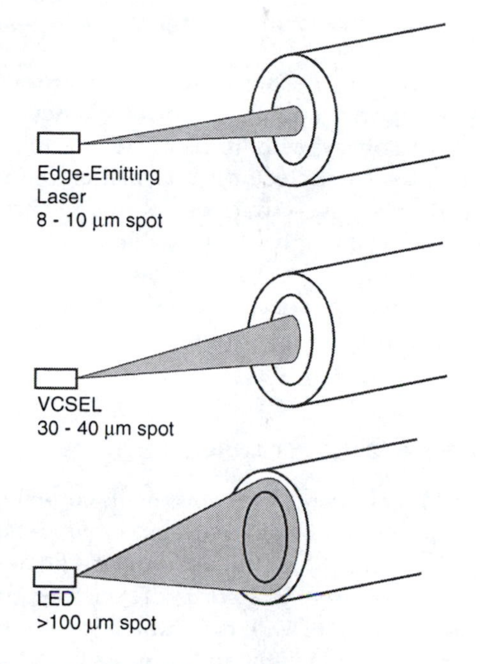

Figure 3–18. VCSELs are well matched to multimode fibers

Resonant Cavity LEDs

Another type of optical source, the resonant cavity LED (RCLED), is even newer than the VCSEL. The RCLED consists of an active region placed in a very thin optical cavity. RCLEDs have a number of advantages:

- An RCLED is smaller, simpler to fabricate, and lower in cost than a VCSEL.
- They have a narrower spectral width and smaller spot size than a standard LED, although not as good as a VCSEL.
- RCLEDs can be modulated at rates of up to about 1 GHz, which is faster than a standard LED, but slower than a VCSEL.
- The power output of an RCLED has very low temperature dependence and is therefore suitable for use over a wide range of ambient temperatures.
- RCLEDs are very efficient and are a good choice for low-power applications.

RCLEDs were first introduced at 650nm (in the visible light range) for use with POF. They are now being commercialized in other wavelengths as well.

Detectors

At the other end of the link is the detector, which serves to convert the optical signal back to an electrical signal. The detector must be able to accept highly attenuated power—down in the nanowatt levels. Subsequent stages of the receiver amplify and reshape the electrical signal back into its original shape. Some detector packages have a preamplifier built in to boost the signal immediately.

Detectors operate over a wide range of wavelengths and at speeds that are usually faster than the LEDs and lasers. The important figure of merit for a detector is its sensitivity: what is the weakest optical power it can convert without error? The signal received must be greater than the noise level of the detector. Any detector has a small bit of fluctuating current running through it; this minuscule current is noise—spurious, unwanted current. The minimum power received by the detector must still be enough to ensure that the detector can clearly distinguish between the noise and the underlying noise. This is expressed as a signal-to-noise ratio or a bit error rate. SNR is a straightforward comparison of the signal level and the noise level, while bit error rate is a more statistical approach to determining the probability of noise causing a bit to be lost or misinterpreted.

Transmitters, Receivers, and Transceivers

Packaged transmitters and receivers offer a convenient, cost-effective method of achieving the electro-optic interface. Typically in small, board-mounted packages, transmitters and receivers include the necessary electronics to permit building of a fiber-optic data link. Figure 3–19 shows 125-Mbps and 155-Mbps transceivers used for high-speed network applications.

A transmitter accepts standard TTL or ECL data, which it converts to light for transmission through the fiber. The receiver accepts the light and converts it back to TTL or ECL signals.

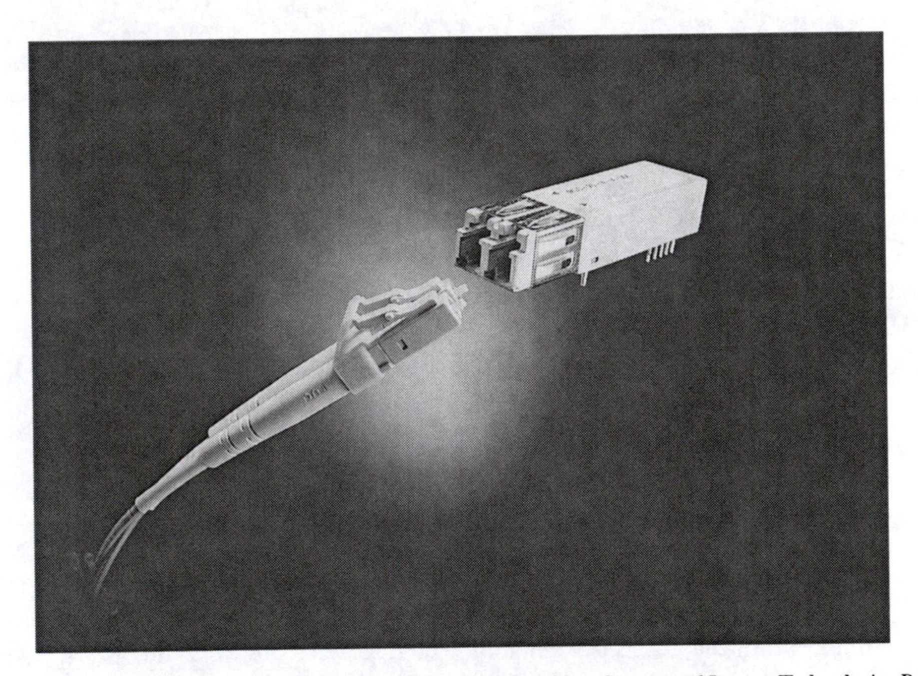

Figure 3–19. Fiber-optic transceiver (right) and cable assembly *(Courtesy of Lucent Technologies Bell Labs Innovations and Methode Electronics)*

Today's packages are easier to use for a number of reasons. First, package dimensions and pinouts have been standardized so that packages from different vendors can be used interchangeably. Second, specifications for using the devices make building links easier. For example, the output power of a transmitter is now given as the power launched into a given type of fiber. Not too long ago, manufacturers listed only the total output power, plus some specs like NA and diameter. The designer was left to calculate and verify the actual power launched into the fiber that was being used. Third, transmitter and receiver packages are offered in matched pairs. That is, you can buy an FDDI set, an ATM set, or an Ethernet set and know that they will meet the published specifications for achieving the speeds and distances allowed by the standards.

Link Power Margin

The difference between the power the source launches into the fiber and the minimum sensitivity of the detector defines the link power margin. Suppose, for instance, that the source launches –19 dBm into the fiber and the receiver sensitivity is –33 dBm. The link power margin is 14 dB. During transmission from end to end, from transmitter to receiver, the signal must not lose more than 14 dB of optical power. Sources of loss include not only fiber attenuation, but losses associated with interconnections at patch panels, wall outlets, and so forth. In addition, a prudent power margin reserves 3 dB for aging of the source.

Most networking and building cabling standards define a power margin. These margins are routinely achievable within the physical limits allowed by the standards. That is, the standards recommend cable distances and the number of interconnections along the path. By following these guidelines, the link should be well within the power margin. Fiber-optic networks typically have a link margin of about 11 dB.

A convenient way to look at a link budget is by graphing the losses along the path. Figure 3–20 shows a simple power margin graph. Fiber attenuation can be estimated by assuming loss is linear

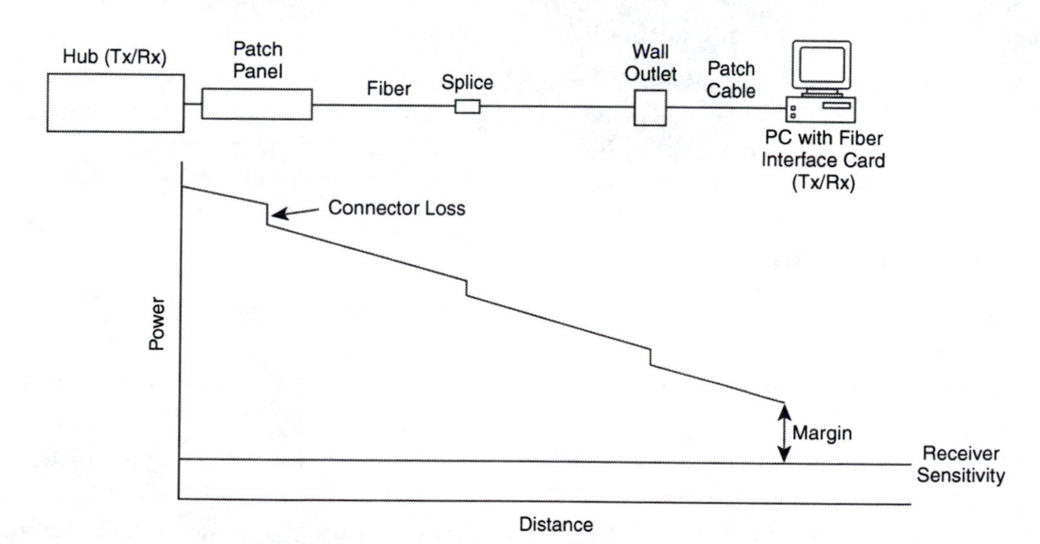

Figure 3–20. Link power budget

per unit length. Thus, if the attenuation is 1.5 dB/km, then the loss is 0.75 dB for a 500-meter run or 0.15 for a 100-meter span. Standards allow 0.75 dB loss for each interconnection. Allow 3 dB for aging. A quick addition of sources of loss in the link will allow a budget to be quickly calculated. In most cases, a measurement of link losses will show losses that are lower than this simple rule-of-thumb addition. One reason is that most interconnections provide well under 0.75 dB of loss, so that your quick addition is for worst-case conditions. If the measured loss is higher than the calculated loss, then there is probably something wrong with the installation.

Efficiency is nice; it is like getting as much optical power as possible from one end of the link to the other. Along the way, of course, there will be interconnections. Each interconnection will contribute some loss. The role of the connector or splice is to provide a mechanical coupling mechanism that minimizes loss.

Alignment: The Key to Low Loss

Interconnection is a critical part of a fiber-optic link. The job of a connector or splice is to provide low-loss coupling of light through a juncture. Low loss is the result of precise alignment of the two fibers being joined. Any misalignment of the two fiber cores increases the loss.

Losses stem from three sources:

1. Tolerances in the connector. As with any device, perfection is in the control of tolerances. A fiber-optic connector has tolerances associated with it that tend to allow some small amount of misalignment.
2. Tolerances in the fiber. The fiber should have a perfectly round core centered exactly within a perfectly round cladding. As these depart from perfection, loss is increased.
3. Different fiber types. Loss is increased if the receiving fiber has a smaller core or NA. In most premises cabling application, the same type of fiber is used throughout the installation, so these losses are not a concern. Such mismatches can contribute considerable loss. For example, coupling light from a 62.5/125-μm into a 50/125-μm can add up to 6 dB additional loss from the mismatches in core diameter and NA.

If tight connector and fiber tolerances are critical to low-loss interconnections, the good news is that both connectors and fiber are made with a high degree of precision. Today's connectors offer an insertion loss of around 0.1 to 0.3 dB—well under the 0.75 dB permitted by network standards. This means that connector performance far exceeds applications requirements.

Flammability Ratings

The NEC® establishes flammability ratings for fiber cables just as it does for copper. There are two different classes of fiber-optic cable:

- Nonconductive—cable that does not contain any metallic or other electrically conductive materials.
- Conductive—cable that contains conductive elements, such as metallic strength members, metallic vapor barriers, or metallic armor.

Figure 3–21 lists the different flammability ratings of fiber cable and the permitted substitutions. The substitutions follow the same hierarchy as for copper cables (plenum cable can be

Cable Type	Description	Permitted Uses	Substitutions
OFNP	Nonconductive plenum	Anywhere, including air ducts and plenums	None
OFCP	Conductive plenum	Anywhere, including air ducts and plenums	OFNP
OFNR	Nonconductive riser	In vertical shafts, and general use	OFNP
OFCR	Conductive riser	In vertical shafts, and general use	OFNP, OFCP
OFNG OFN	Nonconductive general-purpose	General use except for risers and plenums	OFNP, OFNR
OFCG OFC	Conductive general-purpose	General use except for risers and plenums	OFNP, OFCP, OFNR, OFCR, OFNG, OFN

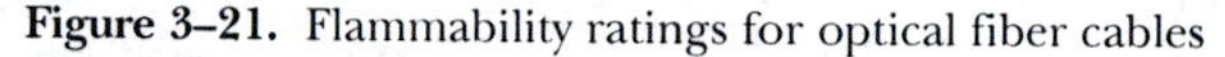

Figure 3–21. Flammability ratings for optical fiber cables

substituted for riser cable, which can be substituted for general-purpose cable) with the additional restriction that conductive cable can never be substituted for non-conductive cable.

Fiber Versus Copper: The Debate

Fiber has become the dominant medium for backbone connectivity. However, in the horizontal subsystem, copper accounts for almost all of the worldwide installed base. For years, a debate raged about whether fiber would take over the horizontal also (often known as fiber to the desk, or FTTD.) This debate has been pretty well settled—in the horizontal cable market, UTP has scored as decisive a victory as Ethernet has achieved in the LAN market. To wrap up this chapter, let's look at this issue in a little more detail.

Pro-Copper Arguments

- Copper-based systems, especially UTP, are less expensive to buy and to install. This issue centers on the cost of the electronics, not the cost of the cable and related hardware. The electronics for fiber-based systems are more expensive than electronics for copper-based systems. The gap is narrowing, but copper retains an edge.
- UTP cables are well understood, with years of proven experience. Both craftspeople and users are comfortable with the technology.
- UTP has proven to be capable of meeting the evolving demands of networks, and there is no reason to believe that it cannot continue to do so. Witness the ability of Category 5e cable to carry gigabit data rates and the emergence of Category 6 and 6A for even more robust performance.
- UTP supports the widest range of applications, from Gigabit Ethernet to ordinary telephones and fax machines.

- Fiber is a niche product. It's best applied in building backbones, in long spans between buildings, and in other situations that clearly require fiber.
- Fiber to the desktop is overkill—too costly and with more speed than most users need. Who needs more than a gigabit to the desktop? And even if you do, you can achieve it with low-cost UTP.
- Fiber has not lived up to its promise of unlimited bandwidth. Indeed, 62.5/125-μm fiber could not handle Gigabit Ethernet at the full 300-meter distances allowed by premises cabling standards. Now it seems the issue is more complex.
- Fiber is tricky stuff to work with.

Pro-Fiber Arguments

- Fiber is clearly superior technologically. Installing fiber ensures performance as even higher speed networks emerge in the future. Fiber optics future-proofs your building.
- There is no assurance that Category 5e UTP can continue to meet emerging needs. Now we are greeted with Categories 6 and 6A—and even 7—in a seemingly endless procession of new cable categories.
- Fiber-based systems are only slightly more expensive to buy and install. But because it is more reliable, the lifetime costs of the cabling system may prove to be lower than with UTP. For example, the costs associated of network interruptions, downtime, and repairs will be much lower because there will be fewer disruptions. Besides, prices for fiber-optic components are steadily declining and should continue to do so.
- Fiber gives you more flexibility in how your network is constructed. While UTP is limited to 90 meters in high-speed runs, fiber can run up to 300 meters. This increased distance allows new network configurations that can save money in equipment costs, in administration and maintenance, and in long-term reliability.
- While Gigabit Ethernet created a small bump in fiber's ability to meet premises cabling needs, it by no means stopped the train. Newer transceiver technology and improved fibers make fiber even more attractive. Newer fiber can carry 10 Gbps over the 300 meters!
- Fiber to the desktop is not overkill. The history of computers and networking demonstrates that demand for higher performance is insatiable. Fiber is the better bet to future-proof your building. Where people thought 10 Mbps to the desktop was perfectly adequate only a few years ago, today's common connection to the desktop is ten times that—100 Mbps.
- Fiber has more room for innovation, for higher performance, and for newer lower cost components.
- At gigabit data rates, transmitter and receiver electronics for copper-based systems are becoming increasingly complex and expensive. The cost differential between copper and fiber is lessening.
- Fiber is no trickier than high-performance UTP to install.
- Category 6 will probably be the last generation of UTP. As we move to higher and higher gigabit data rates, fiber makes more sense economically and technologically.

Who's right? Both sides, of course, have valid and not so valid points when they push their arguments to extremes. For example, fiber hardware, such as a wall outlet, was once significantly more expensive than a copper-based outlet. Today some vendors price both copper and Category 6 hardware nearly identically. Arguments over economics are slippery because they can change rapidly. What is clear is that at Gigabit Ethernet speeds and below, copper has a price advantage, although fiber is finding wider use in backbone and special applications. In high-speed systems of several Gb/s or more, fiber makes more sense. Copper gigabit systems are available, but fiber systems are becoming less and less expensive. The murky area is around 10 Gbps.

SUMMARY

- An optical fiber offers higher bandwidth and lower attenuation than a copper cable, allowing higher data rates over longer distances.
- The two main types of fiber are multimode and single mode.
- Single-mode fiber has incredibly high bandwidth.
- 62.5/125 and 50/125 multimode fibers are the preferred multimode fibers in premises cabling.
- LEDs have slower switching speeds and wider spectral widths than lasers. Hence, LEDs are typically used for lower-bandwidth applications.
- Lasers are required for single-mode fibers.
- A VCSEL is a new type of low-cost laser that is well matched to both single-mode and multimode fibers.
- A resonant-cavity LED (RCLED) is another new type of optical source that is intermediate between a standard LED and a laser.
- Fiber is widely used in the backbones, but not as commonly to the desk (UTP is the dominant media for horizontal cabling).
- Dense wavelength-division multiplexing (DWDM) refers to using many colors (wavelengths) of light to provide different communication channels on the same fiber.
- Great improvements have been made in the performance of plastic optical fiber (POF), but so far, it remains a niche player in the commercial cabling market.

REVIEW QUESTIONS

1. What principal material is used as the conductor in an optical fiber cable?
2. Name the two main parts of an optical fiber. Which part propagates the light?
3. What are the two *main* advantages of an optical fiber over UTP?
4. While there are many types of fiber, we generally distinguish two main categories. What are they?
5. What are the two optical operating windows used by fiber optics in premises applications?

6. What is the link power margin?
7. What are the two sizes of multimode fiber most widely used in premises applications?
8. Why is a VCSEL of great interest in premises applications?
9. Horizontal UTP runs have a maximum length of 90 meters. What is the maximum recommended length for multimode fiber?
10. What is the principal drawback to fiber optics in premises cabling?
11. What is laser-optimized multimode fiber?
12. Explain the difference between OFNP and OFCP fiber cable. Which can be substituted for the other?
13. What is an RCLED and how do its properties compare with lasers and LEDs?
14. What limitations are encountered when using standard multimode fiber to support a 10-Gigabit Ethernet network?
15. Briefly state the main advantage and disadvantage of POF.

CHAPTER 4

Connectors and Interconnection Hardware

Objectives

After reading this chapter, you should be able to:

- List the main types of connectors used in premises cabling.
- Distinguish between the main connecting functions of copper connector and fiber-optic connectors.
- Define the roles of connecting hardware in equipment cross connects and the user area.
- List critical application guidelines for copper and fiber connectors.
- Describe how connectors are used in equipment closets and user outlets.

Connectors and other interconnection hardware play an important role in premises cabling because they are the "glue" that holds everything together. Equally important, they provide the flexibility that allows the moves, adds, and changes that are part of the building's life.

There are several main types of connecting hardware for each major type of cabling.

- For UTP, there are connectors (especially the modular plug and jack), patch panels, punch-down blocks, and cords.
- For fiber, there are connectors, splices, patch panels, and cords.
- For coaxial cable, there are connectors, patch panels, and cords.

Attaching a connector to the end of a cable is referred to as ***terminating*** the cable. In this chapter, we will first discuss methods of terminating UTP, fiber, and coaxial cables, then proceed to discuss the other types of connecting hardware.

Terminating a Copper Cable

The key to terminating a cable properly is to achieve an intimate gastight joint between the connector contacts and the cable conductor. Ideally, the connector will become a transparent

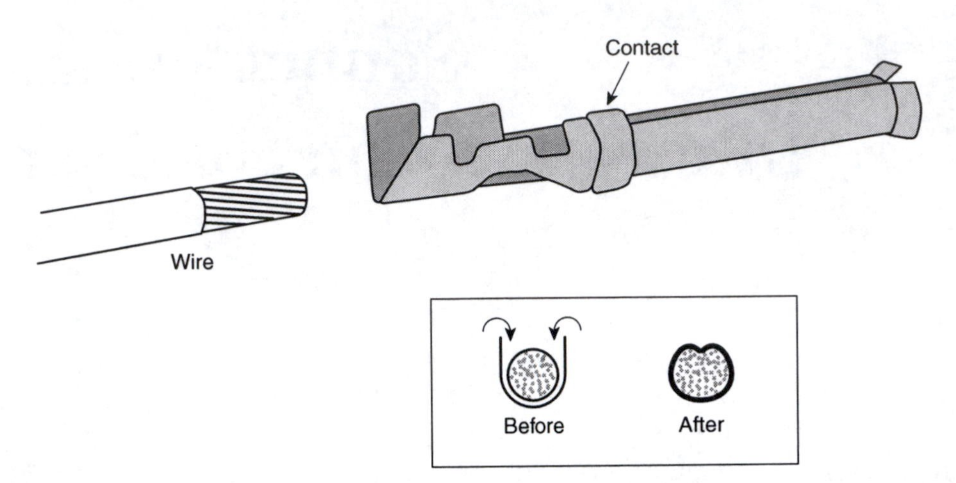

Figure 4–1. Crimp *(Courtesy of AMP Incorporated)*

extension of the cable. In reality, it does not. It increases the resistance a bit, can cause impedance discontinuities, can present a point of mechanical or electrical failure, and can add crosstalk. The gastight joint is important to ensure the long-term reliability of the connector by preventing moisture, gases, humidity, and other environmental effects from degrading the connection. Degradation comes, most often, in the form of increased resistance.

There are two main approaches to terminations applicable in premises cabling: crimping and insulation displacement (IDC).

With both crimping and IDC, it is important to use the correct size and type of wire and the correct tool. A contact, for example, might be designed for 22- to 24-gauge stranded conductors. Using a solid conductor or a 20- or 26-gauge conductor will result in a termination of dubious value—one more likely to fail.

Crimping

In the crimp, a portion of the contact is "crushed" under great pressure around the conductor. The conductor and contact fuse in a cold weld that provides the required electrical and mechanical properties. Crimping tools are designed to apply the proper pressure by closing the dies a fixed amount. An undercrimp results in high resistance and a loose connection. An overcrimp can damage the wire or contact so that they fail mechanically. Figure 4–1 shows the cross section of a crimp.

Insulation Displacement

The insulation displacement contact uses a slotted beam. A wire is driven between the slot, the beams pierce the insulation and deform the conductor. As with the crimp, an intimate gastight joint is formed. The beams maintain a residual spring pressure on the conductor to achieve long-term reliability. One advantage of IDC terminations is that the wire insulation does not have to be removed.

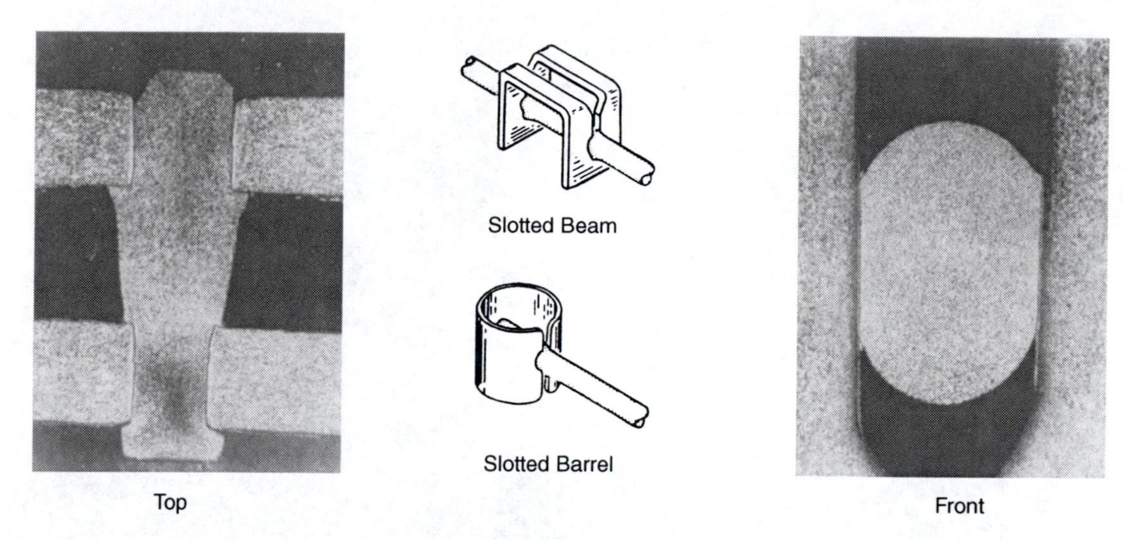

Figure 4–2. IDC termination *(Courtesy of AMP Incorporated)*

An IDC contact can either be a flat formed plate or a barrel. Some barrel-shaped IDC contacts allow more than one wire at a time to be terminated. IDC terminations are the most common in premises cabling applications. Figure 4–2 shows typical IDC contacts, as well as cross-sectional microphotographs of the deformed conductors.

Modular Jacks and Plugs

Modular plugs and jacks (Figure 4–3) are familiar to everyone as the ubiquitous "telephone connectors" that are found on telephone handsets, cords, and wall outlets. Although some residential telephones use four- or six-position modular connectors, for commercial premises, an eight-position connector is used to terminate all four pairs of the UTP cable. The modular plug is typically used on wires and cords, while the modular jack appears on equipment (phones, LAN hubs, and so on) and in telecommunications outlets.

Modular plugs and jacks are often referred to as *RJ* connectors from USOC specifications defining their use. An RJ-11 is a 6-position connector; an RJ-45 is an 8-position connector. As commonly used to indicate simply connectors, both terms are technically wrong. RJ11 and RJ45 refer to specific connectors, wiring patterns, and applications within the telephone system. To the telephone person, they have narrow, precise meanings. However, in the cabling industry, the terms RJ-11 and RJ-45 have been appropriated to refer to any six-pin or eight-pin modular plug/jack, respectively. This terminology has gained wide acceptance because "RJ-45" is a lot easier to say or write than "eight-pin modular connector." In this book, we will often use the RJ-45 terminology, but will seldom refer to RJ-11s because they are not used in commercial premises cabling.

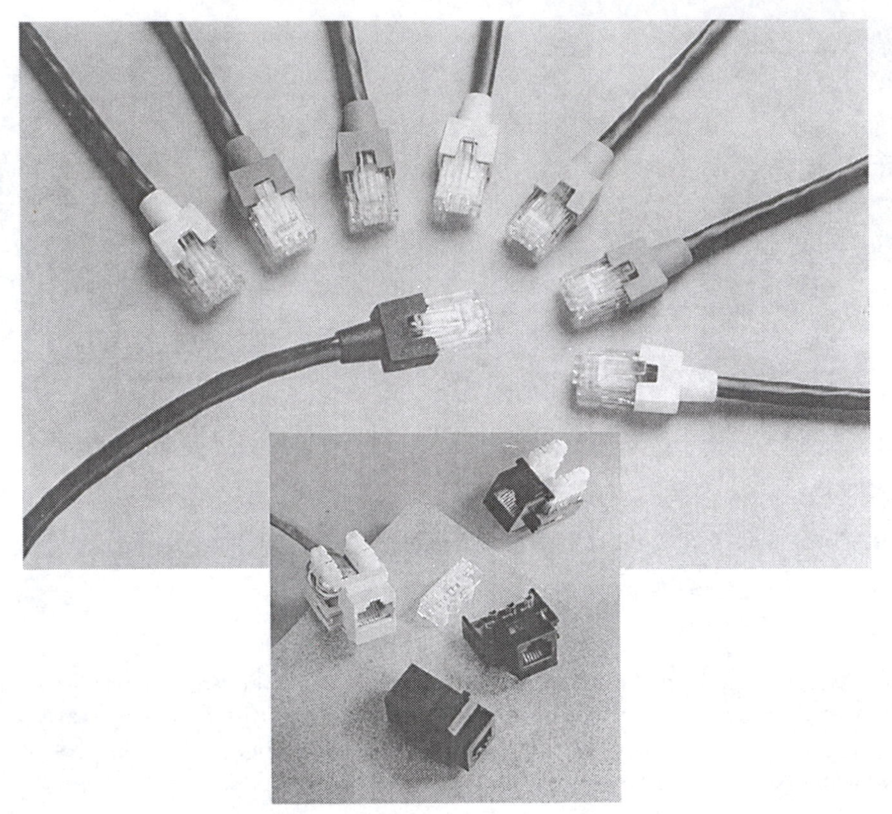

Figure 4–3. Modular plugs and jacks *(Courtesy of AMP Incorporated)*

Modular plugs use IDC terminations

In using modular plugs and jacks, here are some important things to remember:

- You must have plugs and jacks rated to the category of cable you are using. This means that a Category 5e application requires a Category 5e or better plug and jack.
- Plug contacts sometimes differ depending on whether they are used with stranded or solid conductors. Some connectors can terminate both stranded and solid conductors; others can only be used with one type or the other. Do not use the wrong type of plug connector for the cable you are using. While most premises applications use solid conductors, some patch cables use stranded conductors because of the greater flexibility.
- Follow the termination procedure carefully. Especially with Category 5e and higher cables, the proper installation procedures are essential for achieving good high-frequency performance.
- Avoid closely spaced connectors. When designing a cabling system, consideration should be given to the effect of spacing connectors too closely along the cable path. If two connectors of identical design (for example, two mated modular plug/jack combinations) are placed close together in a link, their crosstalk will tend to add in-phase and give worst-case results. If these same connectors are separated by 10 meters or more of cable, the crosstalk does not

tend to add in-phase, and the link/channel performance is improved. For this reason, the use of three or more connectors at one end of a link should be avoided.

Compatibility

The idea of compatibility has taken on increasing importance with Category 6 components. Compatibility falls into two areas:

1. *Interoperability.* Components from one manufacturer must be able to be used with components from another manufacturer and still achieve rated performance. In other words, you should be able to achieve Category 6 performance if you use a cable from company A, a modular plug from company B, and a modular jack from company C.
2. *Backward compatibility.* If you mix different categories of products, you should achieve the performance of the lowest rated category—and not something less. If you connect a Category 5e cable assembly into a Category 6 jack, you should achieve Category 5e performance.

Although it seems simple at first glance, backward compatibility implies several things. First of all, applications that used to work should still work (for example, 10BASE-T). This means that the standard eight-pin modular jack interface must be used at the work area and equipment connections.

Second, if a new Category 6 component (patch cord, for example) is inserted in an existing Category 5e link or channel, it must continue to meet the Category 5e specifications. This is primarily an issue with connecting hardware. Because connecting hardware specifications are given for a mated plug/jack combination, the connector designer has some leeway in determining the crosstalk profile of the plug over the frequency range. The jack must then be designed to cancel most of the plug's crosstalk, leaving the crosstalk of the mated plug/jack within the Category 5e specifications. If the crosstalk profile of the Category 6 plug is substantially different, then the Category 6 jack must have different compensation to cancel the crosstalk to the much more stringent Category 6 mated plug/jack specifications. When a Category 5e plug is connected to this Category 6 jack, unless the design was very carefully done, it will not adequately cancel the crosstalk and the mated combination may only meet Category 3 specifications, for example.

This is not just an abstract theoretical problem, but a very real practical consideration. When Category 6 cabling was first introduced, there was no installed base of data communication equipment with Category 6 jacks built in. Therefore, every Category 6 link was initially connected to a Category 5e jack in the equipment, as shown in Figure 4–4a. If the performance of the Category 5e jack on the equipment and the Category 6 plug on the cord does not meet the backward compatibility specifications, then the operation of the entire channel is jeopardized—adding a Category 6 cable could actually result in a Category 3 system.

If you attempt to get around this problem by using a Category 5e cord, then the exact same problem resurfaces at the telecommunications outlet, as shown in Figure 4–4b.

The same sort of problems can also arise when you use components from different manufacturers. Designs are "tweaked" to adjust the electrical characteristics of the connector, to control capacitance, inductance, crosstalk, and impedance variations, and to achieve the proper level of performance. Traditionally, most of this tweaking was done in the jack. Therefore, any Category 5e plug could be plugged into any Category 5e jack to achieve a Category 5e system. With Category 6, both the plug and jack were modified. But if one manufacturer modified its

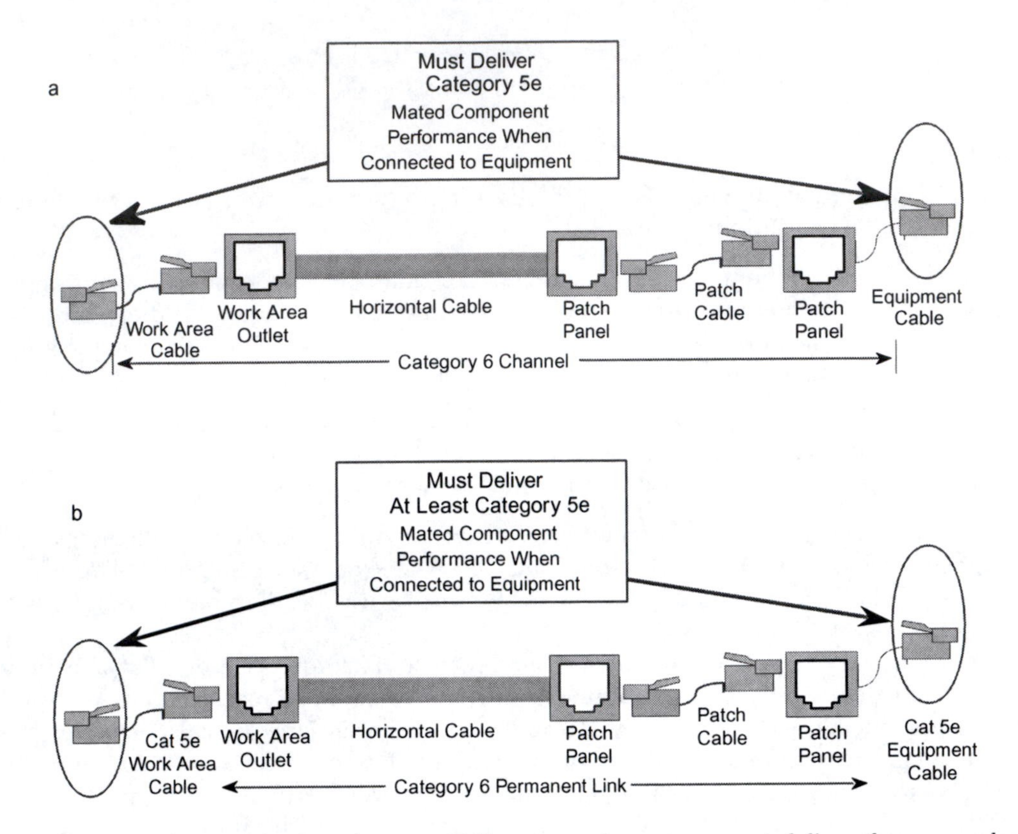

Figure 4–4. The issue of backward compatibility means the system must deliver the proper level of performance in mixed-category systems

plugs and jacks one way and another manufacturer modified its components another way, there was no guarantee that the components would interoperate at the rated level of performance. In the pre-standard days, interoperability was not a major concern—simply having components that met Category 6 performance was. With the official publication of the Category 6 addendum to TIA/EIA-568B.2, interoperability is a requirement for claiming Category 6 compliance.

Of course, manufacturers like to sell *systems,* encouraging users to install only their components throughout the system. This argument is sensible, but self-serving if the manufacturer does not offer backward compatibility and interoperability. Most premises cabling installations use components from a single vendor—or at least those components that determine performance, like connectors and cable. But lack of compatibility locks you into a single vendor. You can't shop around for a patch cable, for example. (Notice that the system can sometimes involve two companies: a connector company and a cable company may join forces to offer the end-to-end system. But we will still consider this situation to represent a single vendor.)

There are many good reasons for sticking with a single vendor. A system's approach simplifies specifying and buying components. There is less hassle and less finger-pointing if things go wrong. Even so, the prudent plan is to insist on compatibility. That way, you're not locked into a single vendor. And you're not left out in the cold if the vendor decides to exit the business.

UTP Connector Performance Requirements

As with cable, connectors are also subject to performance requirements to ensure their suitability to the applications. Performance requirements for UTP connectors are given in TIA-568-B.2 (for Categories 3 and 5e) and TIA-568-B.2-1 (for Category 6). Category 6A connecting hardware requirements will be in TIA-569-B2-10 when it is issued. Figure 4–5 shows NEXT and FEXT loss specifications for Category 3, 5e, and 6 connecting hardware.

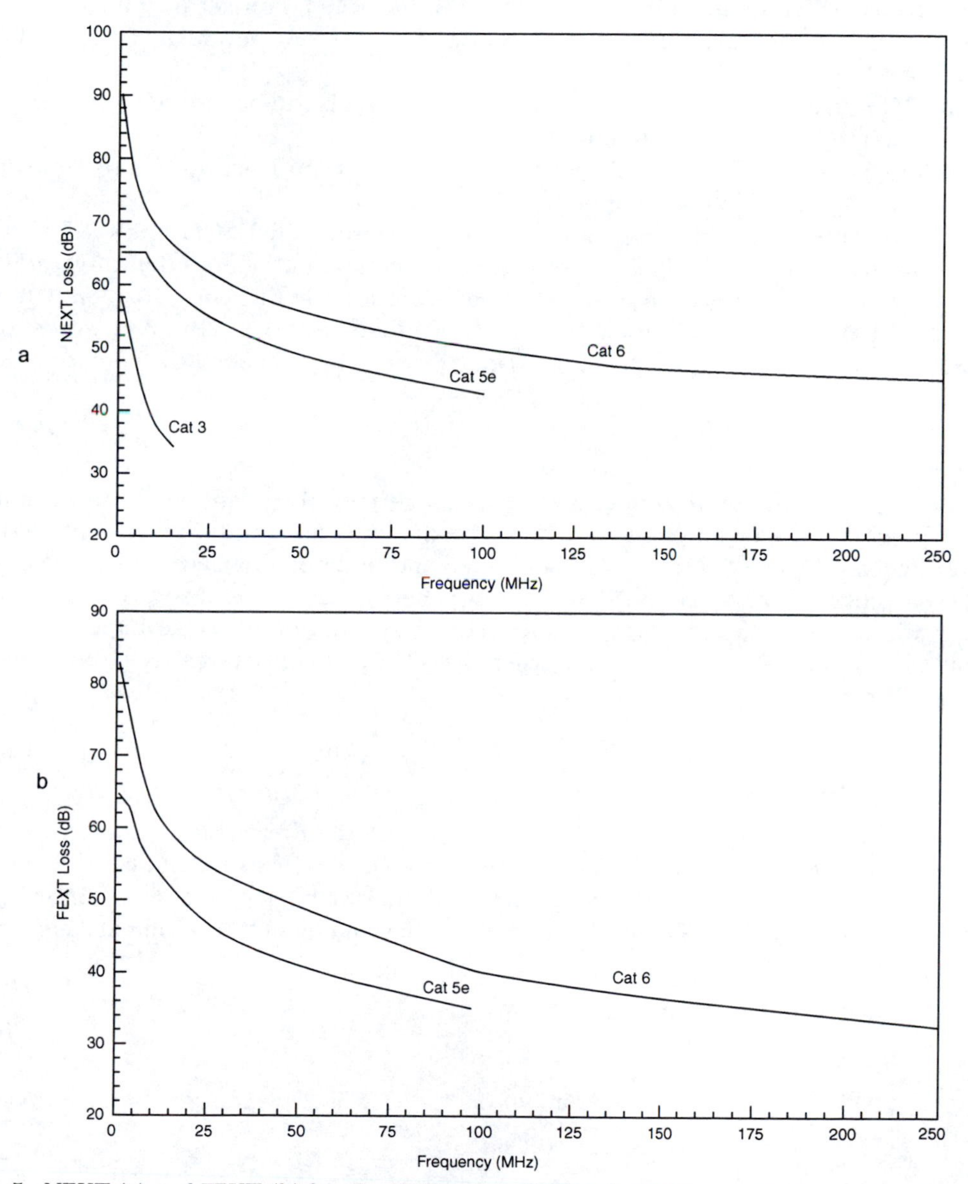

Figure 4–5. NEXT (a) and FEXT (b) loss for UTP connecting hardware

Category 6A NEXT and FEXT specifications will extend to 500 MHz, but will be the same as Category 6 up to 250 MHz.

Modular jacks and plugs are also available in shielded versions for use with ScTP cable.

Pinouts

Connector pinouts refer to the assignment given to each contact in the connector. In other words, which conductor of the UTP goes into which contact of the cable. Actually, there are three standard pinout configurations for UTP and modular jacks, as shown in Figure 4–6.

USOC RJ61X. This configuration allows a 6-position plug to be inserted into an 8-position jack and still make the connection on Pairs 1, 2, and 3.

AT&T 258A/T568B. Originally devised by AT&T, this configuration is also recognized by TIA/EIA-568 as an optional wiring pattern (hence the B).

T568A. This standard represents a compromise between the USOC and AT&T wiring patterns and standards. It is backward compatible with equipment of both the USOC and AT&T style. It is the preferred pattern in TIA/EIA-568B. Notice that the TIA/EIA-568B pattern and the AT&T pattern are identical except for the assignment of pairs. In other words, you can convert from a T568A to a T568B wiring pattern merely by relabeling the contacts.

Keying

Keying refers to methods to ensure that only certain plugs can mate with certain jacks. Some plugs have a key on one side or the other; mating jacks must have a similar cutout to accept the key. Figure 4–7 shows standard keyed and unkeyed interfaces of jacks. Notice that unkeyed plugs will also mate with keyed plugs. Keying can be used in premises cabling to clearly differentiate different types of lines. The most common use of keying is in the Digital Equipment DEC connect system. This system uses key jacks for data ports and unkeyed jacks for telephone ports.

Wire All 8 Positions

Some applications use only two pairs (four wires). But different applications use different connector positions. To achieve a performance-driven cabling system that can be used with any application, it becomes important to terminate all four pairs. For example, 10BASE-T, Token Ring, and TP-PMD (for UTP on FDDI) all use different contacts to terminate cable pairs as shown in Figure 4–7.

10BASE-T:	Pins 1-2 and 3-6
Token Ring:	Pins 4-5 and 3-6
TP-PMD:	Pins 1-2 and 7-8
ATM:	Pins 1-2 and 7-8
100BASE-T:	Pins 1-2 and 3-6

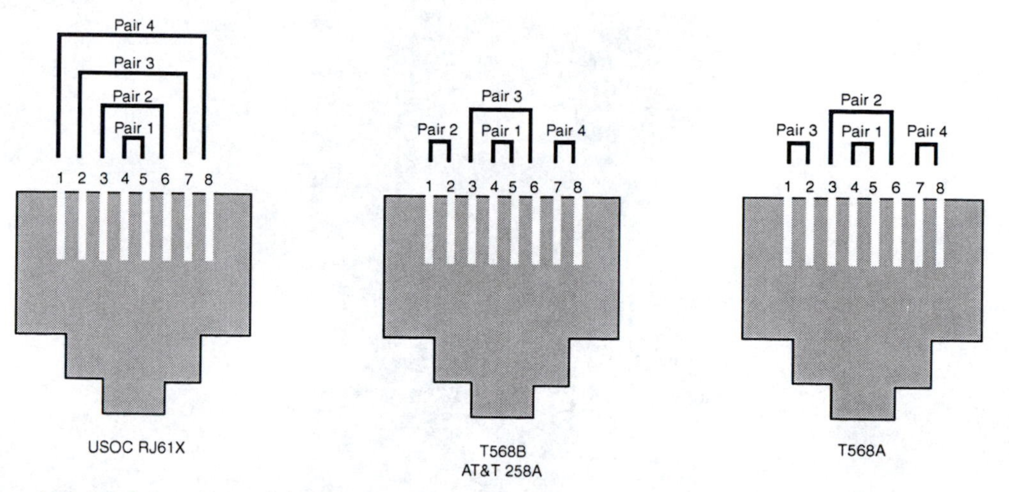

Figure 4–6. Modular plug and jack pinouts

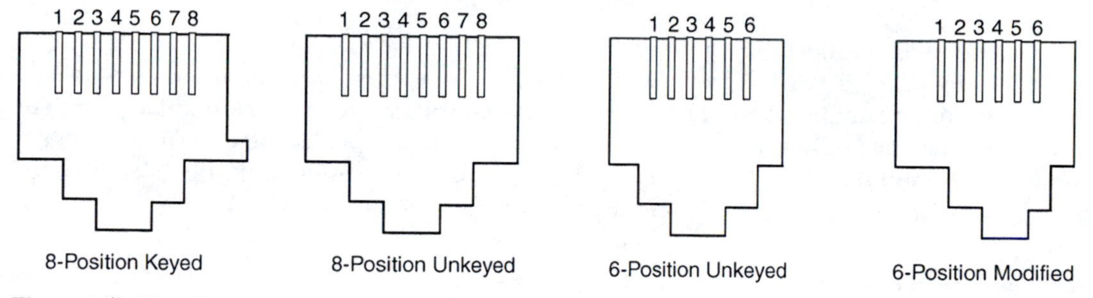

Figure 4–7. Keyed and unkeyed modular jack interfaces

A few years ago, when most applications only used two pairs like the ones listed above, a single UTP cable would sometimes be split among two modular jacks, with two pairs terminated on each jack. Although this violates the TIA-568 standard, you could sometimes get away with it. But modern applications, such as 1000BASE-T, IEEE 1394c, and 10GBASE-T use all four pairs. The bottom line is:

> ***Don't even think of splitting a UTP cable between two jacks! Follow the standard and hook up all four pairs of every jack.***

Data Connectors

Data connectors (Figure 4–8) are 4-position connectors used with Type 1 STP in Token Ring networks. They use a hermaphroditic interface—that is, both halves of the connector have the same mating face.

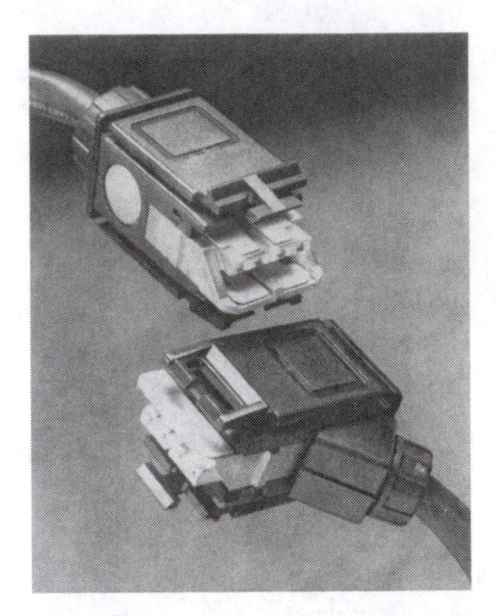

Figure 4–8. Data connector *(Courtesy of AMP Incorporated)*

Data connectors short the transmitter and receiver pins when they are disconnected. This automatically wraps the signal through the connector to preserve the circuit when a cable is disconnected. The connectors use IDC barrel contacts and a stuffer cap to terminate pairs. The cable shield terminates in a metal ferrule. While data connectors are somewhat large and bulky compared to modular jacks, they are rugged and perform well. Although originally designed to operate at Token Ring speeds, newer versions are rated to 300 MHz.

25-Pair Connectors

Because a basic cable unit in telephone systems is the 25-pair cable, it's sometimes practical in premises cabling to run these cables as a unit. For example, a 25-pair cable can be connected directly from a PBX to a cross connect. Some network hubs also allow a 25-pair cable to run from the hub to the cross connect or patch panel.

Figure 4–9 shows a 25-pair cable assembly. The connector is a miniature ribbon connector. The connector uses slotted-beam style IDC contacts to terminate the pairs; the mating face has a tongue-and-groove style interface. The "ribbon" part of the name derives from the shape of the contacts.

Miniature ribbon connectors were originally designed for low-speed voice communications. While they are quite capable of being used in lower-speed applications such as 10-Mbps Ethernet, recent versions are rated to Category 5. Cross connects and patch panels are available that allow 25-pair connectors to plug in directly.

Category 7 Connectors

In selecting a connector for Category 7 cables, the ISO/IEC committee settled on two connectors. The primary choice is compatible with modular plugs and jacks, with backward

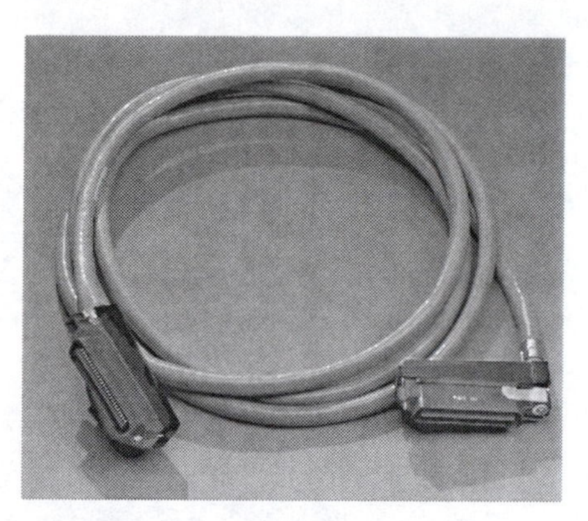

Figure 4–9. 25-pair connector *(SYSTIMAX Solutions graphics are courtesy of SYSTIMAX Solutions™, a CommScope company)*

compatibility with Categories 5, 5e, and 6 hardware. The alternate choice is a connector designed specifically for Category 7 applications.

Figure 4–10 shows the primary Category 7 plug connector. Two pairs of contacts are added to the bottom of the connector. In addition, a retractable stub protrudes from the front of the connector. The stub is extended for Category 7 applications and retracted for other applications. The stub also determines which contacts are active.

In Category 5, 5e, or 6 applications, the top four pairs of contacts are used, just as in a normal modular plug connector. For Category 7 applications, only the upper outer two pairs and the lower two pairs are active. Figure 4–11 shows the active contacts for different applications. This arrangement allows better separation of pairs and high performance. The jack is similarly arranged, with the protruding stub operating an internal switch to indicate which contacts should be active.

The alternate non-RJ connector uses a completely new design, forgoing compatibility. The connector, shown in Figure 4–12, can be used in 1, 2, or 4 pairs.

Even though the design is new, components will snap into the same holes in patch panels and outlets as modular jacks do. This means the jacks can be used in any standard hardware.

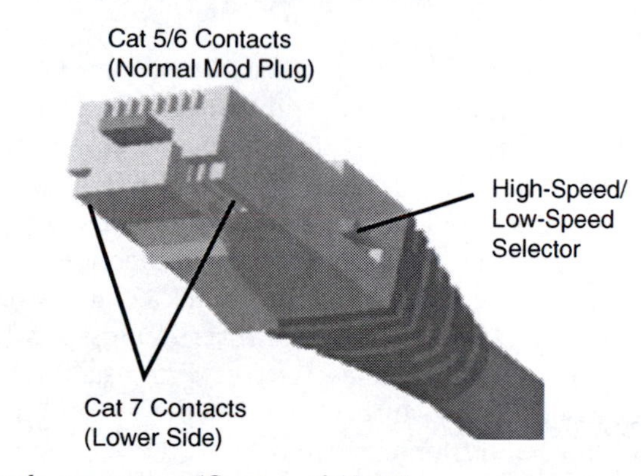

Figure 4–10. Category 7 RJ-style connector *(Courtesy of Alcatel)*

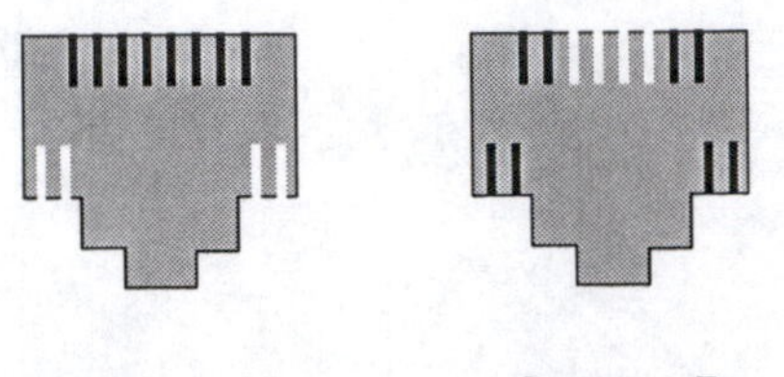

Category 5, 5e, 6
(Upper 8 contacts)

Category 7
(4 upper contacts
4 lower contacts)

Figure 4–11. Active pins in Category 7 RJ-style connector

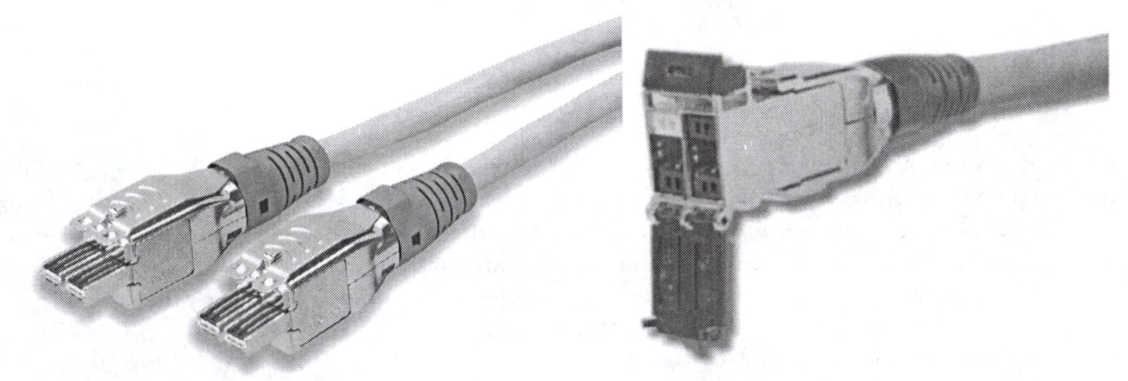

Figure 4–12. Example of a Category 7 plug and jack *(photos courtesy of The Siemon Company)*

Fiber-Optic Connectors

Fiber-optic connectors differ significantly from electrical connectors. Connectors for both copper and fiber have a similar function: to carry signal power across a junction with as little power loss and distortion as possible. But the difference between electrical signals and optical signals is that the connectors are very different. Where the electrical connector forms a gastight, low-resistance interface, the key to an optical connector is alignment. The light-carrying cores of the fibers must be very accurately aligned. Misalignment means that optical energy will be lost as light crosses the junction.

Alignment of the two fibers is a formidable task. Not only are the dimensions quite small—a core diameter of 62.5 µm for multimode fibers and 9 µm for single-mode fibers—tolerances complicate the matter. One hopes that the core is perfectly round and perfectly centered within a perfectly round cladding. And that the cladding is 125 µm, not 125.5 or 126 µm. The same applies with the connector dimensions. Unfortunately, some variations creep into any manufacturing process. These variations represent misalignment that can increase the loss of optical power.

Even so, both fiber and connector manufacturers have evolved processes to create tightly toleranced components. Some years ago, the goal for insertion loss was 1.5 dB. Then 1 dB. Today's connectors offer loss of around 0.3 dB, while premises wiring standards call for a loss of 0.75 dB.

Loss is measured as insertion loss. Insertion loss is the loss of optical power contributed by adding a connector to a line. Consider a length of optical fiber. Light is launched into one end and the output power is measured at the other end. Next the fiber is cut in two, terminated with fiber-optic connectors, and the power out is again measured. The difference between the first and second measurement is the insertion loss.

Most connectors use a 1.25 or 2.5 mm ferrule as the fiber-alignment mechanism. A precision hole in the ferrule accurately positions the fiber. The two ferrules are brought together in a receptacle that precisely aligns the two ferrules through a tight friction fit.

Ferrules are typically made of ceramic, plastic, or stainless steel. Until recently, there were clear differences between the materials. Ceramic was more expensive and offered lower insertion loss. Plastic was the least expensive material, but offered higher insertion loss. Stainless steel fell in between. Today the price and performance of the materials is much closer, so there is little difference. The main exception is single-mode applications, where ceramic is still the preferred material.

Terminating a Fiber

Figure 4–13 shows the termination procedure for an epoxyless connector.

Epoxy has long been a staple for ensuring that the fiber is properly held in the ferrule. And epoxy has long proven that it does its job well. But epoxy has also been seen as a necessary evil. It has several drawbacks. It's an extra step in the process, it's potentially messy, and it requires curing. Curing time can be shortened with a portable curing oven, but that means another piece of equipment to lug around. Finally, when working in a completed building, the possibility of spilling epoxy on an executive's carpet or furniture is an unpleasant prospect.

Epoxyless connectors eliminate the need for epoxy and require only a crimp. Because early attempts at epoxyless connectors were less than successful, some people are reluctant to specify epoxyless connectors. But epoxyless connectors today offer low loss and long-term stability to make them a well-suited alternative to epoxy connectors.

Several approaches exist for eliminating epoxy. The most straightforward uses an internal insert that is forced snugly around the fiber during crimping. The insert clamps and positions the fiber, while the crimp secures the cable strength members.

The hot-melt connector uses preloaded adhesive so that external mixing is not required. The connector is placed in an oven for one minute to soften the adhesive. The fiber is then inserted into the connector; in three to four minutes the adhesive dries. The connector is then polished.

A third variation uses a short piece of fiber factory-assembled and polished into the end of the connector ferrule. The cable fiber butts against this internal fiber and the cable is crimped in place. This approach also eliminates the need to polish.

The main benefit of epoxyless connectors in premises cabling is increased productivity and, hence, lower costs.

For premises cabling applications, there is little performance difference between ceramic, plastic, and stainless steel-based connectors and between epoxy and epoxyless connectors. All can meet the requirements of premises cabling.

Types of Fiber-Optic Connectors

SC Connectors

Nippon Telephone and Telegraph (NTT) originally designed the SC connector (Figure 4–14) for telephony applications. It has subsequently become the preferred general-purpose connector, and is being specified for premises cabling in TIA/EIA-568A, ATM, low-cost FDDI, and other

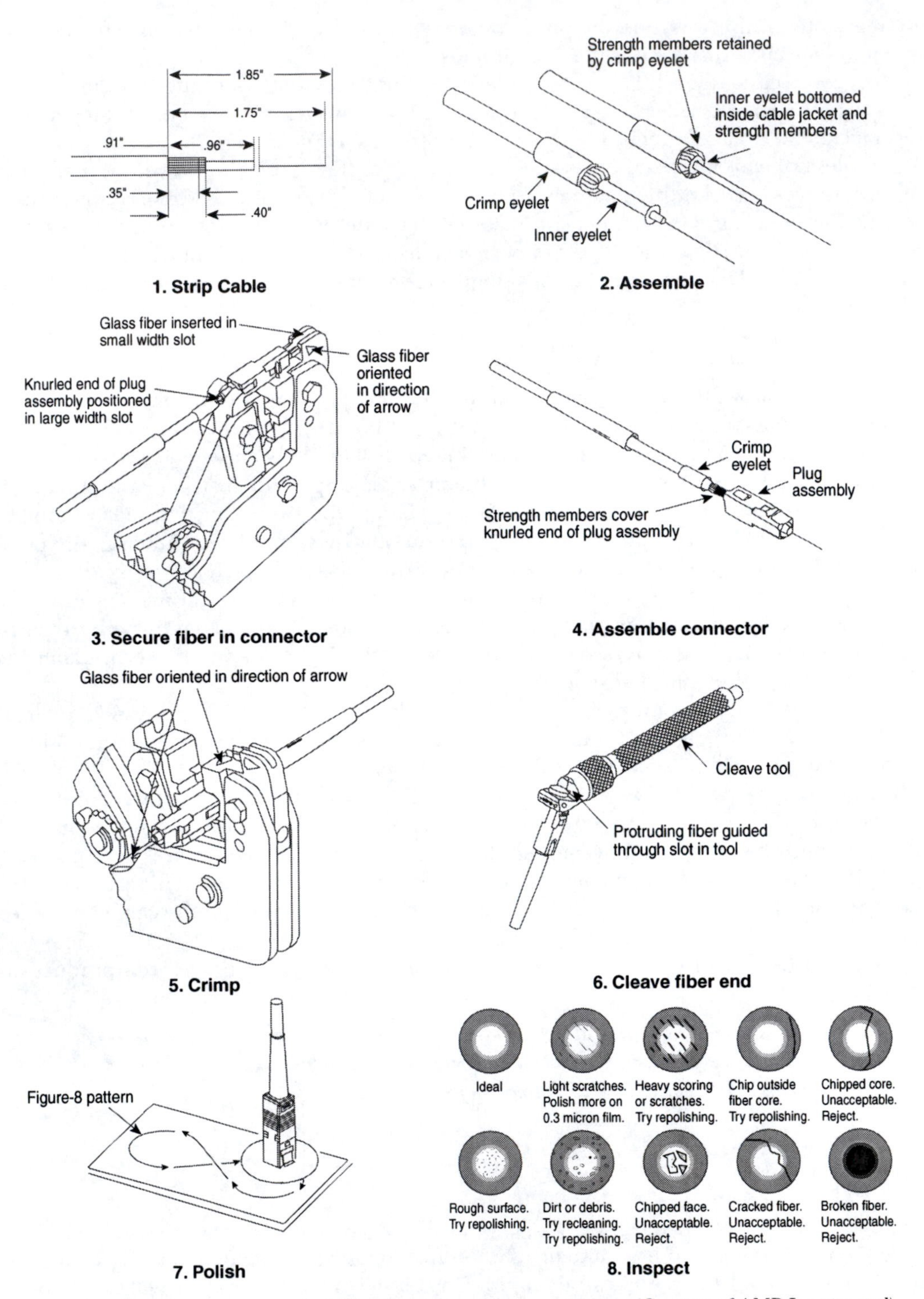

Figure 4–13. Terminating an optical fiber in an epoxyless connector *(Courtesy of AMP Incorporated)*

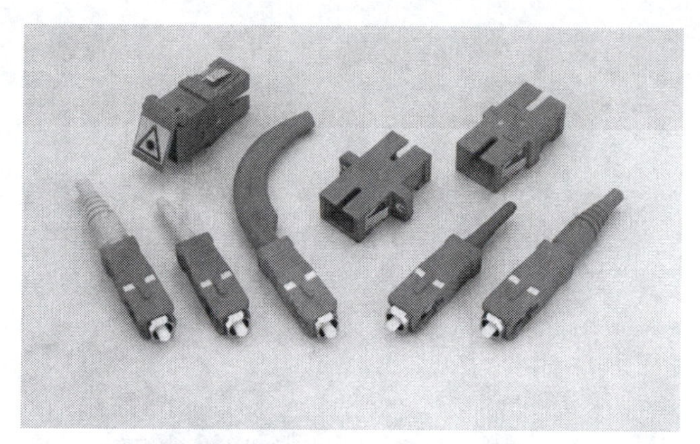

Figure 4–14. SC connectors *(Courtesy of Molex Incorporated)*

emerging applications. The SC is still recognized in TIA/EIA-568B, although the standard no longer specifies a specific connector. Any connector meeting the required performance can be used. Three characteristics make it popular.

First, its design protects the ferrule from damage. The connector body surrounds the ferrule so that only a small part sticks out.

Second, it uses a push-pull engagement mechanism. Finger access is easier and requires less all-around space than a connector with a rotating engagement mechanism like threaded or bayonet-style coupling nuts. This permits closer spacings and higher density designs in patch panels, hubs, and so forth. And third, the rigid body eliminates intermittency problems caused by accidental pulling on the cable at the rear of the connector.

Connectors are easily snapped together to form multifiber connectors. Joining two connectors to form a duplex connector simplifies installation, use, and troubleshooting of the premises cabling plant. Because all links require two fibers, each carrying signals in opposite directions, a duplex cable assembly obviously makes sense. If you have 10 point-to-point links, it's easier to deal with 10 duplex cables rather than 20 individual cables.

TIA/EIA-568A recommends color-coding for SC connectors: beige for multimode connectors and blue for single-mode connectors.

ST Connectors

Invented by AT&T, ST connectors were the connector of choice during the late 1980s and early 1990s, being replaced by the SC. As shown in Figure 4–15, the ST uses a bayonet coupling that requires only a quarter turn to engage or disengage the connector.

ST connectors are still an important force in premises cabling. Even TIA/EIA-568A acknowledges that they must be considered. Many buildings already wired with fiber use ST connectors. Most network equipment uses ST connectors for fiber-optic ports. So even new buildings will need to accommodate ST connectors in one way or the other in the future. One way is cable assemblies with STs on one end and SCs on the other. Another way is hybrid adapters that accept ST connectors on one side and SC connectors on the other side. Such hybrid adapters are available for nearly every combination of connector types.

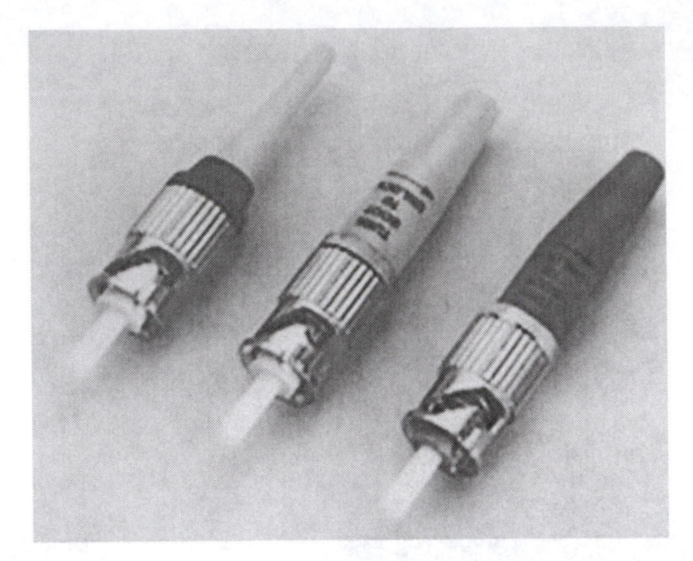

Figure 4–15. ST connectors *(Courtesy of Molex Incorporated)*

FDDI Connectors

FDDI MIC (for medium interface connector) connectors (Figure 4–16) are two-channel devices having two ferrules on 0.7-inch centers. The connectors use a fixed shroud to protect the fibers and a side-latch mechanism for keeping connectors engaged. While the connectors were designed for FDDI networks, they are not restricted to such uses. The connectors accept either factory- or field-installed keying to ensure that the cables are connected to the right FDDI station.

Small-Form-Factor Connectors

Traditional fiber-optic connectors like the SC and ST tend to be large, about twice the size of a modular plug and jack. The number of connectors you can pack into a given area (and especially in a side-by-side row) is called the *port density*. A modular jack has a port density twice that of SC or ST connector pairs. Remember, you need two ST or SC connectors for each port, one for carrying signals in one direction and the other for transmission in the other direction.

The implication of port density is this: in the space you can fit 24 copper modular jacks, you can only fit 12 duplex SC fiber connectors. Put another way, you need twice as much space to fit the same number of SC connectors as you need for modular jacks. One consequence of this lower port density can be seen in network electronics. Consider a network switch having plug-in boards. If one board can hold 24 modular-jack-based ports, you need two boards to achieve 24 fiber ports. This means even higher costs for fiber-based equipment. Similar space savings are seen in equipment rooms for patch panels, where a larger patch panel is needed for fiber ports.

The latest generation of fiber-optic connectors, small-form-factor (SFF) connectors offer the same port density as modular jacks. What's more, they offer a press-to-release latch very similar to those found on modular plugs. The latch means the same familiar easy use for users,

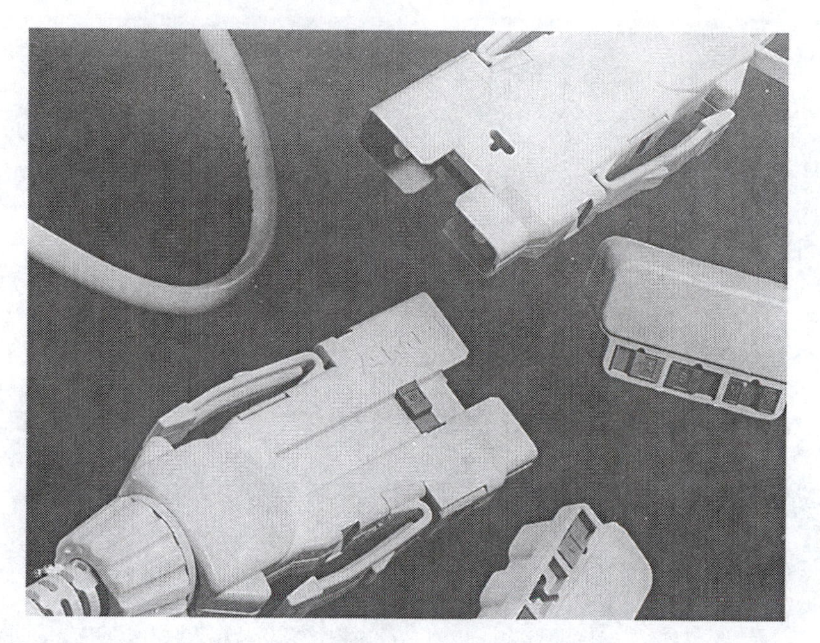

Figure 4–16. FDDI connectors *(Courtesy of AMP Incorporated)*

while the small size means there is no significant penalty in port density. Some fiber connectors, in fact, will fit the same panel cutouts as modular jacks use.

While TIA/EIA-568B recognizes SC and ST connectors, it does not specify an SFF connector. When the standards committee considered SFF connectors, they declined to specify one for premises cabling. Instead, the committee decided to let the market decide which was the preferred connector. There are several SFF connectors vying for market acceptance.

MT-RJ Connector. The MT-RJ connector was designed by AMP, Siecor, and USConec as an alternative to RJ-45 modular telephone connectors. The MT-RJ small rectangular ferrule that holds two fibers. It also uses a press-to-release latch quite similar to that found on the modular jack to make its operation familiar. Figure 4–17 shows an MT-RJ connector, along with the small-form-factor transceiver it uses.

Most fiber-optic connectors use two plug connectors mated through a coupling adapter. While the MT-RJ can be used in this manner, it also offers a true plug-jack interconnection. Designed for very simple termination of fiber, the jack contains two fiber stubs in a polished ferrule. The backend of the stub fibers are in a mechanical splice. To terminate the fiber-optic cable, the fibers are cleaved, slid into the splice, and the splice is closed to form a low-loss interconnection.

LC Connector. The LC connector was designed by Lucent Technologies as a smaller connector for premises cabling and telecommunications applications. Figure 4–18 shows the connector. Based on a 1.25-mm ceramic ferrule, the LC uses an RJ-45-style housing with a push-to-release latch that provides an audible click when the connector is mated in an adapter. Available in both duplex and simplex configurations, the connector doubles the density over existing ST and SC designs. The connectors are color-coded beige for multimode and blue for single mode.

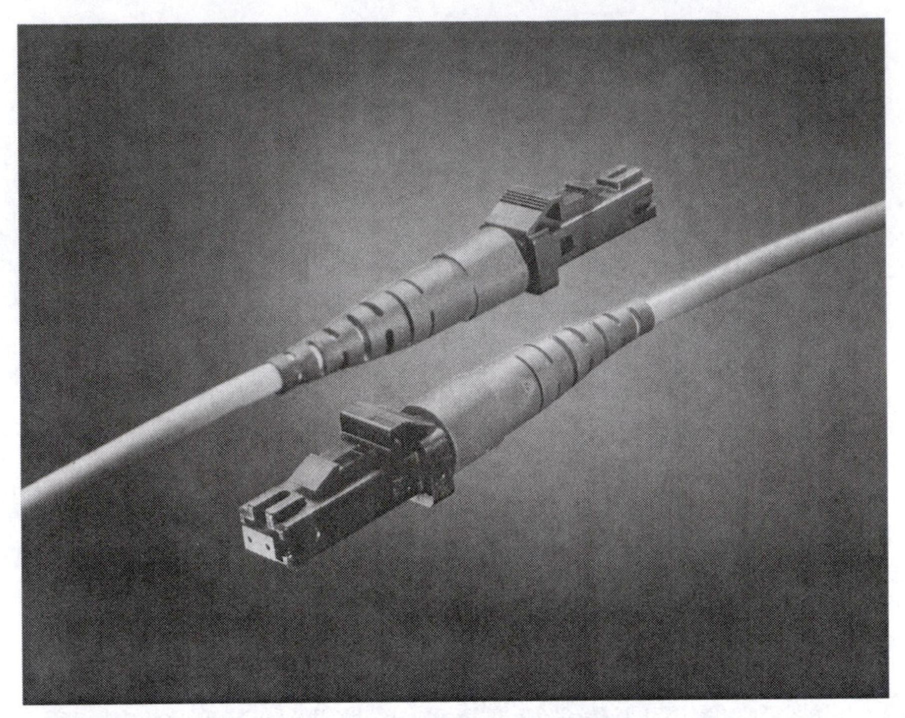

Figure 4–17. MT-RJ connector *(Courtesy of Siecor Corporation)*

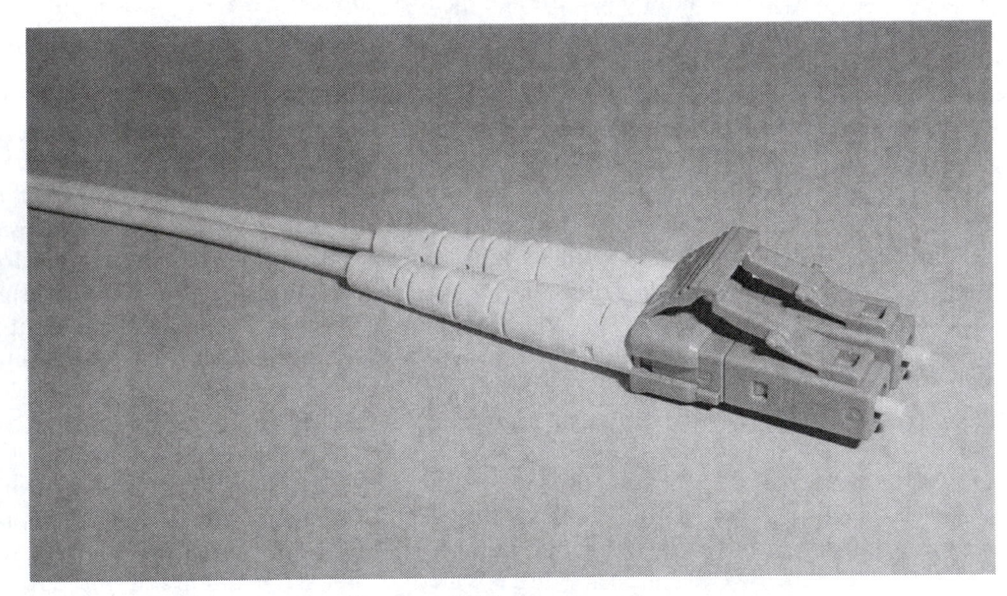

Figure 4–18. Duplex LC connector *(Courtesy of Molex Incorporated)*

Figure 4-19 shows an indication of the relative sizes of standard and SFF connectors. It shows duplex ST and duplex SC connectors alongside four duplex LC connectors.

Volition VF-45 Connector. The VF-45 connector from 3M is a ferruleless connector, as shown in Figure 4–20. The plug connector consists of a fiber holder to secure the fiber, a shroud and boot that secure the connector to the cable, and a hinged door that acts as a dust cover. The mating socket has a fiber holder, a hinged door, and a body that houses V-grooves to align the fibers. When the plug is inserted into the socket, the fibers slide along the V-grooves until they mate with the fiber in the socket (also in the V-grooves). The fibers in the plug bow, creating a forward and downward pressure to optimize the fiber-to-fiber contact. The plug snaps into place when fully inserted and uses a press-to-release latch similar to modular telephone plugs.

The Volition connector has been adopted as a standard connector in Fibre Channel applications.

SMI Connector

The Small Multimedia Interface (SMI) connector is the most recently introduced of all the connectors mentioned in this chapter. SMI connectors are available for both POF and GOF and are illustrated in Figure 4-21.

MU Connector. The MU connector was invented by NTT as a smaller replacement for the SC. Using a 1.25-mm ferrule, the connector is essentially a smaller version of the SC and uses the same push-pull latching mechanism.

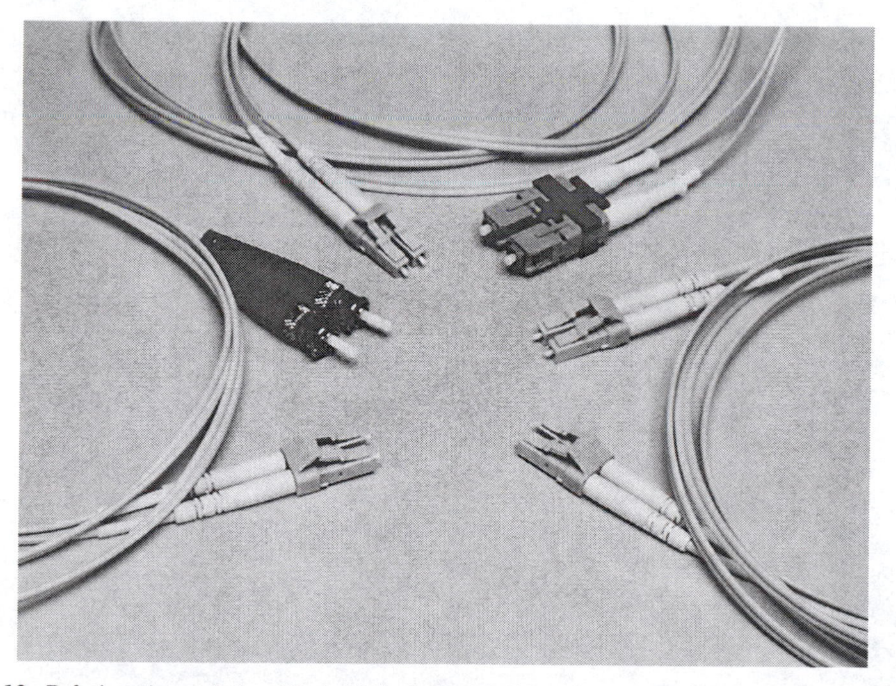

Figure 4–19. Relative sizes of the ST (middle left), SC (top right), and LC connectors *(SYSTIMAX Solutions graphics are courtesy of SYSTIMAX Solutions™, a CommScope company)*

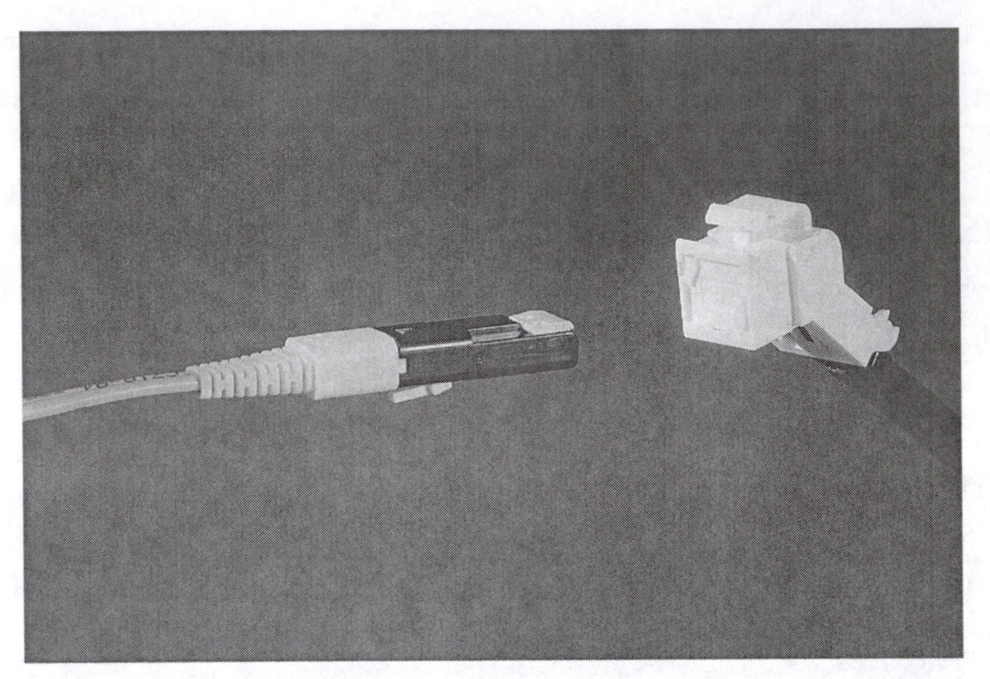

Figure 4–20. VF-45 Volition connector *(Courtesy of 3M)*

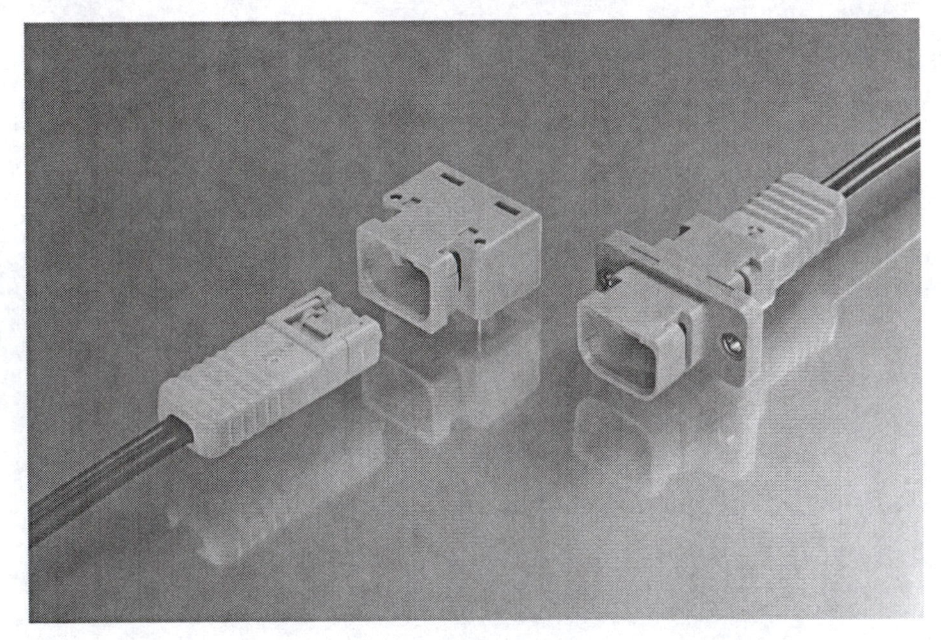

Figure 4–21. SMI connector *(Courtesy of Molex Incorporated)*

OptiJack Connector. The OptiJack connector from Panduit also uses a 1.25-mm ferrule and RJ-45-style latch. Unlike the LC, which is a discrete connector that can be combined for duplex operation, the OptiJack is inherently a duplex connector.

SC-DC Connector. This connector, from Siecor and IBM, is an SC-style connector that holds two or four fibers in a single 2.50-mm ceramic ferrule.

There are two things to mention about SFF connectors in particular, and fiber-optic connectors in general. First, we are concerned with the connector as an interface. Transceivers should also be available with the interface to allow the interface to be used throughout the system. SFF transceivers are the key to higher port densities in equipment.

The second thing to remember is that connector choice doesn't have to be an either/or choice. You can put one type of connector on one end of the cable and another type of connector on the other end. Suppose, for example, you cable your building with the LC interface. Does this mean your network equipment must also feature the LC? What if you buy an Ethernet switch with an MT-RJ interface? No problem. You simply need a cable with a duplex LC connector on one end and an MT-RJ connector on the other end.

Even so, it might be easier and more convenient to have an end-to-end system with the same interface. It will be easier to buy patch cables off the shelf, without special ordering this connector here and that connector there.

Array Connectors

Array connectors are a special category of SFF connectors and are used to connect more than two fibers. An array connector holds more than one fiber within the ferrule. The most popular style is the MT connector invented by NTT in the 1980s for use in telephone equipment in Japan. During the 1990s, connectors based on the MT ferrule and its derivatives have become popular for numerous other applications.

The MT ferrule holds up to 12 single-mode or multimode fibers in an area smaller than the single-fiber SC connector. Most MT-style connectors use a push-pull mechanism to maintain two plug connectors mated in an adapter. The MT ferrule has been adopted for different styles of connectors. The most popular is the industry-standard MPO connector, which is used for wiring data centers and other applications. The MT-RJ is based on the MT-style ferrule, but substitutes a latch for the push-pull locking mechanism.

Although the connectors can be used with a variety of fiber constrictions, ribbon fiber is very popular. Figure 4–22 shows an MPO connector, along with an SC connector for comparison. The SC connector holds one fiber; the MT holds 12 fibers in a connector significantly smaller than the SC—at about the same width and half the height. This size reduction is one reason why MT-style and other small-form-factor connectors are gaining in popularity.

Array connectors are widely used to create structured cabling systems in large-scale computer installations. The connectors are used as backbone or trunk cable running between equipment.

In premises applications, array connectors serve a function similar to 25-pair copper connectors: they carry multiple fibers between two points. Array connectors can be used in a building backbone or in horizontal zone cabling (discussed in Chapter 6).

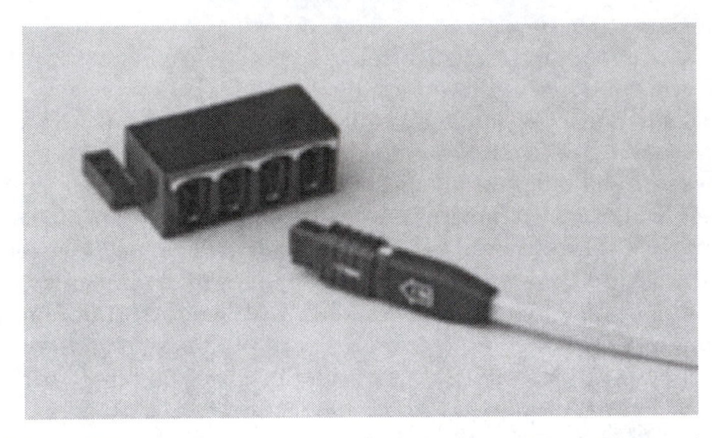

Figure 4–22. A 12-fiber MPO array connector *(Courtesy of Sumitomo Electric Lightwave)*

Splices

Splices are used to join two fibers permanently or semipermanently in cases where the need for connecting and disconnecting is not anticipated. We said "semipermanent" because some splices are re-enterable—that is, the splice is reusable.

The attraction of a splice is that it offers the lowest insertion loss—under 0.2 dB. Because compatibility is not an issue, splice designs tend to be proprietary to a single manufacturer. Fusion splices—which heat and actually fuse two fibers into a single unit—are used by telephone companies because they offer the lowest losses obtainable, as low as 0.05 dB. Fusion equipment is expensive so you need to perform lots of splices before fusion techniques become cost-effective. For premises applications, mechanical splices that simply align two fibers precisely are also used.

Because disconnectable, rearrangeable circuits are central to a flexible premises cabling system, splices are not as common as connectors.

Splices have three main uses in premises cabling:

1. Splices can join lengths of cable together to make very long runs.
2. Fibers can be connectorized by splicing on factory-installed connectors that come with a short fiber pigtail.
3. Broken or damaged fibers can be repaired by splicing in a new section.

Figure 4–23 shows a typical splice.

Coaxial Connectors

Coaxial connectors must terminate both the braid and the center conductor of the coaxial cable.

The most popular coaxial connector types are defined by military standard MIL-C-39012, which provides dimensions, materials, and performance requirements for a variety of connector types. From the military standards, many variations have evolved to provide lower costs, easier application, and greater flexibility.

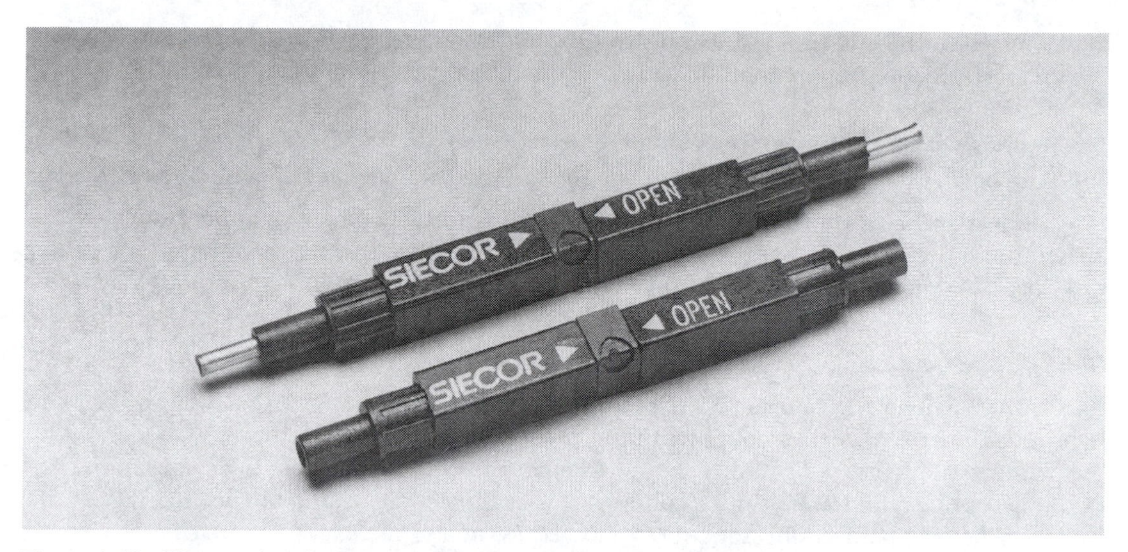

Figure 4–23. Fiber-optic splice *(Courtesy of Siecor Corporation)*

BNC connectors are used with 10BASE-2 thinnet. They use a bayonet coupling and offer a 50-ohm nominal impedance, the same as the cable's. Figure 4–24 shows the BNC. BNC connectors are available in several styles:

- **Dual Crimp,** which is designed to meet rigid military specifications. The connector requires two crimps. First, the center contact is crimped to the center conductor. The connector is then assembled, and the braid is crimped to the connector body.
- **Single Crimp,** which requires only a single crimp cycle to terminate both the center and outer conductors. This significantly speeds the termination process.
- **Commercial,** which is a lighter, more compact, and less expensive single-crimp connector. It still meets the performance requirements of MIL-C-39012.

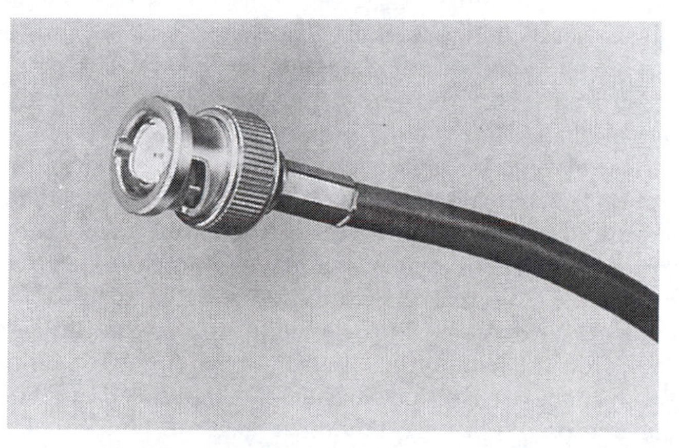

Figure 4–24. BNC connector *(Courtesy of AMP Incorporated)*

- **Consumer,** which is even less expensive than the commercial version. It does not meet all the performance requirements of MIL-C-39012; it does exceed the requirements of network applications.
- **Toolless,** which requires no special tools for termination.
- **Twist-on,** in which the termination is made by twisting a coupling nut onto the connector body.

Toolless and twist-on connectors are not recommended for network applications, except for temporary repairs. They are less able to withstand excessive flexing.

Coaxial connectors consist of a cable-mounted plug and an equipment-mounted jack.

Other Coaxial Connectors

10BASE-5 networks use an N-series connector. Larger than the BNC connectors, the N-connector uses a threaded coupling nut.

Video systems use a 75-ohm F-connector like those used by cable TV. The traditional F-connector does not have a center contact—it uses the center conductor of the cable as the contact. The original F-connectors were designed for analog signals at frequencies to around 216 MHz. Analog video signals can be very forgiving of poor performance from the connector. Consequently, F-connectors were designed more with cost in mind than performance. Traditional F-connectors are generally unsuited to digital data. Recently, however, more rigorous standards have upgraded the F-connector to make it suitable for digital signals and frequencies to 1 GHz.

Some mini- and mainframe computer systems use either twinaxial or triaxial cable. Connectors for these are similar to N-series connectors, but are often larger, bulkier, and harder to use.

Patch Panels

Patch panels provide a method of arranging circuits. For example, an employee who changes offices might want to keep the same phone number. By rearranging circuits at the patch panel, you can have the number ring in any office you want. Similarly, networks are rearranged to add, subtract, and change workstations. Again, people can move around but stay physically attached to the same network by changing the circuits. The patch panel is the place where circuits are connected and reconnected. Figure 4–25 shows a patch panel, while Figure 4–26 shows a typical application.

Many patch panels use a feed-through connector so that a cable can be plugged into both sides. For example, an RJ-45 patch panel has a modular jack that accepts a modular plug in each side. The horizontal cables running to work areas plug into one side of the panel. Patch cables from equipment or from the backbone cable termination plug into the other side. While cables on either side of the panel can be rearranged, it is usually the patch cables that are changed. Think of the back side—the side with the backbone or horizontal cable—as being fixed and unchangeable in most cases. The front or patch side is where circuit reconfigurations take place.

Feed-through patch panels are typically not suited to high-speed operation. UTP panels typically feature IDC contacts on the back and modular jacks on the front. The IDC contacts can be either 110 style or barrel style. This offers better electrical performance to reduce NEXT.

Figure 4–25. UTP patch panel *(SYSTIMAX Solutions graphics are courtesy of SYSTIMAX Solutions™, a CommScope company)*

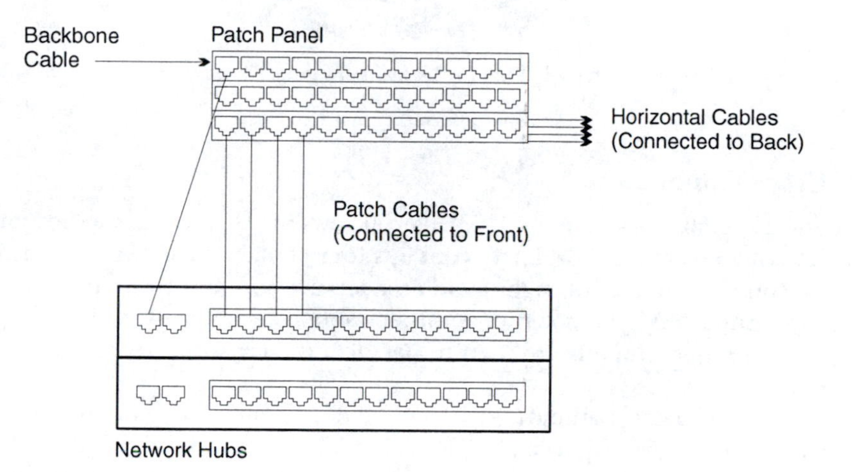

Figure 4–26. Application of patch panels

Fiber-optic patch panels often provide a transition between different connectors. Transitions among SC, FDDI, and ST connectors are common. Figure 4-27 illustrates a typical fiber patch panel.

Punch-Down Blocks

Punch-down blocks are so called because the wire is positioned in the top of an IDC contact and then forced (punched) down into the slot for termination by a tool called the punch-down tool. The tool also trims the wire by cutting off any excess on the end. Standard for telephone installations, punch-down blocks are the most common form of cross connect for twisted-pair cable. They're also called cross connect blocks. Like a patch panel, the purpose of the punch-down block is to provide the connection between two cable segments.

The two main varieties of punch-down cross connects are termed Type 66 and Type 110, which are AT&T designations.

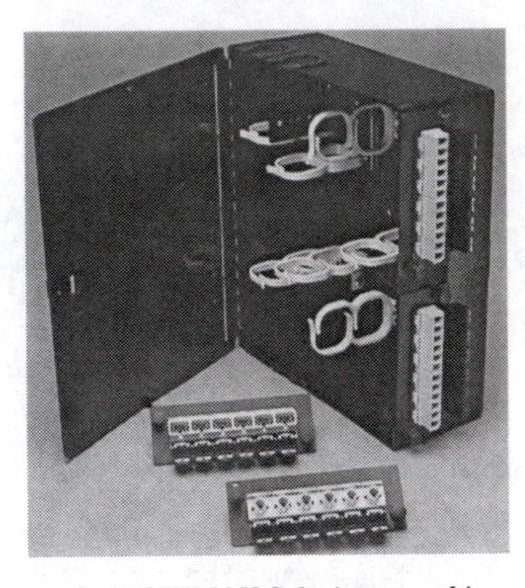

Figure 4–27. A small fiber patch panel *(SYSTIMAX Solutions graphics are courtesy of SYSTIMAX Solutions™, a CommScope company)*

Type 66 Cross Connects

The basic Type 66 block (Figure 4–28) has 50 rows of IDC contacts to accommodate the 50 conductors of 25-pair cable. Each row contains four contacts. The two left-hand contacts of each row are commoned; the two right-hand contacts are commoned. In use, one cable is terminated to the outer left-hand column. Another cable is terminated to the outer right-hand column. The two inner columns are used to perform the cross connect by terminating cross connect cables.

Type 66 blocks represent an older style designed originally for voice circuits. More recent versions meet Category 5e requirements. You should, however, be careful when using older blocks for digital circuits. Some older blocks have high crosstalk that makes them unsuitable even for moderate speeds.

110 Cross Connects

The basic 110 cross connect consists of three parts: mounting legs, wiring block, and connecting block. The mounting legs provide cable-routing management and the means for holding the wiring block. The wiring block consists of molded plastic blocks that position the cable. The block contains horizontal index strips. Conductors are laced into the slots of the index strip. The strip typically has 50 slots to accommodate a 25-pair cable. It is marked every five pairs to help visually simplify the installation and reduce errors. In addition, color coding helps identify the circuits. In application, the wires are laced into position and then punched into place with the impact tool. The wiring block, however, does not terminate the conductors. It simply positions them.

Figure 4–29 shows some examples of 110 wiring blocks, Figure 4–30 shows a close-up of the 110C connecting blocks that plug into the wiring blocks, and Figure 4–31 shows a typical 110 cross connect installation.

After the wires are positioned in the wiring block, the connecting block is seated over the wiring block. The connecting block has IDC contacts at both ends. One set of contacts terminates the contacts in the wiring block; the other set is used for jumper wires that actually

Figure 4–28. Type 66 cross connect blocks *(Courtesy of The Siemon Company)*

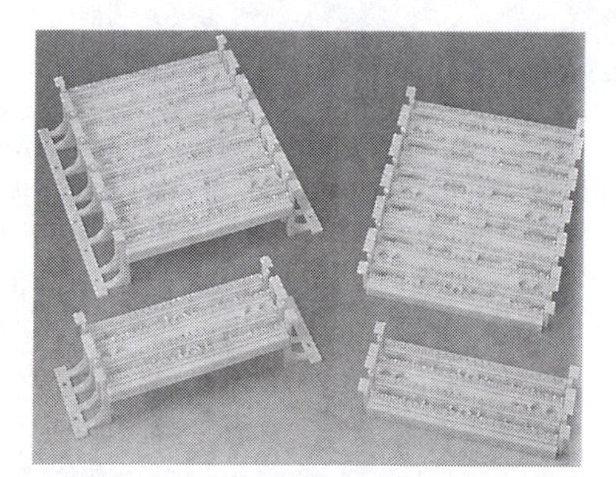

Figure 4–29. 110 wiring blocks *(SYSTIMAX Solutions graphics are courtesy of SYSTIMAX Solutions™, a CommScope company)*

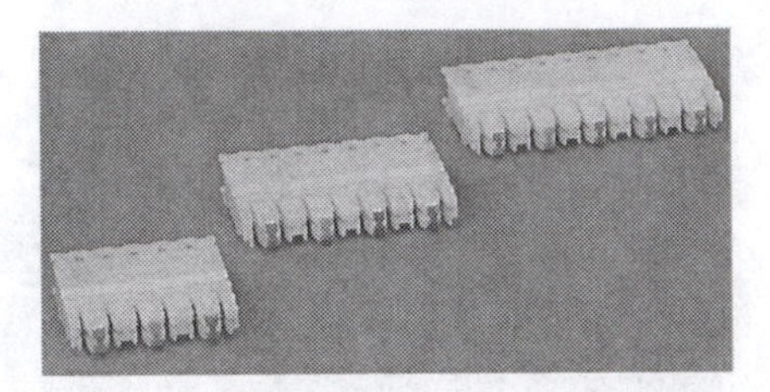

Figure 4–30. 110C connecting blocks *(SYSTIMAX Solutions graphics are courtesy of SYSTIMAX Solutions™, a CommScope company)*

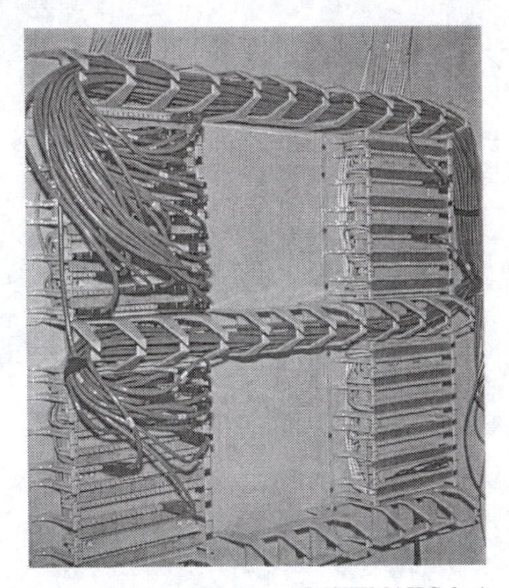

Figure 4–31. Typical 110 cross connect application *(SYSTIMAX Solutions graphics are courtesy of SYSTIMAX Solutions™, a CommScope company)*

perform the cross connect. Once the connecting block is seated, the cross connect wires can be used to interconnect the cables properly.

Wiring blocks are available to accommodate up to 300 pairs. The basic structure is that each horizontal index strip handles 25 pairs; a cross connect uses multiple strips for higher pair numbers. A 100-pair cross connect requires four index strips; a 200-pair cross connect requires eight index strips. The connecting blocks come sized to accept either 3, 4, or 5 pairs. Each block has an alternating series of high and low teeth. Individual conductors of a pair are separated by low teeth; the high teeth separate pairs. The positions for each pair are color coded on the high teeth:

Pair 1: Blue

Pair 2: Orange

Pair 3: Green

Pair 4: Brown

Pair 5: Slate

In addition, each contact for the pair is marked either "T" or "R," for tip and ring. Tip and ring are telephone terms that date back to the days of manual switchboards. The switchboard plug had three parts labeled tip-ring-sleeve. In modern multipair telephone cable, the ring conductor is color coded and the tip conductor is white.

Notice that the color coding of the connecting block matches the color coding of the pairs of a telephone cable. It does not refer to the function of the cable within the hierarchy of premises cabling—such as backbone or horizontal cable. Chapter 10 describes the color-coding system for marking the cross connect positions by color. Don't confuse the conductor color coding of the connecting blocks with the color coding by cable function.

The normal method of cross connection is that the cables to be joined are in different index strips. For example, the telephone backbone might terminate in row 1, the top strip. The horizontal cable might terminate in row 2. The cross connect function is performed by running a cable between rows 1 and 2.

We've just described the basic function and use of a 110 cross connect. As with anything as flexible and widely used, 110 blocks have evolved into a wider line of products. The two areas of interest here are cross connect methods and handling 25-pair cable.

110 blocks typically allow three types of cross connect:

- **Direct IDC.** In the system described above, the cross connect wires are simply run between connecting blocks. They are terminated by punching the conductors into the IDC contacts.
- **Patch Plugs.** (Figure 4–32) These are connectors that plug into the connecting block. This eases rearrangements by simplifying the move, add, or change to a unplug/plug operation.

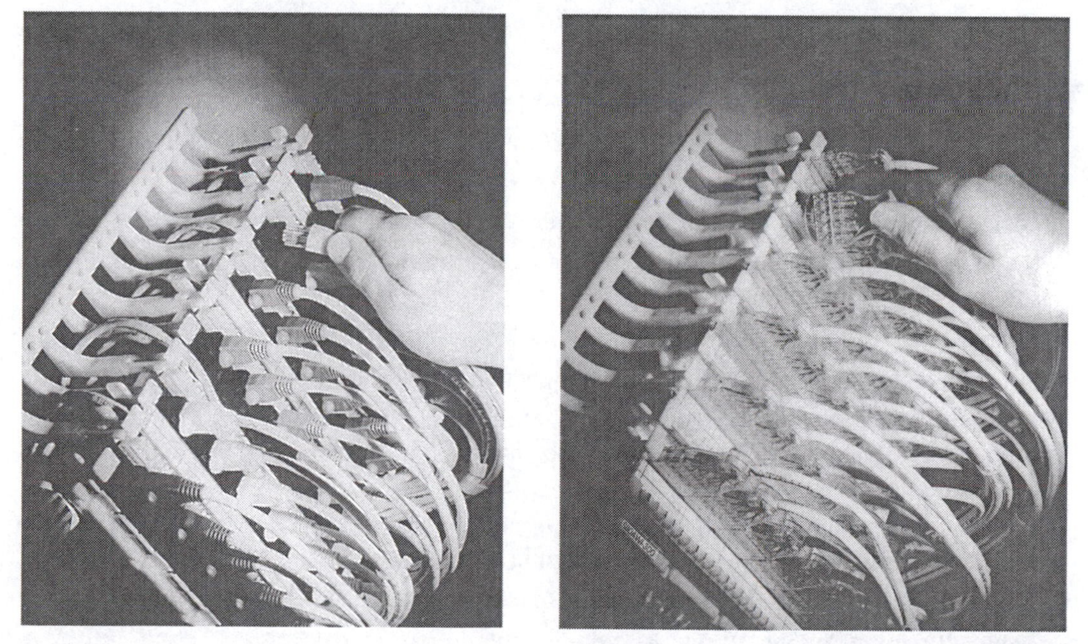

Modular Jack   Patch Plug

Figure 4–32. Modular jack and patch plug 110 styles *(Courtesy of The Siemon Company)*

- **Modular Jacks.** (Figure 4–32) Cross connects can also be supplied with modular jacks to perform the cross connect.

Cross connects can also be supplied prewired for specific applications. One popular variation uses a 25-pair connector. The block is prewired to the connector to allow a 25-pair cable from a device like a hub or PBX to simply plug into the cross connect without having to terminate individual pairs. Other versions terminate the backend cable in the same manner, but offer modular jacks for output.

The choice of cross connect style depends on many factors.

Cross connect blocks offer higher densities and so they require less space than patch panels. They are also less expensive. They can be wall mounted to reduce the amount of space required in a wiring closet, especially floor space, because they don't need a floor-standing rack. Larger systems are, however, rack mounted.

Direct IDC patching of cross connects is the least expensive and most straightforward of the various approaches. It is also the least friendly for making moves, adds, and changes. Skill is involved in removing and rearranging cables. Patch cables are easier, and modular jacks are easiest of all. Ease of use can be a two-edge sword. One edge is the ease and speed with which changes can be made. No skill is needed to unplug a modular plug from a jack and plug it into another jack. Skill *is* needed to reterminate patch wires. The other edge is security: do you want anybody to be able to perform moves, adds, and changes? Any user with access to a closet can make changes. Costs are also a consideration. In small systems, patch panels can make economic sense, but in larger installations the increase in cost may become unattractive. Fiber systems, of course, are patch panel-type systems.

Application Guidelines

Here are some suggestions for successful application of cross connect systems.

- Make sure the components are rated for the application.
- Group similar types of applications together.
- Use the cable management features of the block. Feed cables to be terminated on the right side into the right side of the block.
- Maintain the proper bend radius. Do not kink or crimp the cables.
- Don't strip any more of the cable jacket away than is absolutely necessary.
- Maintain pair twist right up to the contacts. Maintaining twist becomes more important the higher the cable grade. With Category 5e or higher cable, no more than 0.5 inch of a pair should be untwisted.
- Begin at the end that the cable enters. If the cable comes from the floor, start at the bottom and work upward. If it comes from the ceiling, start at the top and work down.
- Always double-check the routing of pairs, making any corrections immediately.
- Label all pairs. (Chapter 10 discusses documentation and labeling.)
- Do not bind cable bundles too tightly together or transmission performance may be impaired.

The choice of patch panels or cross connect blocks is often a personal preference, which can be strongly influenced by your background in the industry. Cross connect blocks grew out of the telephone industry; those with telecommunications background are quite familiar with them. Patch panels are popular with network people, who are comfortable with the ease with which network connections can be made and rearranged. Telephone service has always been a company thing; the tradition of the telephone company or service provider maintaining the complete system, including any moves, adds, and changes, goes back many years. Even for companies committed to maintaining their own systems, the attitude of central control by trained people is strong.

Networks are moving in the other direction. The individualism of the PC carries over into networking. Many small networks were installed by users. Network administrators tend to want to have tighter and more immediate control over the network and its arrangement and favor patch panels. Patch panels are intuitively easy to understand and require no special training to use. Not so with cross connect blocks. Even as networks become a critical part of the company, with the same bureaucratic restraints, the legacy of patch panels remains.

The choice between cross connects and patch panels, of course, is not an either/or situation. Systems can also be mixed: a 110 block for voice, security, and other low-speed voice, and patch panels for high-speed data.

Outlets

Outlets are the transition between the behind-the-wall cable and the work area cables running directly to telephones, computers, and so forth. Again, there are many variations to meet different situations: wall outlets, in-floor flip-up outlets, surface-mount outlets for floors and walls, and outlets for modular office panels.

Modular outlets are increasing in popularity because of the flexibility they offer in creating and maintaining a cabling system. The idea of modularity is to provide a framework that will accept any type of cable and type of connector interface. Figure 4–33 shows one approach to a modular outlet that uses pluggable interfaces. Each interface—modular jack, BNC connector, data connector, and so forth—is placed on a small printed circuit board. This board, in turn, plugs into an edge connector in the outlet. The edge connector connects to the horizontal cable.

What makes the system so flexible is that the interface modules can be changed without changing the horizontal cabling. The four-pair horizontal UTP, for example, always terminates in the edge connector in the same pattern. The wiring of the printed circuit board/modular jack insert can be changed to suit any need, general or application-specific. Even though all edge connectors are wired identically, you can obtain modular jack inserts wired for 568A or 568B patterns. You can also plug in data connector interfaces for Token Ring or coaxial interfaces that include circuitry for converting from coaxial cable to UTP.

Modularity has several benefits:

- It allows both current, emerging, and legacy applications to be easily supported and even intermixed.
- It provides an easy upgrade path for technology without requiring excessive recabling or changes in hardware.

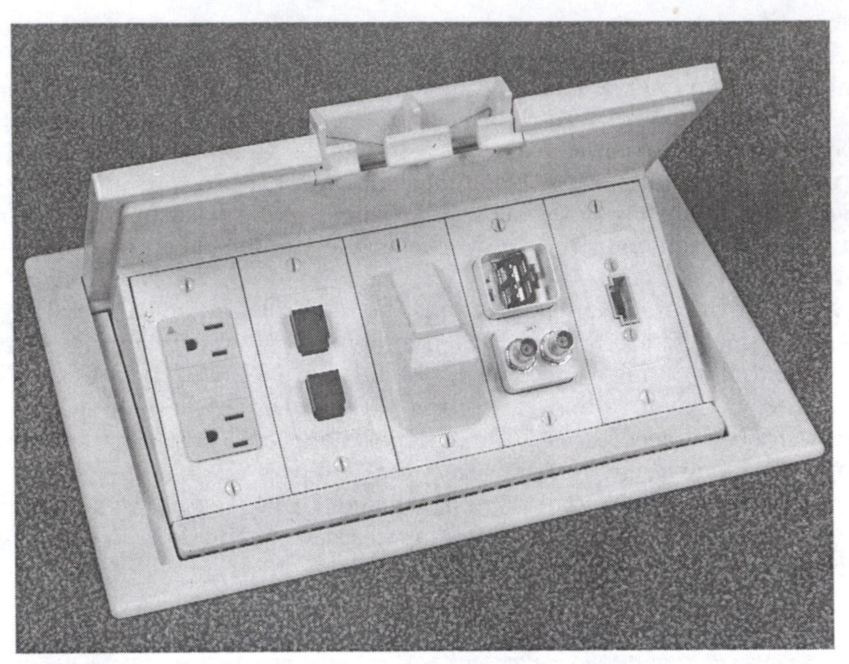

Figure 4–33. Modular floor outlet *(Courtesy of AMP Incorporated)*

- It allows a single outlet to be flexibly wired for different applications. Many outlets accept four inserts. They can be mixed and matched.
- It promotes open systems by allowing both standard and proprietary inserts to be used to achieve vendor-independent cabling.

Other systems use a similar approach, but eliminate the edge connector. The plug-in interface (such as a modular jack) and the horizontal cable termination (such as a 110 block) are all part of a single module. Again, barrel and 110-style IDC termination predominate.

Patch Cords

A patch cord is a unique type of connecting hardware. Most connecting hardware specifications are given for a mated plug/jack combination. A modular patch cord is a hybrid arrangement consisting of two modular plugs (but no jacks) and a length of cable (which can be several meters or more long). Therefore, the specifications given for other types of connecting hardware cannot be directly applied to patch cords. Until recently, there was no sure way to test a patch cord except to insert it into a channel and test the channel. If the channel passed, the patch cord could be assumed to be good. Since patch cords are subject to wear and user abuse, it is desirable to have a means to test them.

This situation was recently rectified with the publication of Addendum No. 4 to TIA/EIA-568-A (TIA/EIA-568-A-4) in December 1999. This document gives specifications and test procedures for testing modular patch cords without inserting them into a channel. This allows cable test equipment to provide patch cord testing functionality in addition to link/channel testing.

SUMMARY

- Connecting hardware is essential to connecting cables to equipment and to other cables.
- The most popular copper connector for premises cabling is the 8-position modular plug and jack.
- 110 cross connects are used for high-density, high-performance cross connect applications in equipment rooms; Type 66 blocks are common in lower-performance applications.
- The 50-pair ribbon connector is used for 25-pair UTP.
- The data connector is used for IBM-style STP.
- Category 7 cable uses both RJ-45-style and non-RJ-45-style connectors.
- No specific fiber-optic connector type is defined by TIA/EIA-568B; any connector meeting the performance criteria can be used.
- The ST and SC connector are traditional fiber-optic connectors used in premises cabling and network applications. The SC was the specified connector in TIA/EIA-568A.
- New small-form-factor (SFF) fiber-optic connectors give the same port density as copper modular jacks.
- There are many SFF connectors, including the LC, SMI, MT-RJ, and VF-45.
- A fiber-optic array connector can terminate up to 12 fibers.
- Careful application of a connector to the cable is critical to achieving the proper performance of the cable assembly.
- Patch panels and telecommunications outlets allow fast, easy connection and disconnection of cables.

REVIEW QUESTIONS

1. What is the most commonly used connector style for copper connectors in premises applications?
2. What is the common name for a termination in which the unstripped conductor is inserted between two opposing beams?
3. How many total contacts does a Category 7 RJ-style connector have? Why?
4. Why should you always wire all eight positions of a modular connector?
5. What is the main advantage of SFF optical connectors?
6. What is the maximum insertion loss allowed for an optical connection? What is a typical range for connectors?
7. For a Category 5e application, what two types of connecting hardware would be typically used in an equipment room?
8. What are two aspects of backwards compatibility?

9. What fiber-optic connector is specified by TIA/EIA-568B?
10. What is meant by port density? If a patch panel is designed to hold 12 modular jacks, how many SC connectors will it accommodate? LC or MT-RJ connectors?
11. Why is backwards compatibility important for Category 6 connecting hardware?
12. What is the functional difference between T568A and T568B jacks?
13. What is the advantage of operating on UTP pins 1-2 and 7-8 as TP-PMD does?
14. What are the four major types of UTP connecting hardware?
15. Why are special test procedures needed for cords?
16. Why do pairs 1 and 2 usually have the worst NEXT performance in a T568A modular plug/jack?
17. Why are coax connectors much simpler than UTP connectors?
18. What is a hybrid cord? In what circumstances would one be needed?
19. What is the difference between a splice and a connector?
20. Why do fiber patch panels use duplex connectors?

CHAPTER 5

Networks

Objectives

After reading this chapter, you should be able to:

- List the four main types of network topologies used in premises cabling.
- Understand the layers in the OSI model.
- Distinguish the main physical topologies used in LANs.
- Distinguish between the different flavors of Ethernet, including data rates and media.
- Discuss the importance of QoS in high-performance networks, especially multimedia.
- Describe the basic difference between a shared-media and switched-media network.
- Discuss the video formats used in the United States and internationally.
- Discuss how video applications differ from typical network applications.
- Identify typical examples of network devices.
- Describe the role of baluns and media filters in premises cabling applications.
- Discuss telephony applications, including VoIP.

Since the earliest days of structured cabling, the trend towards higher-speed networks has driven the performance of premises cabling. Category 5 cable, for example, resulted from the need to support 100 Mb/s networks. This trend has continued unabated. Today, the requirements for 10-Gigabit Ethernet are driving the specifications for Augmented Cat 6 UTP. Although TIA-568b permits the installation of Cat 3 UTP for some telecommunications outlets, almost all new installations are using Category 5e or better UTP (usually Cat 6) to provide flexibility to support new applications. Multimode fiber performance has also increased with the migration from 62.5 μm fiber to 50 μm fiber, and then to laser-optimized 50 μm fiber.

As you saw in Chapter 1, an important feature of building cabling standards like TIA/EIA-568B and ISO/IEC 11801 is that they are application-independent. Rather their recommendations aim to achieve a level of performance that allows any application with the same performance requirements to operate on the cabling system. Ideally a 100-MHz cabling system will accommodate any network up to 100 MHz. Nevertheless, because the cabling system and the network are so intimately related in real life, this chapter provides an overview of local area networks. The perspective is the physical layer—that part that includes the cable and attached equipment.

LANs

A LAN is a sophisticated arrangement of hardware and software that allows stations to be interconnected and pass information between them. The *topology* of a LAN refers to its physical and logical arrangement. Figure 5–1 shows common network topologies.

A *bus structure* has the transmission medium as a central line from which each node is tapped. Messages flow in either direction on the bus. Most bus-structured LANs use coaxial cable as the bus.

A *ring* structure has each node connected serially in a closed loop. Messages flow from one node to the next in one direction around the ring.

A *star* topology has all nodes connected at a central point, through which all messages must pass.

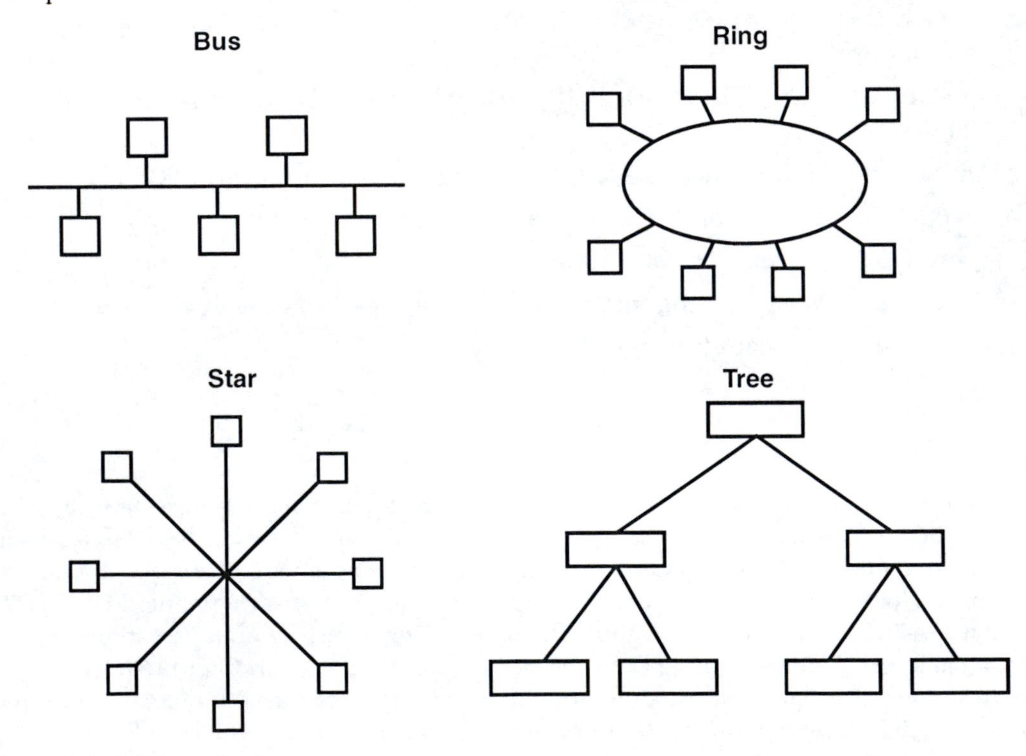

Figure 5–1. LAN topologies

A *tree* topology uses a branching structure. Most often this forms a *hierarchical star* layout such as is used in premises cabling. There are several star points, with each star center feeding a star higher up the chain. A single star occupies the top. Many practical configurations—including networks and premises cabling—use a tree topology.

Hybrid topologies combine one or more of the basic topologies.

It is important to distinguish between a *logical* topology and a *physical* topology. A logical topology defines how the software thinks the network is structured—how the network is "philosophically" constructed. A physical topology defines how the LAN is physically built and interconnected. For example, the Ethernet LAN is a logically bus-structured network. Its physical topology depends on the specific variety of network. Sometimes it is wired like a traditional bus. Other times it is wired as a star or a hybrid configuration. Regardless of how it is wired or interconnected, the LAN appears to the control software as a bus network. Similarly, although Token Ring is a ring network, it is also configured as a star with all stations connecting through a hub.

Star-wired networks are popular because they increase reliability. Cable breaks in ring and bus structures can bring down the entire network. In a star network, only the affected portion crashes; the rest of the network can continue. The cabling topology recommended in TIA/EIA-568B and ISO/IEC 11801 is a hierarchical star configuration.

Network Layers

As data communications become an important part of business, a need arises for universality in exchanging information within and between networks. In other words, there is a need for standardization—a structured approach to defining a network, its architecture, and the relationships between the functions.

In 1978, the International Standards Organization (ISO) issued a recommendation aimed at establishing greater conformity in the design of networks. The recommendation set forth the seven-layer model for a network architecture shown in Figure 5–2. This structure,

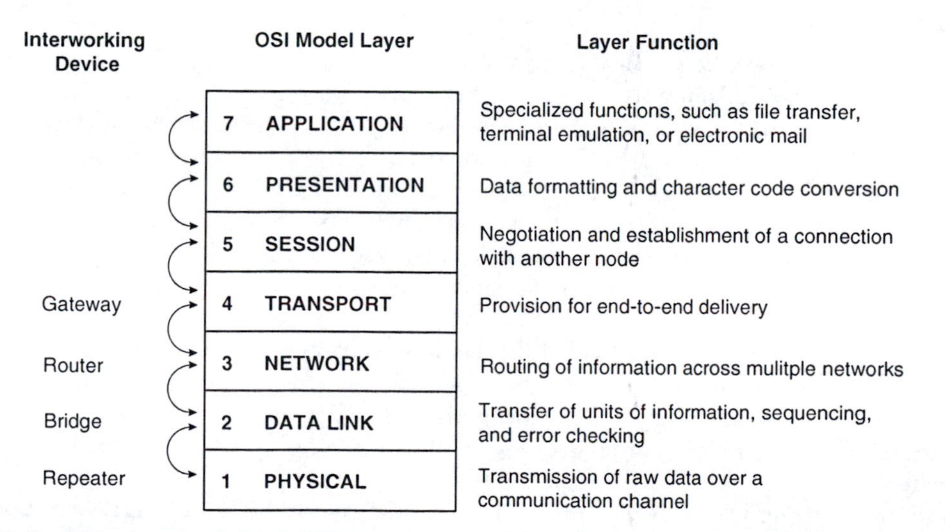

Figure 5–2. OSI network model

known as the Open Systems Interconnection (OSI), provides a model for a common set of rules defining how parts of a network interact and exchange information. Each layer provides a specific set of services or functions to the overall network. The three lower layers involve data transmission and routing. The top three layers focus on user needs. The middle layer, transport, provides an interface between the lower three layers and the top three layers. Unfortunately, each layer has numerous standards that define the layer differently. The network systems in this chapter, for example, define the lower layers differently.

The following is a brief description of each layer.

Physical Layer

The physical layer is where the rubber meets the road. This is where the network actually interacts with the laws of physics. The goal of the physical layer is to communicate bits of information over a single transmission link. For all types of physical layer, the type of transmission media, connector, power levels, bit rate, and noise environment are important considerations. For a copper cable, the relevant concepts also include the current and voltage levels, signal waveshape, cable impedance, encoding technique, etc. For a fiber cable, the physical layer deals with the wavelength, power levels, type of fiber, type of source (laser, VCSEL, or LED) etc. For a wireless channel, the physical layer is concerned with the frequency band, power levels, type of antenna, and the modulation method. The various cable and connector parameters that we discussed in previous chapters all come into play at the physical layer.

One of the most widely known physical layer specifications is EIA-232C (formerly known as RS-232C), which is typically used for the serial port on a PC. EIA-232C specifies the signal definitions, voltage levels, type of connector, etc.

Cabling systems, which are the subject of this book, exist at the physical layer. In order to understand the demands placed on the physical layer, it is helpful to understand the functions of the higher layers.

Data-Link Layer

Once you get above the physical layer, you are concerned with logical concepts like messages, headers, acknowledgements, etc. rather than physical things like volts and ohms. The job of the data link layer is to take the transmission facility provided by the physical layer and reliably transmit information that is free from undetected errors. To do this, the information is usually divided up into frames, to which a header and trailer are added. The header and trailer contain additional information describing the data, such as addresses, sequence numbers, and checksums for error detection and/or correction. For asynchronous data, link layer protocols often include a mechanism for acknowledgement of properly received frames and retransmission of errored frames.

Networks like Ethernet and Token Ring exist on the physical and data-link layers. In a PC network, chips on the PC's network interface card perform the physical and data link functions. Software in the computer performs the functions of the higher layers. From the viewpoint of this book, building cabling and networks are distinguished at the physical and data-link layers.

In networks using broadcast channels (such as Ethernet and Token Ring), as opposed to point-to-point links, the data-link layer is often divided into two sublayers. The lower sublayer (closest to the physical layer) is called the Medium Access Layer (MAC). The MAC provides a method for determining who may transmit next on the shared broadcast channel. Examples of MAC layer protocols are the CSMA method used in Ethernet, the token-passing algorithm used in Token Ring, and the ALOHA protocol used in radio networks.

Where a MAC sublayer is used, the upper part of the data-link layer (nearest the network layer) is often referred to as the Logical Link Control (LLC) sublayer. The LLC provides the remainder of the data-link layer functions as described above.

Network Layer

The network control layer addresses messages to determine their correct destination, routes messages, and controls the flow of messages between nodes. The network layer, in a sense, forms an interface between the PC and the network. It controls where the data on the network go and how they are sent. The Internet uses protocols called transmission control protocol and Internet protocols (TCP/IP) to control communications between computers on the net. IP works on the network layer.

Transport Layer

This layer provides end-to-end control once the transmission path is established. It allows exchange of information independent of which systems are communicating or their location in the network. This layer performs quality control by checking that the data are in the right format and order. TCP exists on the transport layer.

Session Layer

The session control layer controls system-dependent aspects of communication between specific nodes. It allows two applications to communicate across the network.

Presentation Layer

At the presentation layer, the effects of the layer begin to be apparent to the network user. This layer, for example, translates encoded data into a form for display on the computer screen or for printing. In other words, it formats data and converts characters. For example, most computers use the American Standard Code for Information Interchange (ASCII) format to represent characters. Some IBM equipment use a different format, the Extended Binary Coded Decimal Information Code (EBCDIC). The presentation layer performs the translation between these two formats.

Application Layer

At the top of the OSI model is the application layer, which provides services directly to the user. Examples include resource sharing of printer or storage devices, network management, and file transfers. Electronic mail is a common application-layer program. This is the layer directly controlled by the user.

It is often useful to think of the seven layers as comprising two groups. The lower three layers (physical, link, and network) deal primarily with *data communication.* They are concerned with transmission, reception, error checking, routing, and so on. The upper three layers (session, presentation, and application) deal primarily with *data processing*. They are concerned with file formats, record structures, login procedures, user interfaces, and so on. Layer 4, the session layer, is the glue that attaches the data communication layers to the data processing layers.

In practice, data flows from the top layer of one station (the message originator) to the top layer of another station (the message recipient). A message, such as electronic mail, is created in the top presentation layer of one workstation. The message works its way down through the layer until it is placed on the transmission medium by the physical layer. At the other end, the message is received by the physical layer and travels upward to the presentation level. It is at the presentation level that the electronic mail is read.

This layered approach to building a network holds two benefits. First, an open and standardized system permits equipment from different vendors to work together. Second, it simplifies network design, especially when extending, enhancing, or modifying a network. For example, while most LANs use copper wire, either coaxial cable or twisted pairs, adding a fiber-optic point-to-point link involves only the physical layer. Higher levels of the OSI model do not care how the physical layer is actually implemented; they care only that the physical layer follows certain rules in interacting with higher levels.

Bear in mind that the OSI model is a reference model. While the ISO has created protocol standards for each level, they are not widely used. Different systems use different protocols. For example, Novell NetWare and Microsoft Windows networking use different protocols for the different layers. IBM's System Network Architecture (SNA) and Digital Equipment's Network Architecture (DNA) have yet different rules. Nor does a system have to use all seven layers; a system can combine two layers into a single set of protocols.

Since Novell NetWare is one of the most popular network operating systems for PCs, here are the protocols used:

7. Application: NetWare shell
6. Presentation: Network Core Protocols (NCP)
5. Session: NetBIOS Emulator
4. Transport: Sequenced Packet Exchange (SPX)
3. Network: Internet Packet Exchange (IPX)
2. Data Link: Open Data-Link Interface (ODI)
1. Physical: (Specific network type: Ethernet, Token Ring, etc.)

A network can be connected to other networks of the same or different type. Sometimes users purposefully break a large network into several smaller segments to make the network more efficient. There are several ways that these network segments can be connected, including repeaters, bridges, and routers, as illustrated in Figure 5–3.

A repeater (Figure 5–3a) is a physical layer device. Its function is to regenerate an electrical or optical signal. It is used to connect one LAN segment to a similar LAN segment which operates at the same data rate. It enables extended transmission distances. It does not terminate any protocol layers.

A bridge (Figure 5–3b) connects two or more LANs at the data link layer. It terminates the physical layer and the data link layer protocols. A bridge examines the destination address in

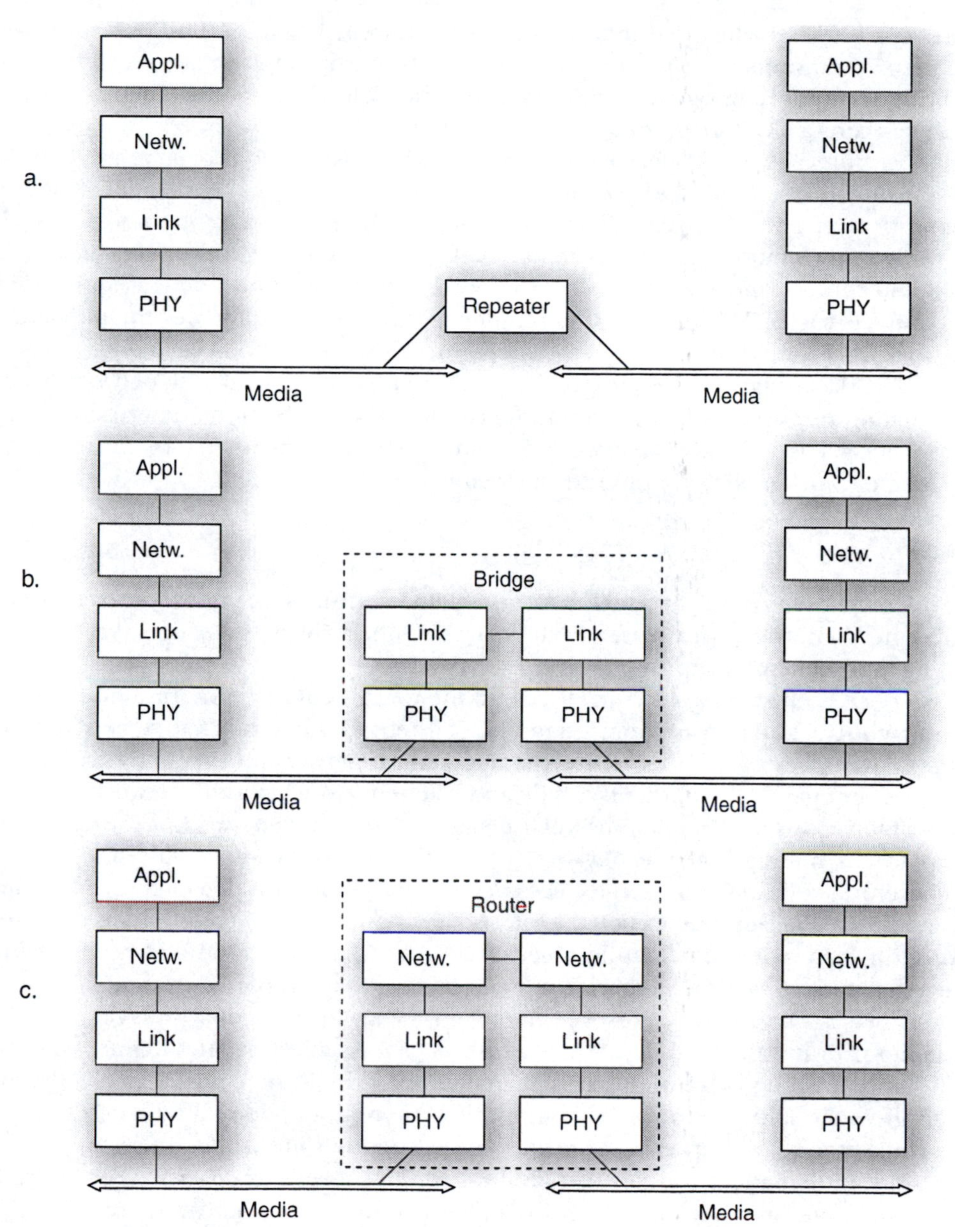

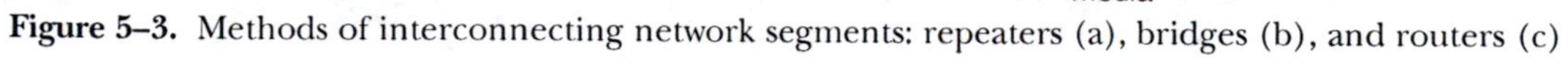

Figure 5–3. Methods of interconnecting network segments: repeaters (a), bridges (b), and routers (c)

each frame. If a frame is addressed to a station on a different LAN, the bridge receives and buffers it, then retransmits it on the destination LAN. If the frame is already on the proper LAN, the bridge ignores it. Bridges are insensitive to network level protocols—they do not care if the frame contains a TCP packet or some other type of data.

A router (Figure 5–3c) is a network level device which connects two or more LANs. It terminates the physical, data link, and network layer protocols. It examines the destination address in each network layer packet to determine how to handle the packet. Routing a packet at the network layer is a much more complex operation than bridging a packet at the data link layer.

Bridges and routers can be used locally or remotely. In a local application, a single bridge or router connects each LAN segment. In a remote application, typically over the public telephone network or private or leased lines, a bridge or router is required at each end.

A gateway works at higher levels, serving as an entry point to a local area network from a larger information resource such as a mainframe computer or a telephone network.

The ISO model is a handy framework for understanding the structure of a network although some network models do not use all seven layers.

Access Method

Access refers to the method by which a station gains control of the network to send messages. Three methods are carrier sense, multiple access with collision detection (CSMA/CD); token passing; and demand priority.

In CSMA/CD, each station has equal access to the network at any time (multiple access). Before seizing control and transmitting, a station first listens to the network to sense if another workstation is transmitting (carrier sense). If the station senses another message on the network, it does not gain access. It waits awhile and listens again for an idle network.

The possibility exists that two stations will listen and sense an idle network at the same time. Each will place its message on the network, where the messages will collide and become garbled. Therefore, collision detection is necessary. Once a collision is detected, the detecting station broadcasts a collision or jam signal to alert other stations that a collision has occurred. The stations will then wait a short, random period for the collision to clear and begin again.

In the token-passing network, a special message called a token is passed from node to node around the network. Only when it possesses the token is a node allowed to transmit.

In demand priority, a central device acts as a broker for stations requesting access to the network. A station wishing to transmit sends a signal to the hub. If the network is idle, the hub immediately gives the station permission to transmit. If more than one request is received, the hub uses a round-robin arbitration to acknowledge each request in turn.

One feature of demand priority is that it allows two or more levels priority. For example, time-sensitive messages like video can be given higher priority than other transfers.

Frames

The information on a network is organized in frames (also called packets)[1]. A frame includes not only the raw data, but a series of framing bytes necessary for transmission of the

[1] *The term* **frame** *is normally used to refer to a data structure at the link layer, while a* **packet** *is a similar structure at the network layer. However, some specific networks may use these terms differently.*

data. Figure 5–4 shows the frame formats for the Ethernet, FDDI, and ATM data transmissions. Notice that ATM has a short, fixed length format of 53 bytes. We will discuss the significance of this later in this chapter when we describe ATM. Other frames can also be used. For example, besides the data frame, a Token-Ring network also uses a token frame for passing the token around the network.

Here is a brief description of the elements of an FDDI data frame.

The *preamble* indicates the beginning of a transmission. The preamble is a series of alternating 1s and 0s—1010101010 . . .—that allows the receiving stations to synchronize with the timing of the transmission.

The *start of frame* is a special signal pattern of 10101011 that signals the start of information.

The *packet ID* identifies the type of packet, such as data, token, and so forth.

The *destination address* is the address of one or more stations that are to receive a message. In a network, each station has a unique identifier known as its address.

The *source address* identifies the station initiating the transmission.

The *data* is the information, the point of the transmission.

The *CRC* or *cyclic redundancy check* is a mathematical method for checking for errors in the transmission. When the source sends the data, it builds the CRC number based on the patterns of the data. The receiving station also builds a CRC. If the receiver CRC matches the transmitted CRC, no errors have occurred. If they don't match, an error is assumed, the transmitter is informed, and the transmission is sent again.

The *end of frame* informs the receiving station that the transmission is over. It also contains the status of the transmission. The receiving station marks the status to acknowledge receipt of the packet.

Preamble	Start of Frame	Destination Address	Source Address	Length	Data	CRC
7 bytes	1 byte	6 bytes	6 bytes	2 bytes	46-1500 bytes	4 bytes

IEEE 802.3 Frame

Preamble	Start of Frame	Packet ID	Destination Address	Source Address	Data	CRC	End of Frame + Status
≥ 2 bytes	1 byte	1 byte	2 or 6 bytes	2 or 6 bytes	0-4486 bytes	4 bytes	≥ 2 bytes

FDDI Data Frame

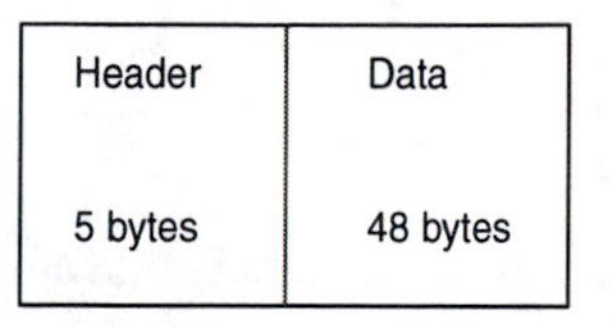

Header	Data
5 bytes	48 bytes

ATM Cell

Figure 5–4. Frame formats: Ethernet, FDDI, and ATM

IP

IP is part of the protocol stack used for the Internet. The TCP/IP (Transport Control Protocol/Internet Protocol) were originally developed in the 1960s and have been evolving ever since. The TCP/IP stack does not strictly follow the ISO model, but omits a couple of layers (session and presentation) and combines a couple of others (data link and physical). IP is the network layer protocol, and TCP is the transport layer protocol. It supports a variety of application layer protocols for e-mail, file transfer, and so on. The purpose of TCP/IP is to allow the interconnection of networks; therefore, it does not specify a protocol at the data link physical layers, but allows the use of any protocols there.

The operation of the TCP/IP protocol suite to interface from one network to another is illustrated in Figure 5–5. In this example, a user at a client PC on a 10-Mbps Ethernet wants to access a server on a 1-Gbps Ethernet. The IP protocol provides addressing information to allow the router to connect the incoming packets to the proper network (the router in Figure 5–5 would normally be connected to several networks; only two are shown for clarity). The operation of the IP protocol is independent of the underlying physical and data link protocols, as long as the router can support the physical data link protocols used by the two networks—in this case 10Base-T and 1000Base-T. The TCP and application protocols operate end-to-end and do not know or care that the two endpoints are on different networks.

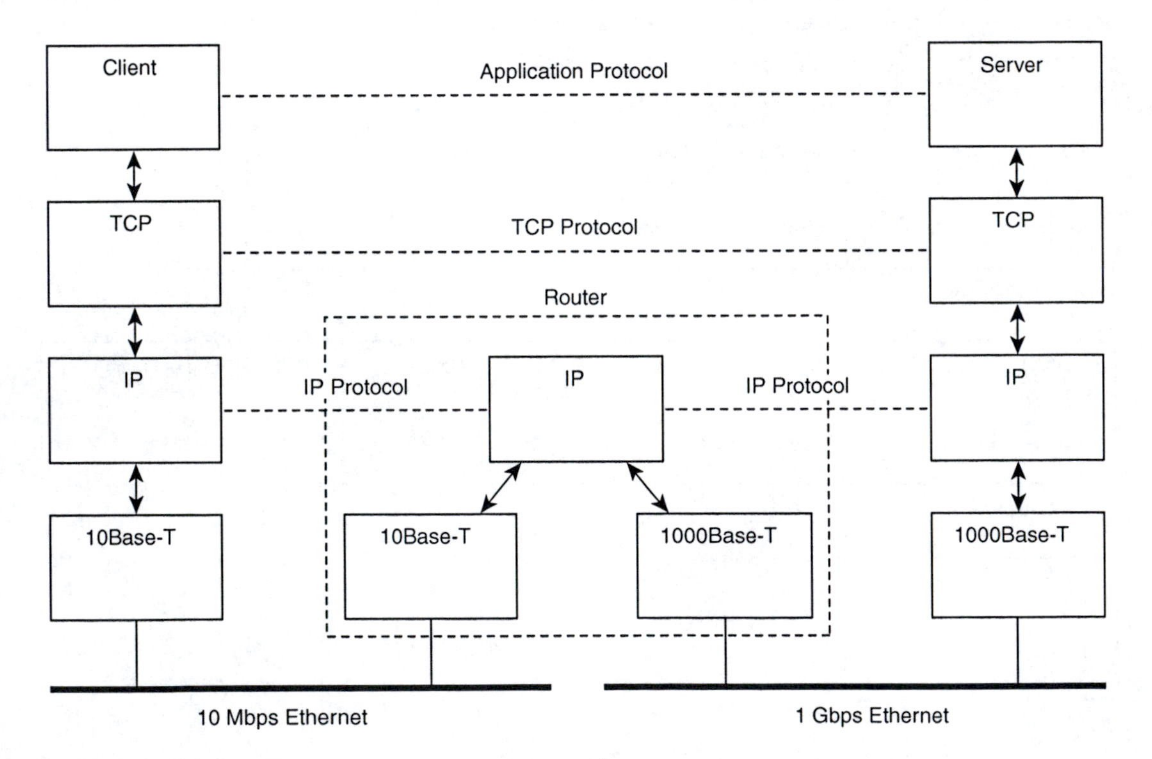

Figure 5–5. Using an IP router to connect two networks

The TCP provides reliable end-to-end communication, using sequence numbers and acknowledgments for each packet. There is another, simpler protocol called UDP (User Datagram Protocol), which provides unacknowledged single-packet data transfer for applications that do not need the complexity of TCP.

IP packets contain a 20-byte header, and have a variable-size data field. The header contains a 16-bit "total packet length" field; therefore, the maximum packet length is 65,535 bytes. This contrasts with the ATM cell, which is always 53 bytes long.

Remember that TCP/IP was designed for packet networks. Because of this, it does not provide support for guaranteed bandwidth connections like ATM does. Applications such as voice or video, which need guaranteed bandwidth to support a fixed-rate bit stream, can be accommodated with an additional protocol such as ITU-T H.323. Standards work is continuing to address this issue, which is often referred to as QoS (Quality of Service), discussed below.

Because IP is the protocol used by the Internet, it is growing in importance as a general-purpose protocol to support a wide range of applications. Support for IP is growing as the Internet becomes a platform for business. Many businesses that used other methods for communications now use the Internet to connect remote offices. Voice-over IP (VoIP) is an area of great interest. In short, both the data networking companies and the telephone companies support IP.

Shared Versus Switched Media

Early networks like Ethernet and Token Ring use a shared-media concept in passing messages. The message goes to every station on the network, with each station checking the address to see if the message is for it.

More recently, switched media has become popular. In a switched-media network, a central device such as a hub contains a switching network. The hub checks the address of the message and then sends the message only to the correct station.

Switched-media networks rely on a star-wired configuration, because there must be intelligence at the star's center to decode the address and perform the switching. Network topologies like rings and busses must be shared media because the intelligence for checking addresses lies in the stations. The message must go to each station.

Figure 5–6 shows the difference between shared and switched media networks. Switched networks make much more efficient use of network resources. The time a message spends on the network is significantly reduced because it passes directly from the originating station through the hub to the destination station. Switching increases network availability and can serve as an alternative to a higher speed network. A 10-Mbps switched Ethernet network (rather than a 100-Mbps Fast Ethernet) may serve as a cost-effective upgrade to a 10-Mbps shared-media Ethernet.

The key difference between shared- and switched-media networks lies in the bandwidth available to each user. In a shared-media Ethernet network, the 10-Mbps bandwidth is shared by the network and attached users. As the number of stations increases, the bandwidth per station drops. In a switched Ethernet network, each station has its own 10-Mbps link. The aggregate or total bandwidth is determined by the central switch, which has a bandwidth many times greater than 10-Mbps.

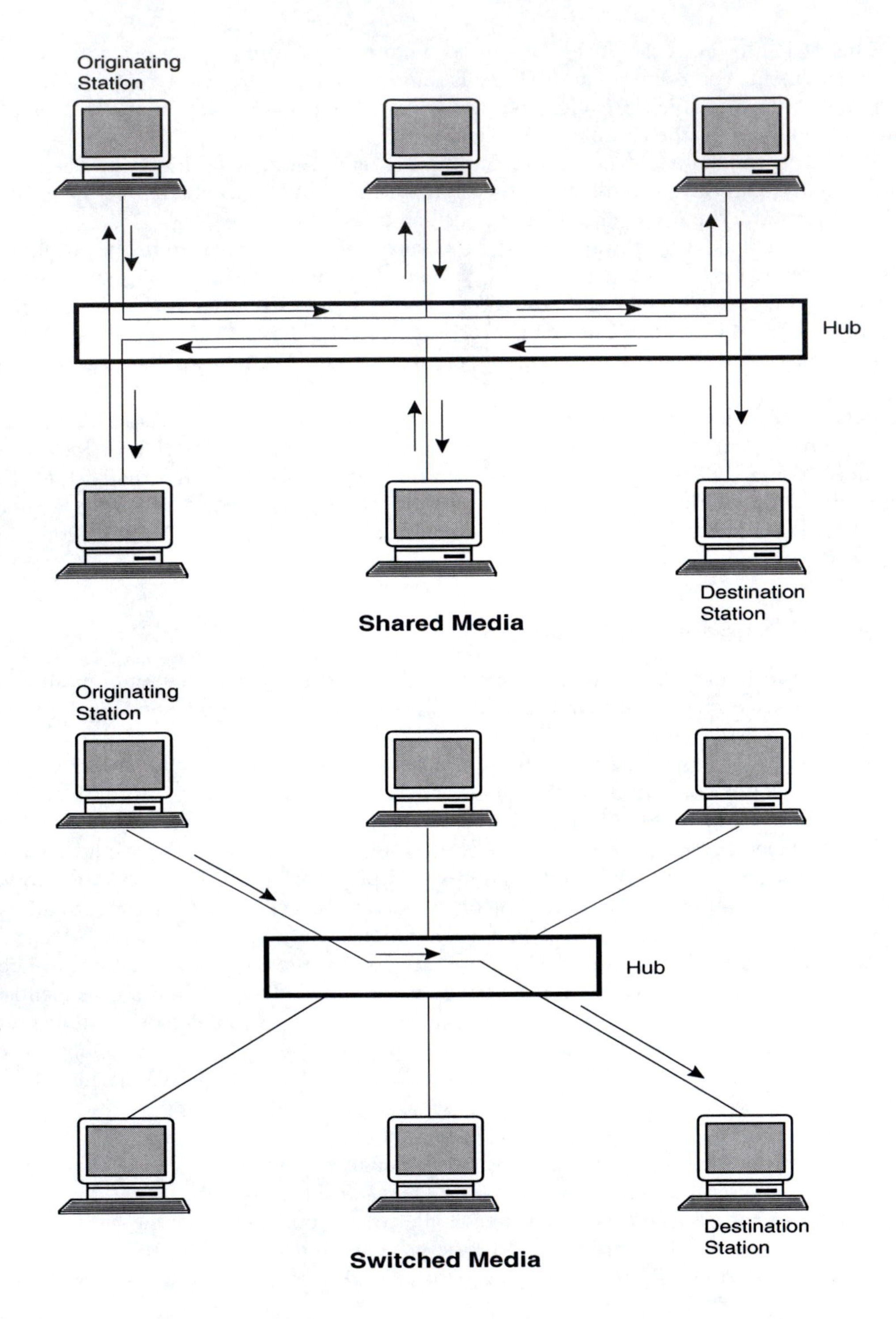

Figure 5–6. Shared versus switched media

Quality of Service

Quality of service (QoS) refers generally to the ability of a network to deliver an application with the performance it requires. In modern usage, QoS refers to providing guaranteed bandwidth with a low and deterministic delay. QoS became of interest as the delivery of video and other time-sensitive, real-time applications became practical. High-quality video requires a steady stream of data. Unless you enjoy jerky, awkward video, the network must be able to guarantee a certain level of delivery. The original networks were incapable of such guarantees. Because the bandwidth was shared among many users who traded large and small files, a steady delivery of data was impractical. It wasn't so much that the network did not have the bandwidth: the bandwidth was there, but there was no mechanism for allocating it among users in a way that *guaranteed* delivery within a specified time.

Although Ethernet has become the dominant LAN technology, it does inherently support guaranteed bandwidth, and therefore cannot guarantee QoS. Ethernet supports isochronous data by means of overprovisioning—that is, providing much more bandwidth than is needed so that delays are negligible. This type of best-effort (as opposed to guaranteed) QoS usually, but not always, avoids problems.

Other types of networks do inherently support isochronous traffic and therefore can offer guaranteed QoS. Asynchronous transfer mode (ATM), which we will discuss later in this chapter, offers a particularly rich approach to QoS with many different levels of service. IEEE 1394 (also known as FireWire) is another type of network that provides inherent support isochronous data streams.

Types of Networks

Many different types of LAN protocols are used in premises networks. Among the most common are Ethernet, FireWire, Token Ring, FDDI, Fibre Channel, and ATM.

IEEE 802.3 Ethernet

IEEE 802.3 is a CSMA/CD LAN commonly called Ethernet, by far the most popular type of network in use worldwide. Today, Ethernet controls more than 90 percent of the market and is actually growing more popular. Nearly all new LANs use some variety of Ethernet. Other networks, such as Token Ring or FDDI, are found in existing applications and are selected primarily to extend or expand existing networks.

Ethernet was originally a LAN defined by Xerox, Digital Equipment, and Intel that used a thick coaxial cable as the transmission medium. Since then, the standard has evolved considerably to be much more flexible, although Ethernet is still widely used to describe this network. Minor differences exist between the Ethernet specification and IEEE 802.3. Still, most people use the two terms interchangeably, and so shall we.

Figure 5–7 lists several flavors of IEEE 802.3. The prefix number defines the operating speed, the middle word defines the signaling techniques, and the suffix defines the transmission medium. For example, 10BASE-T means a network operating at 10 Mbps, using baseband communication, and transmitting over twisted-pair cable. The cable type listed is the type of predominant cable for which the network was designed.

Name	Media	Bit Rate
1BASE5	UTP	1 Mb/s
10BASE-T	Cat 3 UTP	10 Mb/s
10BASE2	Thin coax	
10BASE5	Thick coax	
FOIRL	MMF	
10BASE-FB	MMF	
10BASE-FP	MMF	
10BASE-FL	MMF	
10BROAD36	Coax	
100BASE-TX	Cat 5 UTP	100 Mb/s
100BASE-T4	Cat 3 UTP	
100BASE-T2	Cat 3 UTP	
100BASE-FX	MMF	
1000BASE-T	Cat 5e UTP	1 Gb/s
1000BASE-CX	STP	
1000BASE-SX	MMF	
1000BASE-LX	MMF	
10GBASE-X	MMF, SMF	10 Gb/s
10GBASE-R	MMF, SMF	
10GBASE-W	MMF, SMF	
10GBASE-CX4	STP	
10GBASE-T	Cat 6 UTP Cat 6A UTP Cat 7 STP	

Figure 5–7. Varieties of Ethernet

While IEEE 802.3 is a logical bus topology, it is constructed as either a bus or a star, depending on the type of cable used. Coaxial cables typically use a physical bus topology: the backbone cable runs serially from workstation to workstation. Twisted-pair and fiber use a logical star, with workstations connecting to a hub (also called a multiport repeater). Each hub contains several ports that permit attachment of a workstation or another hub.

10BASE-5

10BASE-5—thick Ethernet—was the original version of the network, using a thick, rigid coaxial cable and N connectors, as shown in Figure 5–8. It permits up to 100 stations over 500 meters for each segment. Up to three segments can be connected by repeaters, for a maximum of 1500 meters. 10BASE-5 is seldom encountered today.

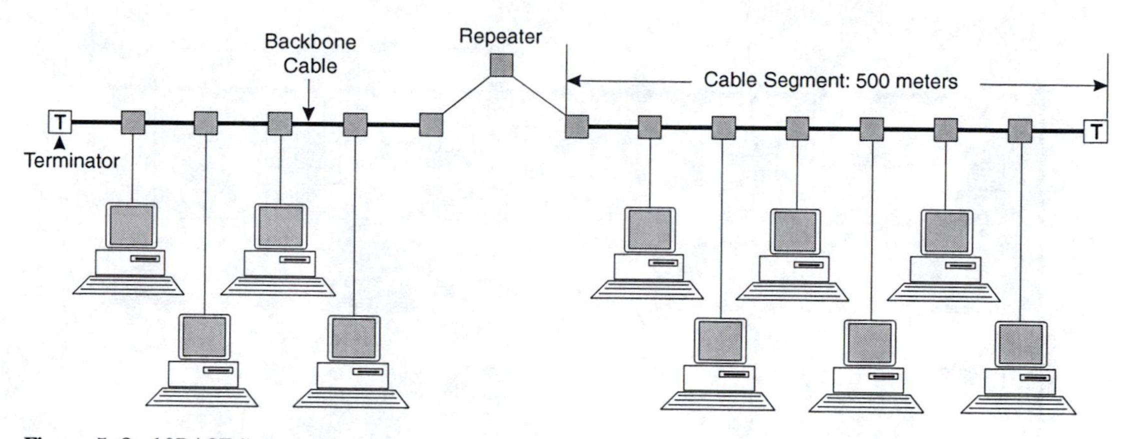

Figure 5–8. 10BASE-5 network

10BASE-2

10BASE-2 is thin Ethernet, using a flexible coaxial cable and BNC connector or thinnet taps. Each segment can be up to 185 meters long, with up to five segments per network. The distance from the backbone bus to the station is 50 meters maximum. Figure 5–9 shows a typical 10BASE-2 configuration.

10BASE-T

10BASE-T is Ethernet on UTP. 10BASE-T departs from the physical bus structure of 10BASE-5 and -2 to use a hub-based star-wired configuration as shown in Figure 5–10. (Hubs have also been made with coaxial ports to make a star-wired 10BASE-2 network.)

The hub is a repeater, so that each signal transmitted from any port on the hub has been amplified and retimed before transmission. 10BASE-T hubs are often called multiport repeaters.

10BASE-T networks can operate on Category 3 or better cable. The maximum recommended distance from switch or hub to station is 100 meters. While distances of 150 meters can be achieved with Category 5 cable, exceeding the 100-meter recommendation is not recommended. Because the cable distance is from station to hub or switch, two stations can be as far apart as 200 meters when they are connected to the same hub or switch. Hubs and switches, too, can be interconnected to extend the distance between two stations. Again, the recommended distance between two hubs or switches connected over UTP is 100 meters. Many 10BASE-T devices also allow fiber backbones that extend the distance between devices to 2 km. You can see that the 100-meter station-to-hub switch limit is really flexible and not so limiting as it first appears.

A standard port acts as a transceiver for the type of cable being used; one function of the transceiver is to ensure that the signal is converted to the form required by the cable.

For example, a UTP cable uses two pairs, one for transmitting and one for receiving, and a modular-jack telephone connector. A coaxial cable uses only a single cable, allowing transmission in both directions over a single wire, and either BNC or N-type connectors. A fiber-optic cable requires two fibers, one for transmitting in each direction, and usually SC or ST connectors.

Another type of port is the attachment unit interface (AUI). The AUI port is a 15-pin subminiature-D connector (similar to the 25-pin connector used in serial ports on personal

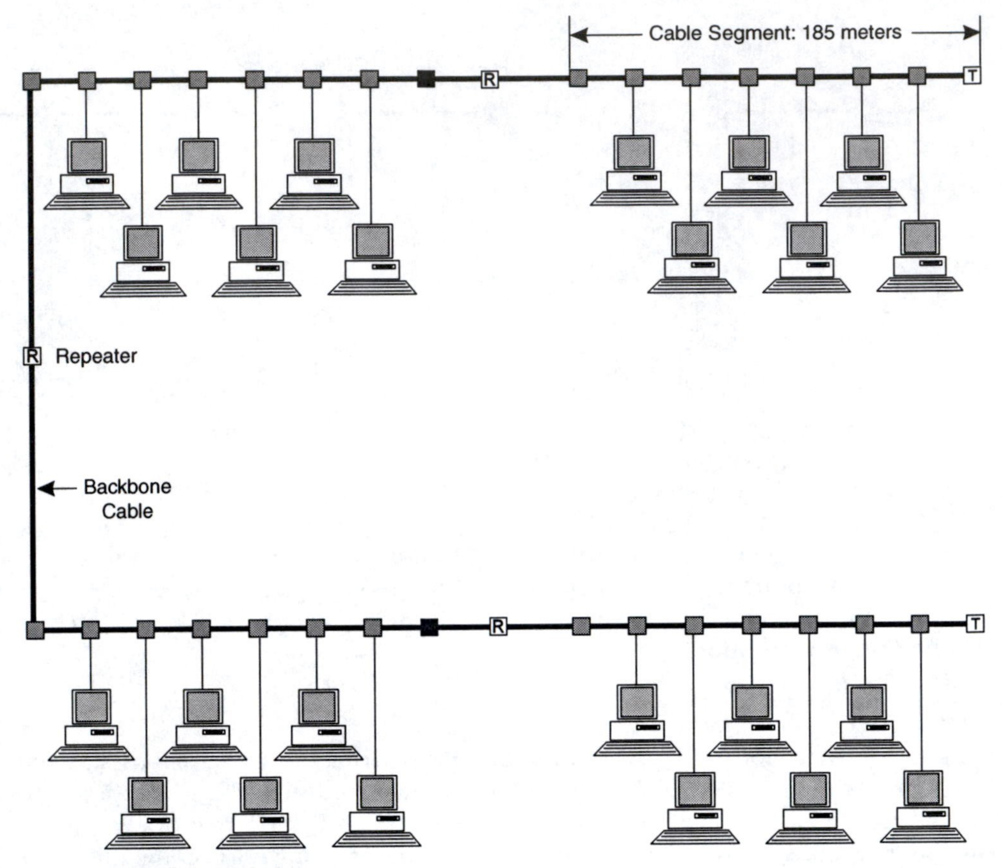

Figure 5–9. 10BASE-2 network

computers) that does not include the transceiver function. An AUI allows different transceivers to be used. For example, consider a network adapter board in a workstation. You can buy a board specific to the type of cable you are using: a 10BASE-T board for twisted-pair cable or a 10BASE-2 board for thin coaxial cable. Alternatively, you can buy a board with an AUI port. The AUI port allows you to plug in a transceiver for whatever cable you wish to use: twisted pair, thick or thin coaxial, or fiber optic. Some high-speed networks call the AUI port a media-independent interface or MII.

10BASE-T networks are prone to cabling errors. Here's why. In the UTP cable, one pair carries signals in one direction; the other pair carries signals in the other direction. The transmitter at one end is connected to the receiver at the other end.

Connecting the transmitter to the receiver can be done in one of two ways as shown in Figure 5–11. The standard way is to have the ports at each end wired differently. Here, the NIC card has a standard port: the transmitting pins are 1 and 2. The receiving pins are 3 and 6. The hub or switch to which the NIC is connected is wired as a crossover port: the receiving pins are 1 and 2 and the transmitting pins are 3 and 6.

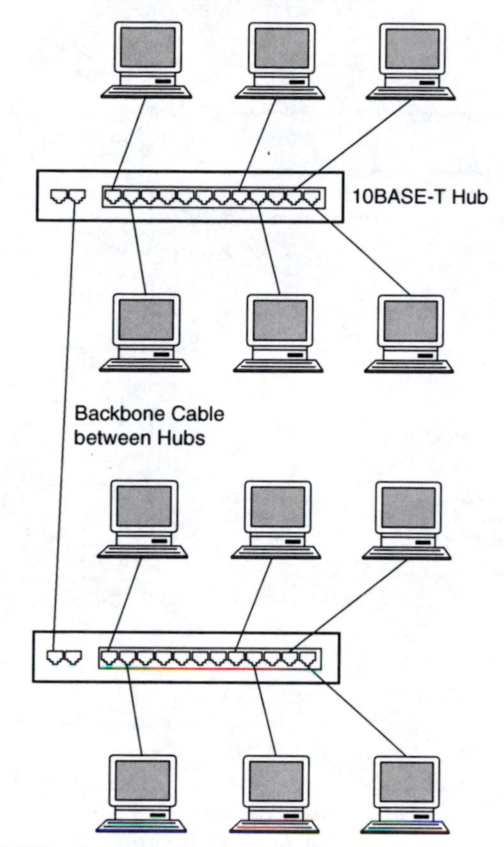

Figure 5–10. 10BASE-T network

This means the cable runs straight through. Pins 1-2 on the transmitting NIC connect to wires 1-2 on the cable and pins 1-2 on the receiving hub. The transmitting lines on the NIC card are thus connected to the receiving lines on the hub.

Instances occur, however, when you want to connect two identical ports: say ports on two different hubs or switches. If you use the same cable you used to connect the NIC to the hub, you are suddenly connecting the receiver to receiver and transmitter to transmitter. This doesn't work. Instead, a crossover cable is required. A crossover cable connects pins 1 and 2 at one end to pins 3 and 6 at the other end. This works. In most installations, crossovers will only be required in patch cables connecting hubs.

One hub can connect to another hub so long as 802.3 wiring rules are followed. The most important is the four-repeater rule. A signal passing from the originating station to the receiving station can pass through no more than four repeaters. Because a typical 10BASE-T network can easily have many hubs, care must be taken to ensure that no signal paths violate the rule. Figure 5–12 shows two methods of minimizing the number of repeaters in a path. One way is to use a single hub to feed multiple hubs. In this case numerous hubs can be used, but only three hubs are ever involved in the path. The second way, which has gained popularity in recent years, is

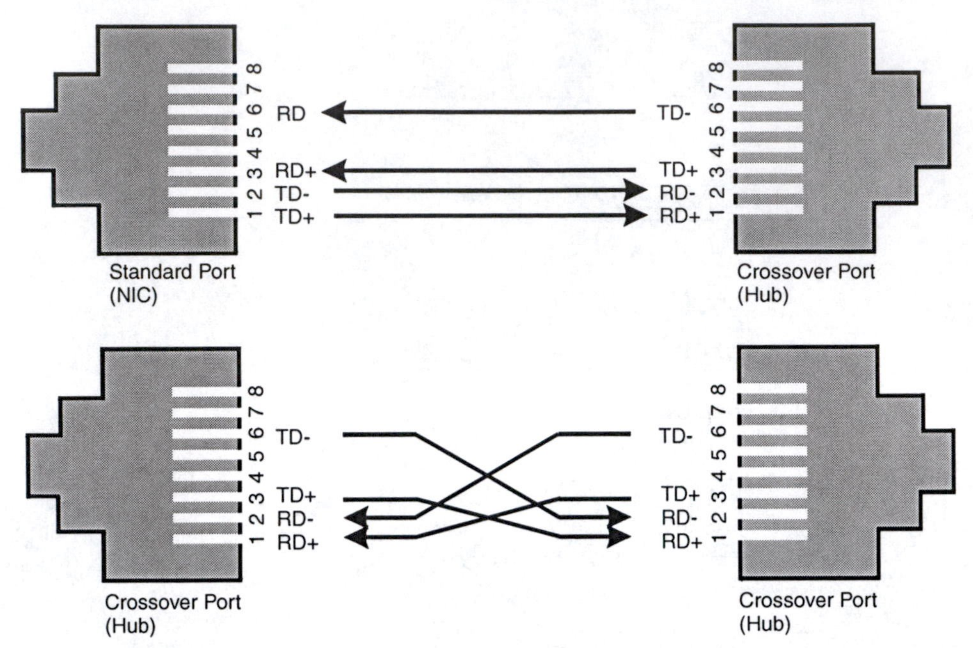

Figure 5–11. Standard and crossover ports and cables

stackable hubs. Stackable hubs are hubs designed so that when they are connected together, they appear to the network as a *single logical hub.* A signal can still pass through four logical hubs. The stackable hub shown in Figure 5–12 stacks four individual 12-port hubs to electrically connect them into a single logical hub having up to 48 ports. Each stack of hubs is treated as a single hub. Stackable hubs greatly simplify the task of avoiding wiring errors in interconnecting hubs.

10BASE-F

The 10BASE-F specification adds a fiber-optic alternative for use in IEEE 802.3 networks. The specification actually lists three different variations: a backbone cable (10BASE-FB), a passive star-coupled network (10BASE-FP), and a fiber-optic link between hub and station (10BASE-FL). Most applications use 10BASE-FL, a simple point-to-point link between two devices. The requirements for a 10BASE-FB and a 10BASE-FL are fairly straightforward. The backbone is simply a point-to-point connection between hubs. The FL version defines connection between a hub and station or between hub and passive star coupler. Notice that the hub is a multiport repeater that contains electronics to regenerate the signal. Each point-to-point link can be up to 2 km long.

IEEE 802.3 Fast Ethernet

Fast Ethernet brings a 100-Mbps upgrade path to Ethernet networks. Also known as 100BASE-X, Fast Ethernet uses the same frame format and medium access-control mechanism as its 10-Mbps sibling. It allows the same applications and network software, but at a tenfold increase in transmission speed.

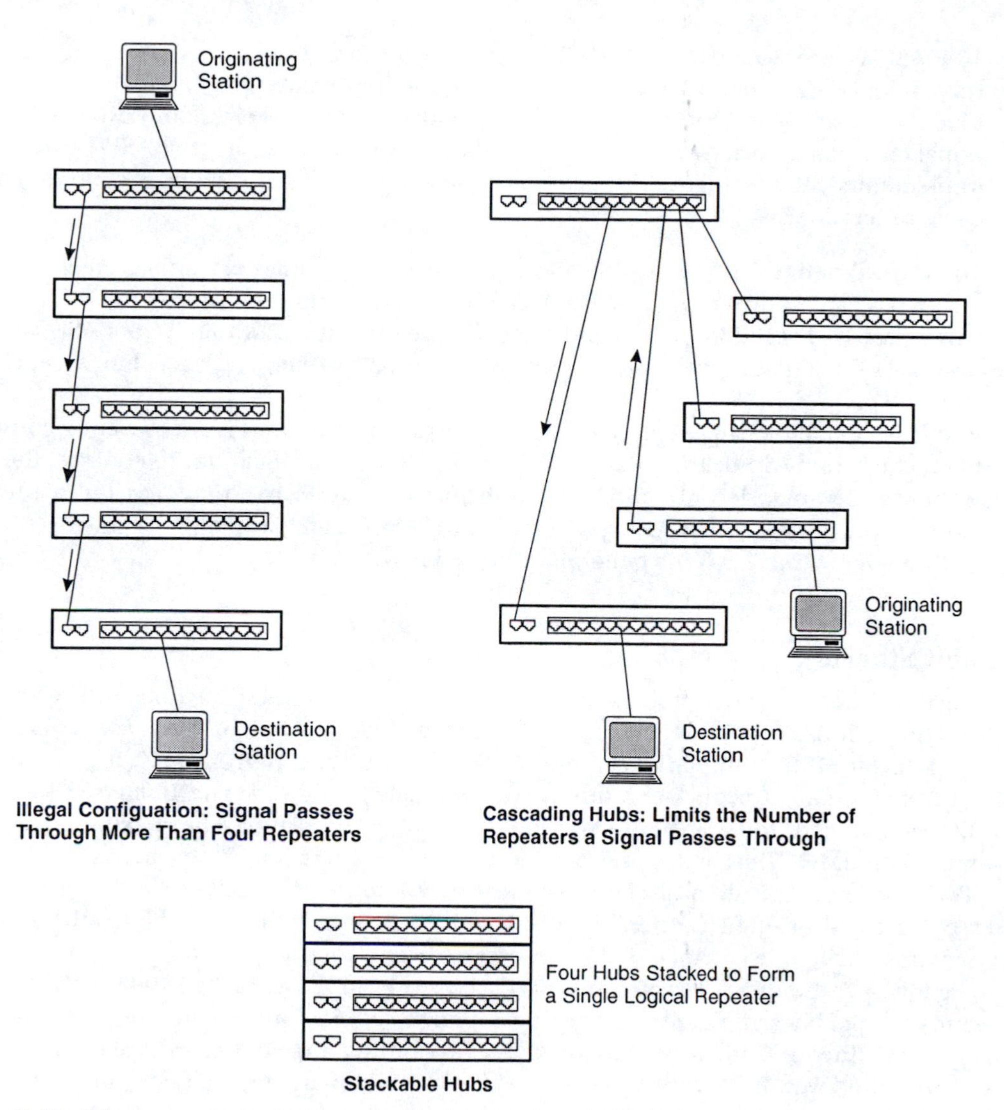

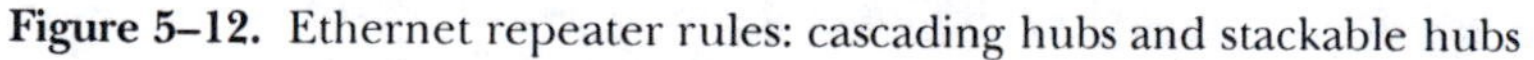

Figure 5–12. Ethernet repeater rules: cascading hubs and stackable hubs

Currently, Fast Ethernet recognizes four variations:

- 100BASE-TX uses Category 5 UTP and STP. The specification for this is derived from the TP-PMD originally created for FDDI on copper. Like 10BASE-T, 100BASE-TX does the crossover inside the hub.
- 100BASE-TX, however, does not use the same RJ-45 pinouts as TP-PMD. To preserve compatibility with existing 10BASE-T wiring, it uses the same pinout. But, from TP-PMD, it uses MLT-3 signaling. For STP, it uses the same pinout as TP-PMD.
- 100BASE-FX defines a fiber-optic link segment.

- 100BASE-T4 uses four-pair cable to allow transmission over Category 3 cable. The standard recommends Category 5 connecting hardware for connectors, patch panels, punchdown blocks, and so forth. Of the four pairs, one transmits, one receives, and two are bidirectional data pairs. Because Category 5 cable has proven so popular, 100BASE-T4 was never implemented. It was originally specified to allow the existing installed base of Category 3 cable to handle Fast Ethernet.

100BASE-T defines two classes of hubs. Class I hubs can connect unlike media, so that segment types can be mixed. A Class I hub can have ports accepting two-pair UTP (100BASE-TX), four-pair UTP (100BASE-T4), and fiber. (Remember that, although 100BASE-TX only uses two pairs for signaling, it is usually wired with four-pair cable.) A Class II hub accepts only one media type.

Fast Ethernet allows 100-meter cable runs. In this case, the limit is set by timing considerations for the round-trip delay of a signal. It is built into the specification. In contrast, the 100-meter limit recommended for 10BASE-T is artificial in the sense that it is not a physical limitation of the network. With Category 5 cable, 10BASE-T can easily support 150-meter cable runs, although 100 meters is the generally accepted maximum.

Gigabit Ethernet

Gigabit Ethernet (often abbreviated as GbE) was first defined on fiber in IEEE802.3z-1998. The 1000BASE-LX system specifies 50μ MMF because of its superior bandwidth, although 62.5μ MMF (and SMF) can optionally be used. Like 100BASE-FX, it recommends the duplex SC connector, although others are optionally permitted. The 1000BASE-LX signal is 8B/10B encoded and transmitted via optical NRZ at 1310 nm. The maximum range is 550m on 50μ MMF. There is also a short-wavelength version (-SX) which operates at 850 nm.

The 802.3z specification also includes a variety known as 1000BASE-CX, which is also referred to as "short-haul copper." It uses balanced shielded jumper cables with non-RJ45 connectors to achieve a distance of 25m. It has not been widely deployed.

Gigabit operation on UTP was introduced a year later in IEEE 802.3ab-1999. To handle the demands of gigabit signaling, Category 5e cabling was introduced, although 1000BASE-T usually works fine on Category 5 cabling also. Transmitting a gigabit of information for a distance of 100m over UTP cable presents some very challenging technical problems, and 1000BASE-T introduced a number of innovations to solve them. These innovations include bidirectional transmission on all 4 pairs and the use of echo and NEXT cancellation. Let's look at the 1000BASE-T PHY in a little more detail.

In contrast to 10 and 100BASE-T, 1000BASE–T uses a very sophisticated encoding algorithm known as 4D-PAM5. This algorithm applies a 5-level code to all 4 pairs simultaneously. In effect, 250 Mb/s are transmitted in both directions on each pair, as illustrated in Figure 5–13.

The top diagram in Figure 5–13 shows a 10/100BASE-T system where there is unidirectional transmission on a single pair in each direction. In this simple case, attenuation and pair-to-pair crosstalk are the most important transmission impairments. The bottom of the figure shows the situation with 1000BASE-T transmission, where 250 Mb/s of data are being sent bidirectionally on all 4 pairs. In addition to attenuation, the following transmission parameters become important:

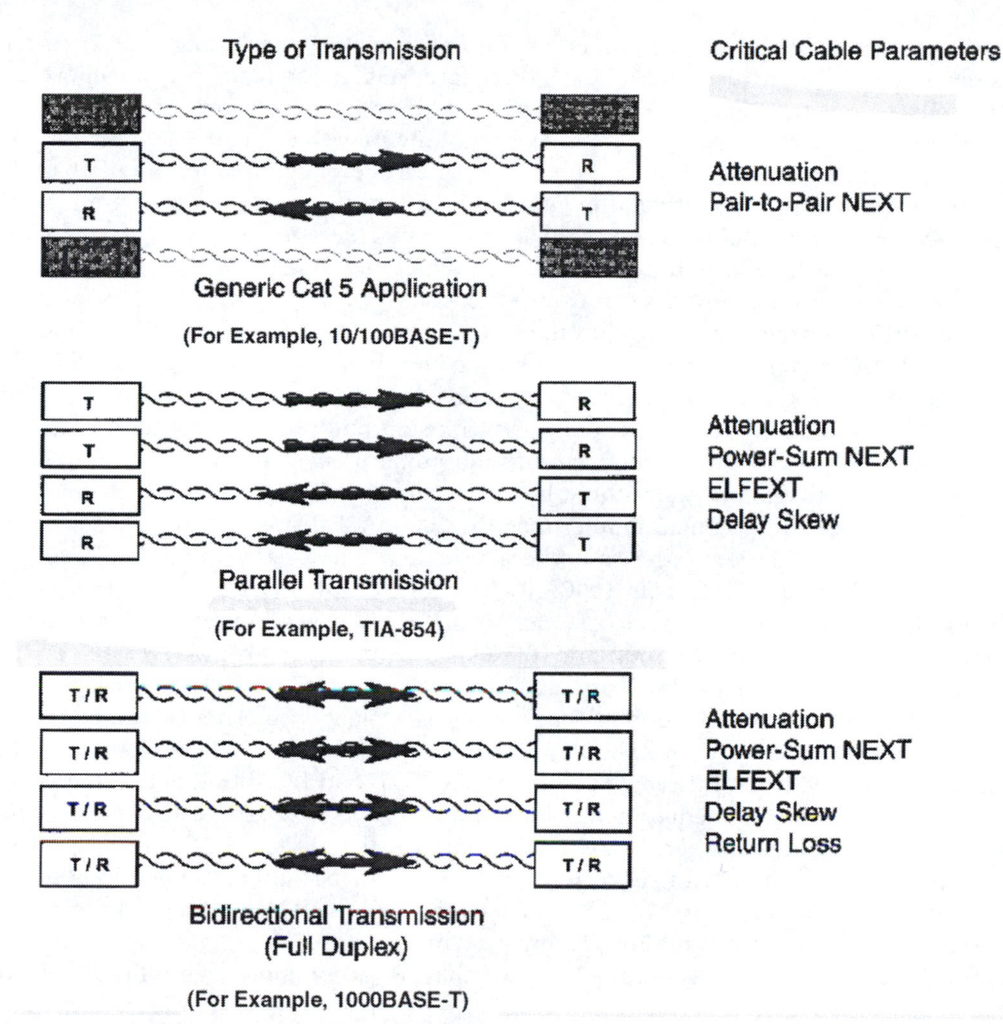

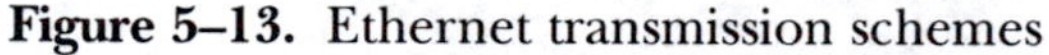

Figure 5–13. Ethernet transmission schemes

- Because there are multiple transmitters in each direction, we are concerned with PSNEXT rather than just pair-to-pair NEXT.
- Because of the bidirectional transmission, crosstalk at either end of the cable is significant, so that ELFEXT is now a consideration.
- Return loss is an important parameter, also because of the bidirectional transmission. In a bidirectional system, reflected signal is not just wasted power; it becomes added noise to the signal coming in the other direction.
- Delay skew is now an issue because, in a multipair transmission scheme, the bits on the various pairs must be kept in synchronization.

The 1000BASE-T PHY uses a number of innovative features including hybrid circuits (which separate out the two directions in the bidirectional data streams), NEXT cancellers, and echo cancellers. The echo and NEXT cancellation is done by large arrays of DSP circuitry which represents the largest and most complex part of the interface.

The PAM5 signal which is transmitted on each pair is a 5-level signal represented by +1, +½, 0, -½, and –1 volts. Before transmitting, the PAM5 signal is put through a smoothing algorithm ($V_i = \frac{3}{4}V_i + \frac{1}{4}V_{i-1}$) which results in a 17-level signal on the line (every ⅛ volt from +1 to –1.) When the link does not have any data to transmit, it sends a 3-level idle code (+1, 0, –1). After shaping, this becomes a 9-level signal (every ¼ volt from +1 to –1).

It is interesting to note that a Gigabit Ethernet system optimized for Category 6 cable was defined in TIA/EIA-854. Due to the superior high-frequency performance of Category 6 UTP, this system transmits two 500 Mb/s signals in each direction, as shown in the middle diagram of Figure 5–13. Eliminating the bidirectional transmission and half of the transmitters and receivers greatly reduces the number of transmission impairments compared to 1000BASE-T. The TIA-854 PHY is able to operate without hybrids, echo cancellers, or NEXT cancellers.

Although this interface is much simpler than the 1000BASE-T interface, it has never been deployed in products. This is another case where a single solution has won out and the other solutions effectively disappear (as with 100BASE-T2 and 100BASE-T4).

Gigabit Ethernet has won rapid acceptance in the marketplace. About 7 million Gigabit Ethernet ports were shipped in 2002 and there has been rapid growth ever since. It is estimated that GbE shipments will surpass Fast Ethernet in 2005.

Many networks use a combination of Ethernet speeds. Gigabit Ethernet is used mainly for connections to servers and for backbone connections between hubs. Connections to the desktop are typically with Fast Ethernet in newer applications and 10-Mbps Ethernet over older hubs and switches. Most newer equipment—hubs, switches, and NICs—have autosensing ports that will adjust the speed between Fast Ethernet and 10-Mbps Ethernet, depending on the speed of the port at each end. In short, Fast Ethernet is fast becoming the general-purpose solution for business-level Ethernet. Still, the widespread base of 10-Mbps Ethernet means that there are still many computers and hubs running at this slower speed.

Figure 5–14 shows a typical multispeed network that also accommodates multiple protocols (Ethernet and ATM).

10-Gigabit Ethernet

The 802.3ae-2002 specification introduced 10Gb/s Ethernet operation on fiber. It defined four major varieties of 10GbE:

- 10GBASE-S that operates at 850 nm on multimode fiber
- 10GBASE-L that operates at 1310 nm on singlemode fiber
- 10GBASE-E that operates at 1550 nm on singlemode fiber
- 10GBASE-LX4 that operates in WDM mode on four fibers (singlemode or multimode) in bands ranging from 1269 to 1356 nm

All of the varieties except LX4 have both LAN and WAN versions that are denoted by the suffixes R and W, respectively. In the commercial premises environment, the most common varieties will probably be 10GBASE-SR and 10GBASE-LX4. The maximum transmission

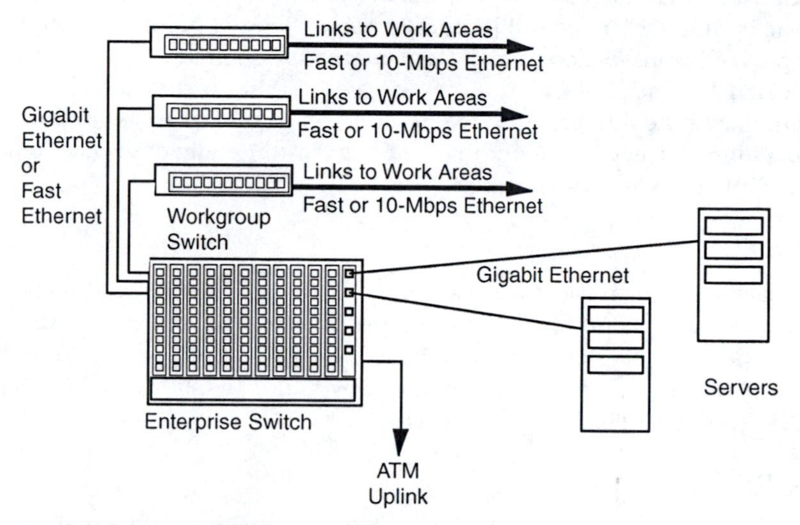

Figure 5–14. Multispeed, multirate network

distances for 10GBASE-SR are 300m over laser-optimized 50µm MMF, and very short (from 26m to 82m) over other types of 50µm and 62/5µm MMF. The LX4 variety can operate over about 300m of most types of MMF.

In March 2003, the IEEE approved a project to develop a 10GBASE-CX4 standard. The purpose of this standard is "to provide a lower-cost option for interconnection of closely located equipment (within ~15 m of cable), typically within a stack or between equipment racks within a room." It is an extension of the 1000BASE-CX concept using a bundle of 4 shielded cables instead of 1, and is aimed at data center applications. It was approved in February 2004 and issued as IEEE 802.3ak-2004. It remains to be seen whether 10GBASE-CX4 will achieve more success in the marketplace than 1000BASE-CX.

In January 2004, the IEEE approved project P802.3an to develop a 10GBASE-T standard. Obviously, there are many technical challenges to supporting a 10Gb/s data rate on UTP cable. The objectives of this project are:

- Preserve the Ethernet frame format and max/min frame sizes.
- Support full-duplex operation only.
- Support autonegotiation.
- Provide compatibility with Power over Ethernet (IEEE 802.3af).
- Support star-wired point-to-point links using structured cabling topologies.
- Define a single 10 Gb/s PHY that will support the following links with a BER of 10^{-12}:
    - 100m on 4-pair Augmented Category 6 UTP.
    - 55m on Category 6 UTP.
    - 100m on 4-pair Category 7 cable.

Most of the technical details have not been worked out at the time of this writing. It appears that the coding technique will be a higher-order type of pulse amplitude modulation. The most important transmission impairment at the frequencies used seems to be alien crosstalk, both ANEXT and AFEXT.

The preliminary schedule calls for the 10GBASE-T standard to be issued in mid-2006. There is quite a bit of uncertainty about this schedule, and it could easily take longer if significant technical difficulties are encountered.

Ethernet Summary

Over the years, more than two dozen varieties of Ethernet have been developed. However, some of them are legacy systems which are rarely encountered anymore and others were never widely implemented. The varieties most commonly encountered in the field are listed in Figure 5–15. Note that the table does not include any of the 10 Gb/s varieties, as they are still (as of mid-2005) too new to be widely deployed.

Power Over Ethernet

A method of delivering power over UTP-based Ethernet systems is defined in IEEE 802-3af-2003. The purpose of this standard is to "Define methodology for the provision of power via balanced cabling to connected Data Terminal Equipment with 802.3 interfaces. The amount of power will be limited by cabling physics and regulatory considerations." The net effect is to eliminate the wall transformers (a.k.a. "wall warts") which are occupying AC outlets all over the world. Key applications that can take advantage of this technique include IP telephones, wireless access points, security cameras, webcams, etc.

There are two alternatives for connecting Power Sourcing Equipment (PSE) to Powered Devices (PD), as illustrated in Figure 5–16.

The most desirable configuration, shown at the top of the figure, is for the hub to supply power. This can be done with 10/100/1000BASE-T equipment.

Data Rate	Ethernet Variety	Media	Max. Distance
10 Mb/s	10BASE-T	Cat 3 UTP	100m +
	10BASE-FL	62.5μm MMF	2000m
100 Mb/s	100BASE-TX	Cat 5 UTP	100m
	100BASE-FX	62.5μm MMF	2000m
1 Gb/s	1000BASE-T	Cat 5e UTP	100m
	1000BASE-LX	50μm MMF (62.5μm MMF and SMF optional)	550m

Figure 5–15. Common varieties of Ethernet

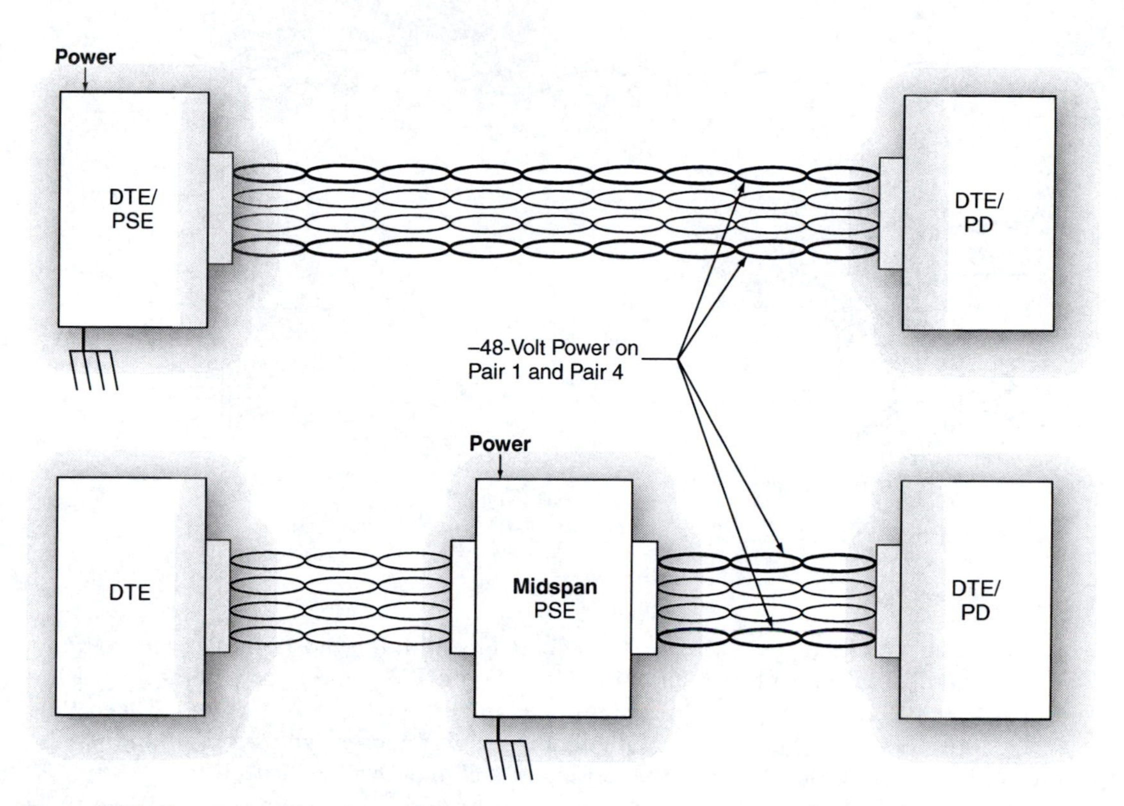

Figure 5–16. Power over Ethernet architecture

The midspan PSE configuration, at the bottom of the figure, is used with existing hubs that do not supply power for 10/100BASE-T only. The midspan PSE puts power onto the unused pairs (pins 4/5 and 7/8). It cannot be used with 1000BASE-T because there are no unused pairs.

In either configuration, –48v power is sent on the signal wires using the technique shown in Figure 5–17. This technique has been used for decades in telephony, and is often referred to as "phantom power" because the power is sent without requiring any extra wires. To prevent damage to non-Ethernet equipment, the PSE applies power only after detecting a particular impedance signature on the pairs.

The PoE standard defines five power classes that depend on the amount of power the endpoint requires. Powered devices can source up to about 13 watts from the Ethernet cable. This will allow for a range of new applications, such as miniature hubs that mount on the telecommunications outlet and are powered via the outlet cable from a hub located at the DD. Other, less obvious applications will also be invented. A new effort in IEEE 802.3 has been started to define PoE-Plus, which can supply much more power, probably up to around 30 or 40 watts. This will allow the remote powering of monitors, speakers, musical instruments, etc.

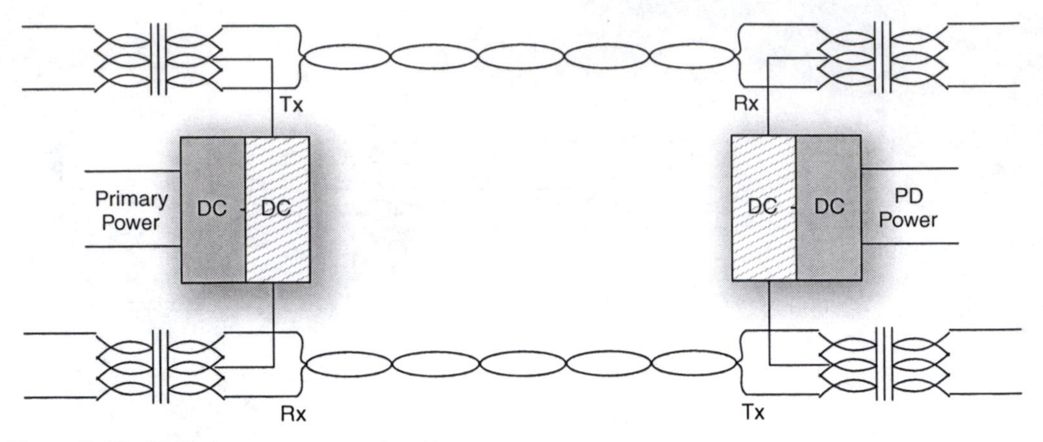

Figure 5–17. PoE phantom power circuit

Token Ring

Token Ring is a token-passing network originated by IBM. It is most popular with so-called IBM shops, companies with a heavy commitment to larger IBM computers. Operating at 4 and 16 Mbps, Token Ring fell behind other high-speed networks. Recently, switched Token Ring and 100-Mbps Token Ring have offered a migration path for users who need additional performance. Even with these advances, Token Ring is declining in popularity.

Like Ethernet, Token Ring is applied today as a star-wired network. If you actually trace the path of the signals through the hub, you would indeed find a ring, with a signal passing from station to station. Token Ring originally used the large data connectors and STP; today UTP and modular plugs and jacks predominate because of the lower system costs they allow and because of their compatibility with structured cabling systems.

IEEE 1394 (FireWire)

IEEE 1394 (also known as FireWire™ and iLink™) is a very different type of network from Ethernet. It was originally developed as a high-performance serial bus that was optimized to support both isochronous streams (like digital audio and video) as well as asynchronous data and has since grown into a full-fledged LAN. It is frequently seen in home networks, but is starting to see many commercial applications as well. It is often used for audio applications such as public address systems, music distribution in stores and restaurants, etc. FireWire is covered in more detail in Chapter 11.

FDDI

The Fiber Distributed Data Interface—FDDI—was the first local area network designed from the ground up to use fiber optics. Compared to its copper-based counterparts at the time, its performance is quite impressive: a 100-Mbps data rate over a 100-km distance and having up to 1000 attached stations.

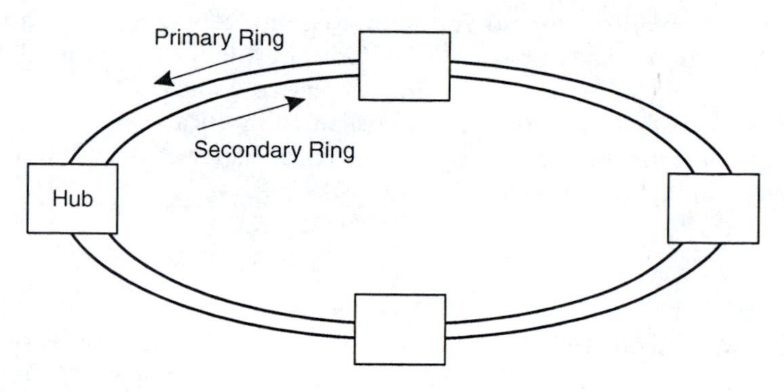

Figure 5–18. FDDI dual-ring topology

While the FDDI standard recommends 62.5/125-μm fiber for multimode applications, it also allows 50/125-μm, 85/125-μm (now obsolete), and 100/140-μm fibers, as long as these do not exceed the power budget or distort the signal beyond the limits allowed by the specification. The power budget does not include optical loss at the interface between the source and connector. The specification calls for a minimum power to be launched into the fiber. This simplifies loss calculations. The power budget between stations must only consider the fiber and any interconnections along the path.

FDDI is arranged as a token-passing ring topology that uses two counterrotating rings as shown in Figure 5–18. The primary ring carries information around the ring in one direction, while the secondary ring carries information in the other direction. The reason for two rings is redundancy: if one ring fails, the other is still available. Redundancy lessens the likelihood of network failure. Further protection lies in the fact that each station is attached only to adjacent stations in the ring. If a station fails or a single point-to-point link fails, the network still functions. If a cable break occurs between two stations, a station can accept data on the primary link and transmit on the secondary link. This wrap-around function makes FDDI highly reliable.

Although FDDI was the first 100-Mbps network and offered a robust, highly reliable system, it did not achieve widespread use. FDDI, even with copper and fiber options, was expensive and tended to be applied only in applications that required the highest levels of reliability.

Even though FDDI is declining in popularity, it did have one important influence. Its specification for operating over copper cable is called TP-PMD (twisted-pair physical medium dependent), an ANSI specification for 100-Mbps operation over UTP or STP. TP-PMD was originally devised for FDDI over copper. It has, however, since been adopted for 100-Mbps ATM and Fast Ethernet. TP-PMD uses 4B/5B, MLT-3 encoding to achieve a 100-Mbps data rate and 125-Mbps transmission rate. The information is also scrambled to spread the frequency content of the signal. So Fast Ethernet, in particular, owes its electrical characteristics to FDDI.

Fibre Channel

Fibre Channel is a network technology designed for transmitting data between computing devices (including computers and peripheral devices) at high data rates. Currently, it operates at data rates of 1, 2, 4, and 10 Gb/s. Transmission distances can be as long as 10 km.

Fibre Channel supports three topologies: point-to-point, arbitrated loop, and a switched (or mesh) network. It uses a 5-layer protocol that does not follow the ISO model. It supports several classes of service, including a circuit-switched service (which is seldom used), connectionless service with and without acknowledgements, and a multicast service.

Due to its high performance and support for redundancy and fault-tolerance, it has become the defacto standard for Storage Area Networks (SANs.)

ATM

Asynchronous transfer mode (ATM) is a network that remains a niche contender against Ethernet in local area networks. Although ATM has not succeeded as a major LAN, it is widely used in telecommunications and Internet applications. It is also used as a backbone in larger networks.

First, its basic operating speed is 155 Mbps. It can also work at 100 Mbps, using the same chipsets signaling techniques as FDDI. For runs to the desktop, additional versions allow 25 Mbps and 51 Mbps for connections (although neither option is widely used).

Equally important, ATM scales upward to higher speeds and is compatible with SONET. SONET is the fiber-optic scheme used by telephone companies for high-speed long-distance communications. SONET speeds start at 51 Mbps and scale upward in multiples of the basic speed to 10 Gbps and above. SONET does not have a top end defined, but today's SONET systems top out around 10 Gbps. The attractiveness of this capability is that both LAN users and the telephone companies can adopt ATM. Because wide-area networking is becoming evermore important, ATM provides a straightforward way to use the telephone system as part of your network. Standards are being set to allow ATM to operate at 622 Mbps and 2.48 Gbps (both of which are SONET speeds). Such speeds, however, will require some reconsideration of building cabling. At these speeds, fiber will be the preferred medium, with single-mode fiber required for long backbone interconnections.

Second, ATM is a circuit switching network. ATM can establish a virtual circuit between two stations, allocating and guaranteeing the bandwidth required for the applications. This virtual circuit is very similar to a telephone call. The telephone company switches your call to the other end. As far as you and the other party are concerned, you have a dedicated circuit for as long as the phone call lasts. When you hang up, the circuit is broken. The difference between the voice call and the ATM virtual circuit is that ATM can guarantee exactly the amount of bandwidth required. As many users require varying amounts of bandwidth, ATM hubs must allocate the bandwidth dynamically. It can distinguish between time-critical and other requirements to prioritize communications between stations.

Third, ATM offers compatibility with existing Token Ring and Ethernet networks through emulation techniques. It is possible for existing low-speed networks and high-speed ATM to coexist. Like FDDI, ATM can serve as a high-speed backbone connecting other types of LANs.

Fourth, ATM offers the most robust quality of service available, with many options for setting requirements for ensuring delivery of services as required by the application.

53-Bytes

Token Ring, Ethernet, and FDDI use variable packet sizes. An Ethernet frame, for example, can range in length from 64 to 1518 bytes, while an FDDI frame can be from 20 to 4506 bytes long. One drawback to variable-length frames is that it becomes difficult to determine how long a frame takes to transmit on the network. In contrast, ATM uses a fixed length 53-byte cell. The cell is divided into two sections called the header and the payload. The 5-byte header carries addressing information; the 48-byte payload carries the information—voice, data, or video.

ATM did not find the widespread use in local networks that early enthusiasts predicted. In effect, the fast arrival of Gigabit Ethernet provided a high-speed migration path using familiar Ethernet. ATM is a complex network that required additional investment in training. Early adoption of ATM was also slowed by delays in specifications. ATM is a very rich and varied network with a range of specifications to cover all manner of applications needs.

Video

Historically, video signals have usually been analog and unidirectional, unlike the full-duplex digital data normally carried by LANs. In a commercial office environment, video is often used in point-to-point mode rather than being broadcast. Nevertheless, structured cabling systems have increasingly been used to distribute analog video signals.

In recent years, digital video has become much more widespread. In this section, we will briefly consider both analog and digital video.

Analog Video

There are two fundamental types of video signal—baseband and broadband. The most basic type is a baseband signal, which is illustrated in Figure 5–19. This consists of the luminance (brightness) and chrominance (color) signals for a single channel multiplexed together with audio and synchronization information to form a rather complex composite video signal.

There are three different specifications for baseband video signals—NTSC, PAL, and SECAM. The NTSC (National Television System Committee) format is the oldest and simplest to implement. It was developed in the United States and first broadcast on January 23, 1954.

It is based on 525 lines per frame, 30 frames per second, and a 4:3 aspect ratio (i.e., the picture is 1⅓ times as wide as it is tall). It is currently used throughout North, Central, and South America as well as Japan, Korea, Taiwan, and the Philippines. It employs a simultaneous amplitude and phase modulated subcarrier to separate luminance and chrominance information and uses a bandwidth of 6 MHz.

PAL (Phase Alternate Line) was developed in Europe and first broadcast in Germany and the United Kingdom in 1967. It is currently used in most of Europe, Asia, Australia, and Africa. It uses 625 lines per frame, 25 frames per second, and a 4:3 aspect ratio, and about 8 MHz of bandwidth. PAL uses a subcarrier similar to NTSC, but alternates the phase of the color difference signals (hence the name PAL) to minimize distortion between chrominance and luminance.

SECAM (Sequential Couleur a Memoire) was developed in France, and also first broadcast in 1967. The line and frame rates, aspect ratio, and bandwidth are the same as SECAM. It employs low frequency modulated subcarriers representing color difference signals that are alternated in time from one successive line to the next. SECAM has never really caught on in the marketplace. It is currently used in France and a few eastern European countries. There is a general trend toward migrating to PAL.

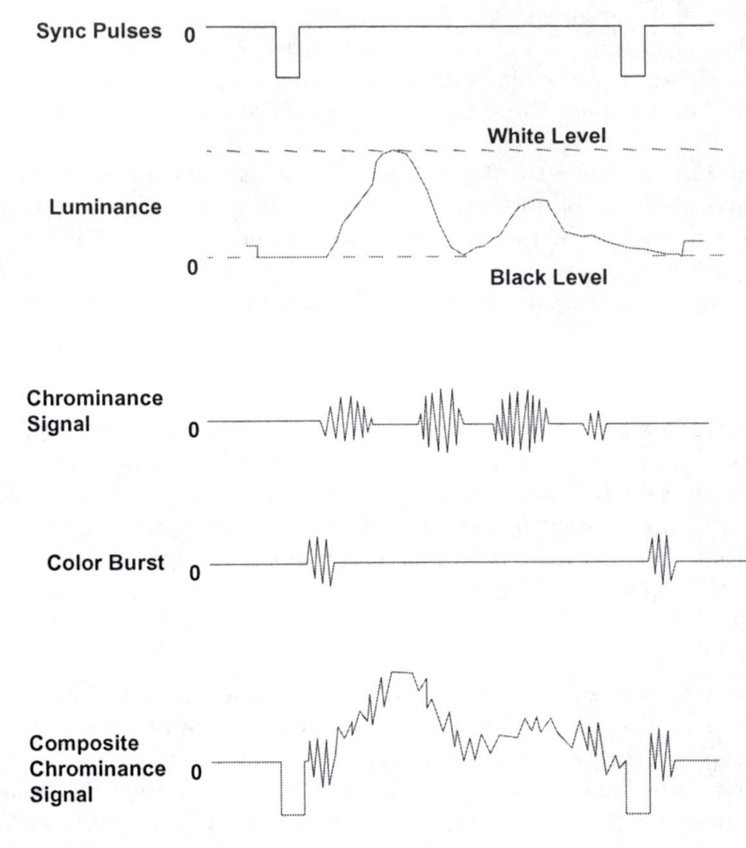

Figure 5–19. Composite video signal

Broadband video consists of a bunch of baseband video signals modulated up to carrier frequencies and multiplexed together. This is the type of TV signal that is broadcast through the air or over cable TV networks. Figure 5–20 gives some of the channel frequency assignments in the United States. Channel 2 is at the lowest frequency, from 54 to 60 MHz—remember that each NTSC channel takes 6 MHz of bandwidth. The UHF and cable TV channels extend up to hundreds of MHz in frequency.

Band	Channels	Frequency Range (MHz)
VHC	2–13	54–210
UHF	14–83	470–890
CATV	2–78	55–547

CATV frequencies vary from system to system. Some channels may be reserved for nonbroadcast use.

Figure 5–20. Typical U.S. TV channel frequencies

Video signals are normally transmitted over coaxial cable. Transmitting baseband video over UTP can be accomplished if a few technical problems (such as balance, impedance matching, and filtering) can be overcome. This is normally done with a video balun (discussed later in this chapter). Balance is an issue because the signal transmitted on coax is unbalanced, and UTP requires a balanced signal. Impedance matching is necessary because coax-based video equipment expects to see a 50-ohm impedance, and the UTP system has a 100-ohm characteristic impedance. Fortunately, both these issues can be handled with a transformer that balances the signal while simultaneously converting the impedance. Finally, the signal must be filtered to prevent high-frequency noise from being radiated by the UTP cable. Bandwidth is *not* a problem, because the single-channel baseband system only requires 6 or 8 MHz, depending on the signal format.

Broadband video signals can also be transmitted over carefully designed UTP systems, even at frequencies up to 500 MHz or more. This works because the TV signal is unidirectional, so there is no crosstalk, and ACR is not an issue. The limiting parameters for the UTP system are the attenuation and EMC performance. A video balun similar in concept to the baseband video balun is used for this application.

Digital Video

The video formats illustrated in Figure 5–19 are all based on analog representation of both the audio and video signals. Digital video refers to digitally encoding both the audio and video and transmitting as a digital bit stream. This has many advantages, including higher resolution, use of DSP-based compression techniques, better signal-to-noise ratio, etc. It also makes the video signal compatible with storage in digital memory (such as a disk drive) and transmission over a digital network, including both home networks and the Internet. Finally, it takes advantage of the declining cost curve of digital electronics.

There are two main types of digital television (DTV)—Standard Definition (SDTV) and High Definition (HDTV). SDTV has image quality equivalent to a DVD and the same 4:3 aspect ratio as analog TV. HDTV has much better resolution and image quality—roughly equivalent to a 35 mm camera, and audio quality similar to a compact disk. HDTV also has a 16:9 aspect ratio—like a movie screen. Figure 5–21 illustrates the difference in width between SDTV and HDTV screens of the same height. It also shows how much an HDTV image must be reduced to fit on an HDTV screen.

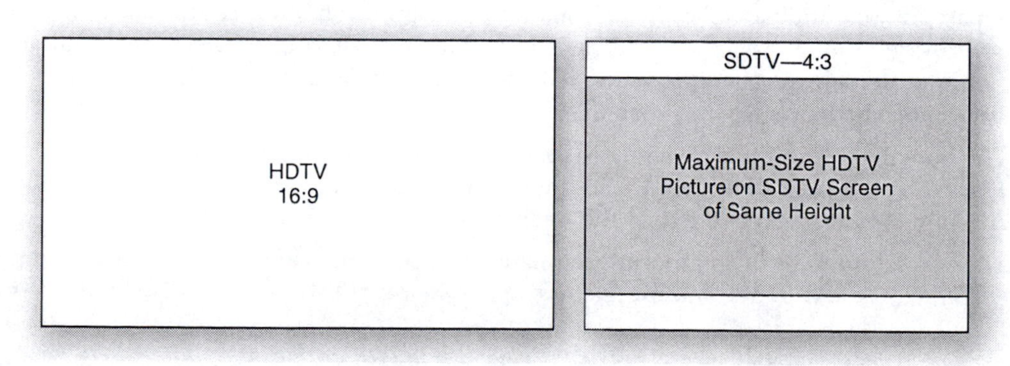

Figure 5–21. SDTV and HDTV aspect ratios

Format	Scanning[2]	Aspect Ratio	Pixels/ Line	Lines/ Frame	Pixels/ Frame	Megapixels Per Second
SDTV	60i	4:3	640	480	307,200	9.216
	30p	4:3	640	480	307,200	9.216
HDTV	30p	16:9	1280	720	921,600	27.648
	60i	16:9	1920	1080	2,073,600	62.208

Figure 5–22. DTV Formats

The Advanced Television Systems Committee (ATSC) is an international, non-profit organization that is responsible for developing digital television standards. The ATSC has defined 18 different video formats that differ in the following parameters:

- Frame rates[2]—24p, 30p, 60p, and 60i
- Aspect ratios—either 4:3 or 16:9
- Spatial resolution—from 640×480 to 1920×1080 pixels

The characteristics of some of the most important formats are summarized in Figure 5–22.

Figure 5–22 lists some of the most common formats—there are 14 others as well. Note that an HDTV picture represents a lot of digital data. The 1920×1080 format, for example, contains over 2 million pixels, each of which may be encoded as multiple bytes of data. So, while a picture may be worth a thousand words, it is also worth several million bytes!

Even HDTV is not the ultimate format. Higher-resolution formats, such as ultra-high-definition video (UHDV) and holographic TV, are currently in the experimental stage.

Video Encoding

Video signals are typically separated into red, blue, and green components (referred to as R, G, and B, respectively.) The RGB components are processed by a mathematical algorithm to yield a format known as YCrCb, where Y represents the luminance (or brightness) and Cr and Cb represent the red and blue chrominance (color) signals, respectively. There are a couple of advantages to the YCrCb format.

- Separating the luminance and chrominance is the traditional way of transmitting a composite video signal, as shown in Figure 5–19. This format is backwards compatible with existing monochrome monitors and televisions that just respond to the luminance and ignore the chrominance information.
- The human eye responds much more strongly to luminance than it does to chrominance. Therefore, the bandwidth needed to transmit the signal can be reduced by sampling the chrominance at a lower rate than the luminance.

There are a number of commonly used standards for digital video, but the most common ones are those issued by the Motion Picture Experts Group (MPEG). MPEG is a branch of

[2] *"i" indicates an* interlaced *scan, where the odd lines are scanned in one frame and the even lines in the next. "p" indicates a* progressive *scan, where all the lines are scanned sequentially in each frame.*

ISO/IEC that is responsible for developing international standards for motion pictures, including encoding, compression, etc. They have issued a series of standards, of which MPEG-2 is the most important. It is widely used for both DVDs and digital TV broadcasting. MPEG-2 uses five different compression techniques to reduce the bit rate by a factor of 25 to 50. It uses intra- and inter-frame compression to remove redundancy both within frames and between successive frames. The MPEG-2 algorithm is asymmetrical—it uses a very complex encoder and a relatively simple decoder. This is a good tradeoff, because decoders are much more numerous.

MPEG-2 has many different options, and can run at data rates as low as 2 Mb/s or as high as 100 Mb/s. For SDTV signals, about 4 to 8 Mb/s is more typical, while for HDTV signals, the data rates are about twice as high.

MPEG-4 is a much more advanced standard which, in addition to providing efficient video encoding for transmission and storage, addresses multi-media presentation. It uses object-based encoding, which allows computer generated graphics, animations, synthesized voices, etc. to be efficiently integrated with the video. The processing required is more than three times as complex as MPEG-2, but the bit rate can be up to 50% lower. Some applications are starting to use MPEG-4, but MPEG-2 is very firmly entrenched in the market and a massive conversion has not started yet.

Another common format is ITU-R BT.601, which specifies parameters for uncompressed video streams that are normally used in television studio equipment. It provides a reference against which to compare other encoding techniques. For SDTV, the output can be encoded as either 8 or 10 bit samples with a resulting bit rate of 216 Mb/s or 270 Mb/s, respectively. ITU-R BT.709 is a similar specification for HDTV formats. Data rates start at about 1.2 Gb/s and go to over 2 Gb/s.

Note the tremendous range of bit rates for digital video, which is summarized in Figure 5–23. Supporting the transmission of video at these rates is one of the major forces behind the continual drive to higher-speed networks and higher-performance cabling systems.

Some Network Devices

We've looked at some of the types of networks and have seen how star-wired systems are at the heart of most networks. Thus, hubs and switches become a central player in the network. These devices can range from large enterprise-level devices to workgroup-level devices. The differences are a matter of scale. Enterprise hubs and switches are typically chassis-based systems that accept

Type	Encoding	Data Rates	Typical Rate
SD	MPEG-2	2–100 Mb/s	4–8 Mb/s
	MPEG-4	1–50 Mb/s	2–4 Mb/s
	BT.601	216 or 270 Mb/s	N/A
HD	MPEG-2	4–200 Mb/s	8–16 Mb/s
	MPEG-4	2–100 Mb/s	4–8 Mb/s
	BT.709	1.2–2.0 Gb/s	1.6 Gb/s

Figure 5–23. Digital video bit rates

plug-in boards. Many enterprise switches or hubs are multiprotocol devices that can handle different types of networks. For example, a switch may contain many plug-in modules for Fast Ethernet links to PCs, one or more Gigabit Ethernet links to servers, and even ATM links to the WAN.

Workgroup hubs and switches (Figure 5–24) are more modest, having fewer ports. They are designed for managing smaller networks or subnetworks. Workgroup hubs might still have installable modules to allow different types of uplinks (such as Fast Ethernet or Gigabit Ethernet, copper or fiber). Often the station ports are copper based, while ports used to link switches over greater distances are fiber.

To connect to a network, a PC must have a network port. Most often, this is provided by a plug-in network interface card (NIC). The NIC in Figure 5–25 accommodates both copper and fiber Ethernet.

Figure 5–24. Workgroup switch *(Courtesy of 3Com)*

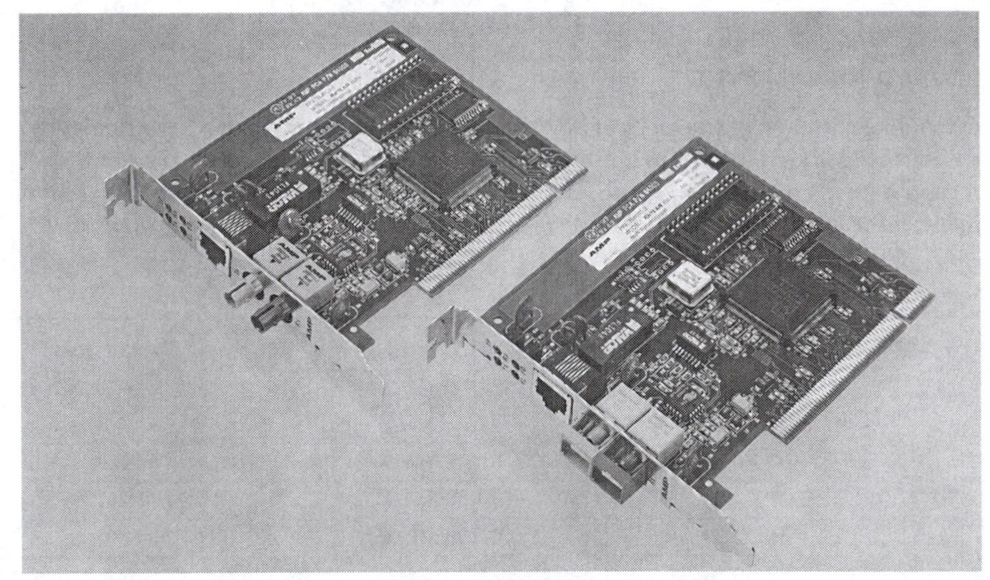

Figure 5–25. NIC, capable of connecting to either copper or fiber cables *(Courtesy of Sun Conversion Technologies)*

One impediment to widespread use of fiber in LANs is the installed base of copper components. The decision between copper and fiber is often presented as an either/or choice. Considering that companies may have hundreds or thousands of PCs with copper NIC cards and lots of hubs with copper ports, there may be some resistance to fiber. In addition, more and more computers are equipped with network ports and some do not use a plug-in card: a copper network port is built directly onto the motherboard. One solution is the media converter, an example of which is shown in Figure 5–26. The media converter performs one main task: it converts signals between optical and electrical. Plug copper in one end and get optical out the other end. Media converters can be a relatively inexpensive way to allow both copper and fiber to coexist. For example, you can wire your building with fiber, buy fiber-based hubs and switches, and use PCs with either copper or fiber NICs. Fiber-based PCs can plug directly to the building wiring. For copper-based PCs, a copper cable from the PC to the media converter saves the investment in PCs.

The wider availability of media converters, fiber NICs, and fiber ports on hubs and switches brings greater options in premises cabling systems, our subject in the next chapter.

Minis and Mainframes

A building cabling system must also accommodate the cabling requirements of minicomputers and mainframe computers. Before the days of open systems, LANs, and PCs, vendors of minicomputers and mainframe computers used proprietary methods to connect terminals to the computer system. While there are many variations, the two most important are the IBM 3270 system and the IBM AS/400 and S3x system. The discussion of how these systems are connected and how they can be integrated into a structured cabling system can easily be extended to other systems.

Figure 5–26. Media converter *(Courtesy of Sun Conversion Technologies)*

The IBM System Network Architecture is IBM's version of the seven-layer OSI model. It provides the layers of protocols to allow IBM equipment to operate together, including a scheme to allow the myriad of mainframe, minicomputer, and even PCs to connect.

At the heart of IBM's system is the 3270 family of products. Many, but not all, of the products begin with the digits "327." Here are some representative models:

3278 and 3279	Terminal
3279 and 3274	Terminal cluster controller
3299	Multiplexer
3245	Communications controller
3705 and 3725	Front-end processor

In a typical configuration, several terminals connect to a terminal cluster controller, which acts as a gathering point for messages between the computer and terminals. Cluster controllers are typically located locally to the terminals.

Groups of cluster controllers, in turn, connect either to a communications controller or front-end processor (FEP). These devices further gather, process, and communicate between the cluster controllers and mainframe. The cluster controller is often remotely located from the communications controller or FEP, communicating with them over the telephone network.

The reason for all this gathering and processing is that mainframe computers are very fast, while communications to the terminal are very slow. You don't want the mainframe wasting time doing "housekeeping" and waiting for slow communications. The communication controller and front-end processors form a bridge between the high-speed computer and the low-speed terminals and cluster controllers.

AS/400 and Systems 34, 36, and 38 refer to IBM midrange computers. In the original configuration, up to seven devices can be daisy-chained together serially. Each daisy-chain connects to a communication channel on the host computer. Because the daisy-chain system is not easily adaptable to other cabling schemes, a star configuration was developed. How terminals connect to the host computer depends on the system used and the type of wiring. Figure 5–27 shows several examples of how AS/400 systems are interconnected.

STP: Up to seven terminals connect to a device called a loop-wiring concentrator. This device forms a star-wired daisy chain. The concentrator has eight ports, seven of which connect to terminals. The seventh port connects to one of the host computer's communication channels. The system is twinaxial from the terminal to the wall plate, STP for the horizontal cable to the telecommunications closet, and a mixed STP/twinaxial from the concentrator to the host computer.

UTP: An active star hub allows the use of UTP. A balun (described below) allows UTP from the terminal to the hub and from the hub to the host computer. Each terminal must be treated as the last device in a daisy chain. One port must be terminated in its characteristic impedance. Allowable distances are:

Terminal to hub:	900 feet
Hub to host computer:	1000 feet
Terminal to host computer:	1000 feet

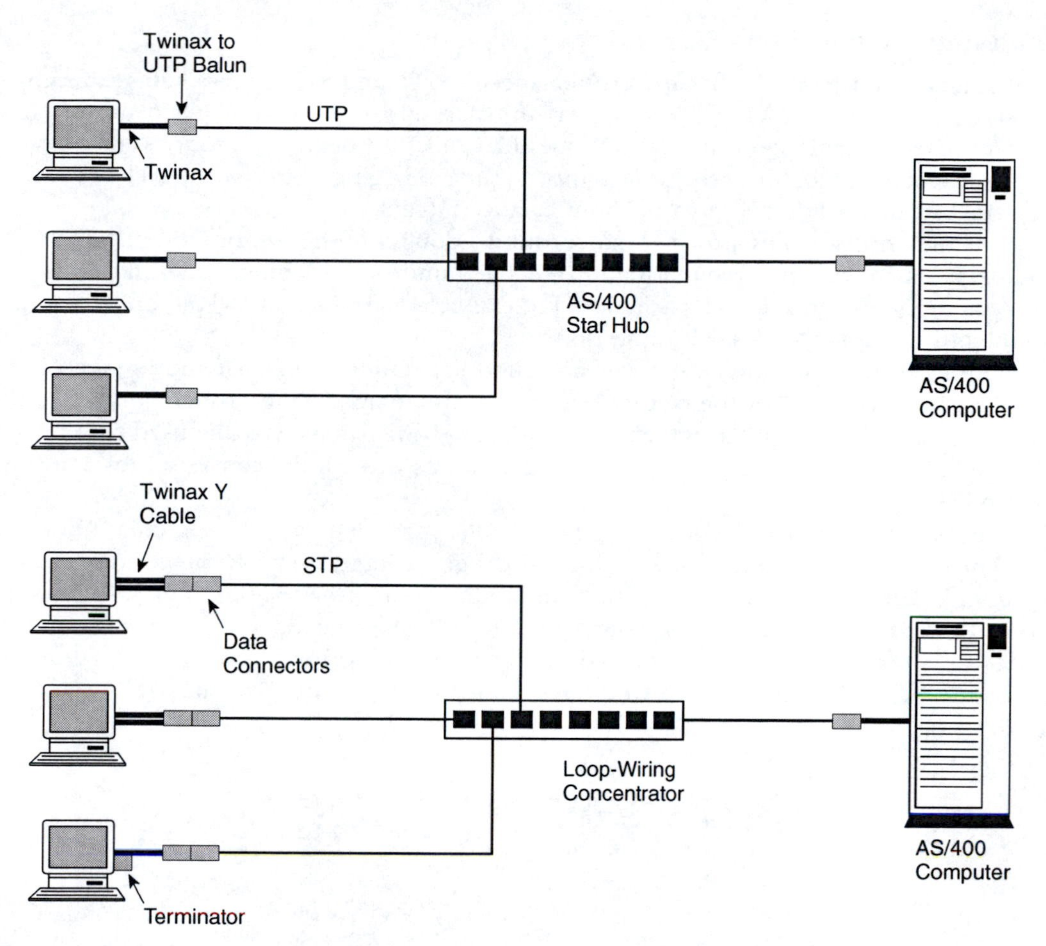

Figure 5–27. AS/400 system interconnections

All three distance conditions must be met, so the total allowable distance is 1000 feet, not 1900 feet. If the hub is located near the host computer, the terminal can be located up to 900 feet from the hub. If the hub is over 100 feet from the host computer, the distance from hub to workstation must be shortened accordingly. If, for example, the hub is 600 feet from the host, the terminal can be only 400 feet away from the hub.

The IBM systems were originally designed to use coaxial or twinaxial cable:

3270 systems:	93-ohm coaxial cable
System 3x and AS/400:	100-ohm twinaxial cable

A twinaxial cable is similar to a coaxial cable, except that it has two center conductors in the dielectric.

To convert these coaxial/twinaxial-based systems to use UTP or STP requires a balun.

Baluns and Media Filters

Balun is a shortening of "balanced/unbalanced." UTP and STP require balanced signals, as described in Chapter 2. Coaxial cables carry unbalanced signals. A balun performs two functions: it converts signals between balanced and unbalanced modes and it provides any required impedance matching between the two circuits. In short, in going from coax to UTP, it converts a 93-ohm unbalanced signal into a 100-ohm balanced signal.

A balun consists of a transformer built around a doughnut-shaped toroid. Wires from the two interfaces are wrapped around the toroid. The number of windings on each side is proportional to the impedance to be achieved. For example, one side is a 100-ohm modular jack and the other side is a 93-coaxial connection.

Baluns can be built into connectors and modular outlet inserts. The advantage of this over discrete baluns is that the connectors and inserts maintain the proper polarity of the signals throughout the cabling system. Figure 5–28 shows baluns typically used for coax-to-UTP conversion. Figure 5–29 shows a balun that allows video to be converted for transmission on UTP.

Two related devices are the impedance-matching adapter and a media filter. An impedance-matching adapter allows cables of different impedances to be connected. For example, Token Ring hubs are available with either 100-ohm or 150-ohm ports. If you wish to connect a 100-ohm cable to a 150-ohm port, an impedance-matching adapter is required to provide a conversion between the two impedances without undue signal distortion and reflections. A media filter removes high-frequency components that may be part of the signal.

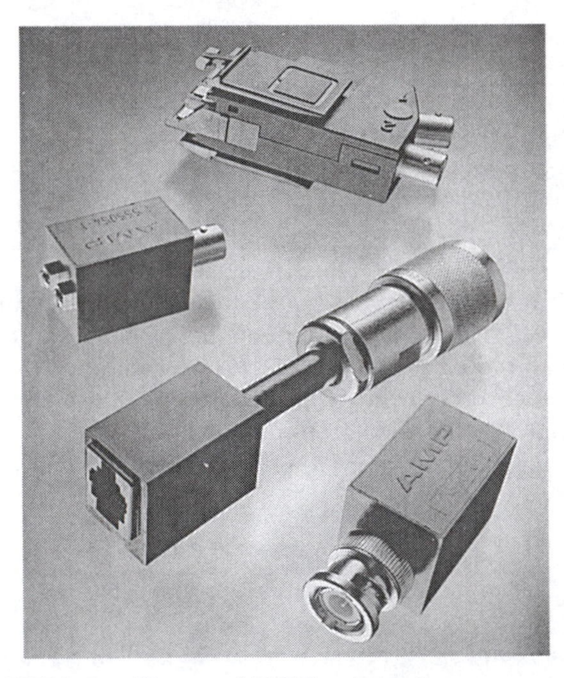

Figure 5–28. Coax-to-UTP balun *(Courtesy of AMP Incorporated)*

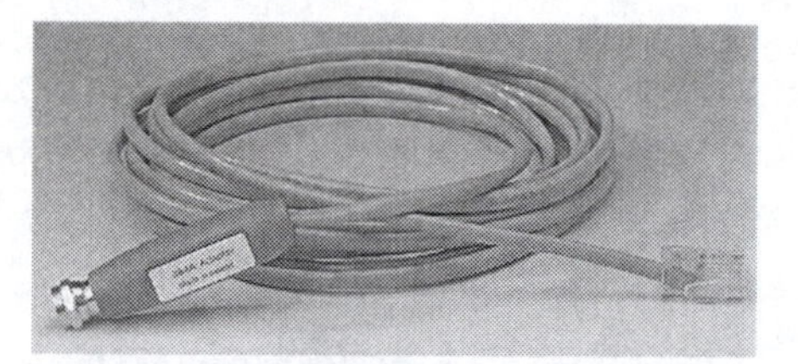

Figure 5–29. Video balun *(SYSTIMAX Solutions graphics are courtesy of SYSTIMAX Solutions™, a CommScope company)*

PBXs and Telephone Cabling

Telephone wiring is also a form of network, although operating at slower speeds that present fewer problems in cabling. The slow speeds allow long runs on lower-grade cable. Even so, TIA/EIA-568B minimizes the differences between voice and data lines, distinguishing them by data rates (or frequency) rather than by applications. Most new installation and renovations use Category 5e or better for all horizontal cabling.

The heart of most telephone systems is the PBX or private branch exchange. The PBX is also known as a PABX (private automatic branch exchange) or CBX (computerized branch exchange). The PBX is an on-premises telephone switching system. It forms an interface between the telephone central office and the telephones in a building. In simplest terms, incoming calls go first to the PBX, which then routes them to the proper telephone. Modern PBXs, of course, are much more sophisticated, offering such features as voice mail, call forwarding, conference calls, and call accounting.

A PBX consists of a cabinet containing one or more shelves. Each shelf accepts plug-in boards that connect to a backplane in the back of the PBX. Different types of boards are available for different needs and levels of capability. Common types of boards in a PBX are:

- **Both-Way Trunk.** These are outside lines that allow both incoming and outgoing calls. Each trunk line connects to a port on the board. Thus an 8-port board can accept eight trunks.
- **Direct Inward Dial Trunk.** These boards accept incoming calls only and allow dialing of individual extensions from within the PBX.
- **Universal Trunk Circuit.** These boards mix both-way and direct-inward-dial capabilities.
- **Tie Line.** Tie-lines connect two PBXs without the need to place an outside call. Businesses use tie lines to connect different buildings without having to go through an outside line.
- **Digital Telephone.** These boards connect to digital telephones directly. Each board typically has 8 or 16 ports.
- **Analog Telephone.** Analog boards handle communications with analog phones. Like the digital telephone board, analog boards typically have 8 or 16 ports.

Telephone companies are increasingly using digital signals for carrying calls. With newer digital telephones and PBXs, it is possible to achieve end-to-end digital communications. The desktop telephone encodes your voice into a digital signal before transmission to the PBX. The PBX retains the digital form for transmission to the central office. Digitation can also occur in the PBX or at the telephone company's central office, so that portions of the transmission are analog.

Voice over IP (VoIP)

Voice over IP, also known as Internet telephony, involves making phone calls over the Internet rather than the traditional voice network. It has been billed as potentially the most revolutionary development in telecommunications since the migration from analog to digital transmission. It allows an almost unlimited variety of new services to be rolled out on the same general-purpose IP network.

Data networking is dominated by IP. VoIP allows voice telephony to share the same networks that have been built for IP traffic. This convergence of voice and data traffic onto a single IP network is widely expected to represent the future of the communications industry.

One of the main attractions for users of VoIP is that IP networks have not historically charged users based on either traffic or distance. Regardless of where a user is located or how long a session lasts, it does not cost the end user any more to access a Web site in Tokyo than one in Chicago. This is completely different from traditional telephony, where billing for long-distance calls is based on both time and distance.

Traditional telephony consists of circuit-switched connections between endpoints whose physical location is known. In other words, your phone number corresponds to a particular geographical location. Signaling is done by means of loop currents, ringing voltages, touch-tones, etc.

When using VoIP, the voice signal is digitized and sent as a stream of packets. There is no longer a fixed path from source to destination, nor is there any correlation between the logical address (IP address) and geographical location of the endpoints. Because of this, it is no longer possible to use currents and tones to perform signaling.

There are two primary methods of signaling in a VoIP network: H.323 and Session Initiation Protocol (SIP).

- H.323 is an ITU-T standard that was approved in 1996 for transmitting multimedia communication (such as audio and video) over a packet-switched network that does not guarantee quality of service (QoS). H.323 defines four major elements—terminals, gateways, gatekeepers, and multi-point control units.
- SIP is an IETF specification (RFC 2543) for initiating, modifying, and terminating multimedia sessions. It is a request/response protocol, with the requests being sent via any transport protocol, such as UDP or TCP. SIP is used for telephony, conferencing, instant messaging, and other applications.

Because VoIP represents a convergence between the voice telephony and data networking worlds, it is not surprising that one of these methods (H.323) came from the telephony industry, while the other (SIP) came from the IETF. H.323 was the first protocol to ensure interoperability among VoIP equipment from different manufacturers and has a significant installed base. Recently, however, SIP has been gaining momentum, as it is generally considered to be simpler and more flexible. Both protocols will probably coexist in the marketplace for some time.

VoIP has been quietly put into service in long haul networks for several years now. Many long-distance calls pass through a VoIP network without the users' knowledge. The voice quality achieved in these networks is good enough that no one notices any difference. In modern premises networks, an IP-based PBX often extends VoIP service all the way to the end user.

Early implementations of Internet telephony often used a headphone attached to a PC. The voice digitization and packetizing were done in the PC and the voice stream shared the PC's broadband Internet connection. This provided telephone service that was essentially free (i.e., no additional cost other than the broadband Internet connection.) However, the voice quality was so low that it was often considered unacceptable, even for free. There are several reasons for this, including delay, packet loss, and echo.

VoIP traffic is *extremely* sensitive to delay. Even a few tens of milliseconds of delay, which would scarcely be noticed in a data connection, can wreak havoc with the perceived quality of a VoIP stream. Corporate VoIP networks generally use a dedicated IP network (or a VPN) that is specifically engineered for low latency.

Packet loss can also have a significant effect on VoIP quality. Studies have shown that a packet loss rate of as little as 3–5% can result in a noticeable drop in users' perception of the quality of the call[3]. Unlike many IP applications, VoIP is inherently bidirectional, so the uplink bandwidth is just as important as the downlink bandwidth.

Due to the extra delay involved in packetizing the voice samples, echo is much more of a problem with VoIP than it is with analog telephony. A small amount of echo that would not be noticeable on an analog connection can significantly degrade the quality of a VoIP connection.

Convergence

Convergence is one of the overarching trends in the communications industry. It refers to the merging of all types of communication and entertainment services, including voice, data, and audio/video, into a single digital bit stream, as indicated in Figure 5–30. VoIP, which merges the world of analog telephony into the packet-switched domain of the Internet, is a perfect example of convergence.

This type of network provides tremendous advantages to both providers and consumers of services. Service providers will gain economies of scale in digital transmission and switching and be able to introduce new digital services quickly. Customers will gain access to many new services (such as video on demand) and integration between features that were not available before (such as the ability to attach a voice message to an email.) Underpinning all these capabilities are the type of high-performance networks and cabling systems discussed in this book.

[3]*J.H. James, et. al., "Implementing VoIP: A Voice Transmission Performance Progress Report,"* IEEE Communications Magazine, *July 2004, pp. 36–41.*

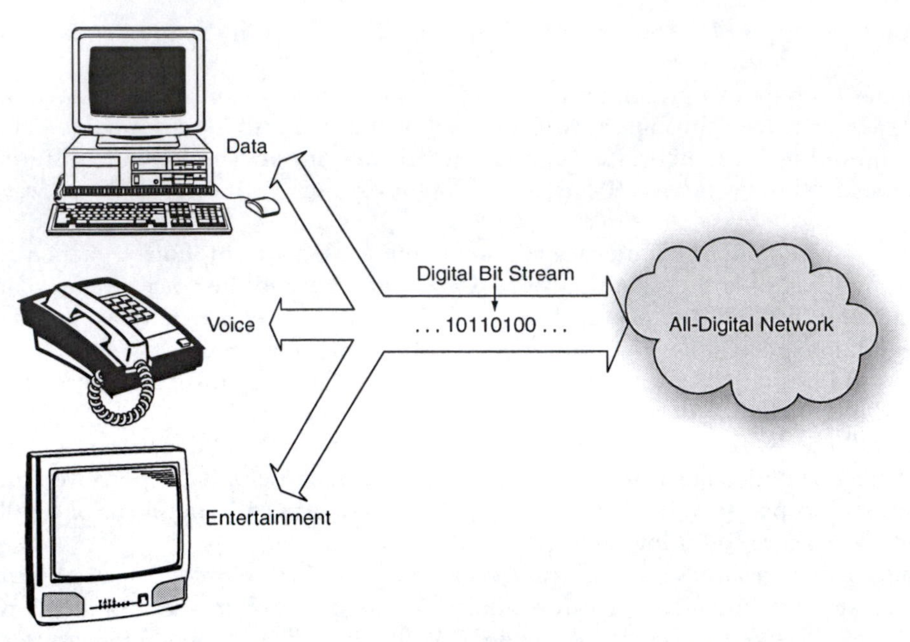

Figure 5–30. Convergence of voice, data, and entertainment to a single all-digital network

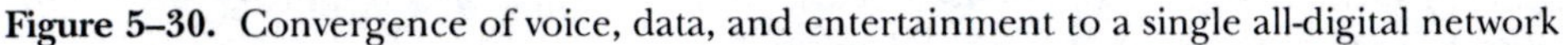

SUMMARY

- Data networks are the applications that require the highest performance in a cabling system.
- Networks are distinguished by access method, frame type and size, and quality of service.
- Premises cabling systems are concerned with the physical layer of the OSI model.
- Quality of service (QoS) is important to delivering real-time applications such as video.
- Ethernet is far and away the most popular type of network.
- Ethernet operates at 10Mbps, 100 Mbps, 1 Gbps, and 10 Gbps.
- Power over Ethernet (PoE) provides the capability to send power to endpoints without a separate power cable.
- IEEE 1394 (also known as FireWire) provides support for isochronous streams and is often used for audio and video networks.
- FDDI and Token Ring are two network types declining in popularity.
- ATM is a high-speed network with excellent QoS; it is widely used in WAN applications, but not in local area networks.
- NTSC is the analog video format used in the United States. It requires 6 MHz of bandwidth for each channel.
- Digital video is becoming increasingly popular. The major formats are standard definition (SD) and high definition (HD).

- There are many methods of encoding digital video. Currently, the most widely used method is MPEG-2.
- Voice over IP (VoIP) involves sending digital voice packets over the Internet rather than using a conventional circuit switch.
- Convergence refers to the merging of voice, data, and video communication into a single digital bit stream.

REVIEW QUESTIONS

1. Define the operation of CSMA/CD in controlling access to a network.
2. Describe the difference between a shared-media network and a switched-media network.
3. What is the most popular network today? At what speeds does it run?
4. Name the new transmission parameters that became important because of the 4-pair bidirectional data transmission used by 1000BASE-T.
5. What innovations were incorporated into the 1000BASE-T PHY to help deal with these parameters?
6. What type of UTP cabling is required to support 1000BASE-T and 10GBASE-T?
7. What applications commonly make use of Power over Ethernet (PoE)?
8. How much power can PoE deliver to an endpoint?
9. Name the three most common varieties of Ethernet for both UTP and fiber.
10. What is the major advantage of IEEE 1394?
11. What are the two major differences between ATM and IP?
12. What are the maximum packet sizes that can be sent with ATM and IP?
13. What is the purpose of quality of service?
14. What device allows you to connect UTP cable to coaxial cable?
15. What is the purpose of a media converter?
16. What network layer is premises cabling concerned with?
17. Describe the two main types of analog video signals.
18. What are the major baseband video formats and where are they used?
19. What technical problems must be overcome when transmitting a video signal over UTP?
20. Why can broadband video be transmitted over UTP at much higher frequencies than are possible with LAN signals?
21. What are the major differences between standard and high definition video?
22. What is the most commonly used encoding technique for digital video, and what is the typical bit rate for a standard definition signal using this technique?
23. What are the major advantages of VoIP?
24. Name three types of network impairments that have an adverse impact on VoIP voice quality.
25. Name at least two advantages of a converged network that provides integrated transport for voice, data, and video transmission.

CHAPTER 6

System Considerations

Objectives

After reading this chapter, you should be able to:

- Define the differences between a permanent link and a channel.
- List the seven classes of ISO/IEC 11801 and relate them to TIA/EIA-568B cable categories.
- List the three main cabling architectures and describe the differences between them.
- Tell why the centralized architecture is well suited to fiber optics.
- Define fault tolerance and tell why it is important.
- Describe various approaches to fault tolerance in intrabuilding and inter-building applications.
- Understand how to interpret cabling system specifications and compare specifications from different manufacturers.

We've looked at the main components of a premises cabling installation. This chapter looks at some additional issues involved in a building cabling system.

Installed Cabling Configurations

It is very important to be able to test the performance of cable after it is installed and to unambiguously specify what configuration was measured. For this purpose, both TIA-568 and ISO/IEC 11801 define two configurations known as the permanent link and the channel, as illustrated in Figure 6–1.

In early editions of TIA-568 and ISO/IEC 11801, there were significant differences in the link configurations. However, these have been harmonized, with one minor exception—ISO/IEC 11801 also defines a CP link for installations that include a consolidation point.

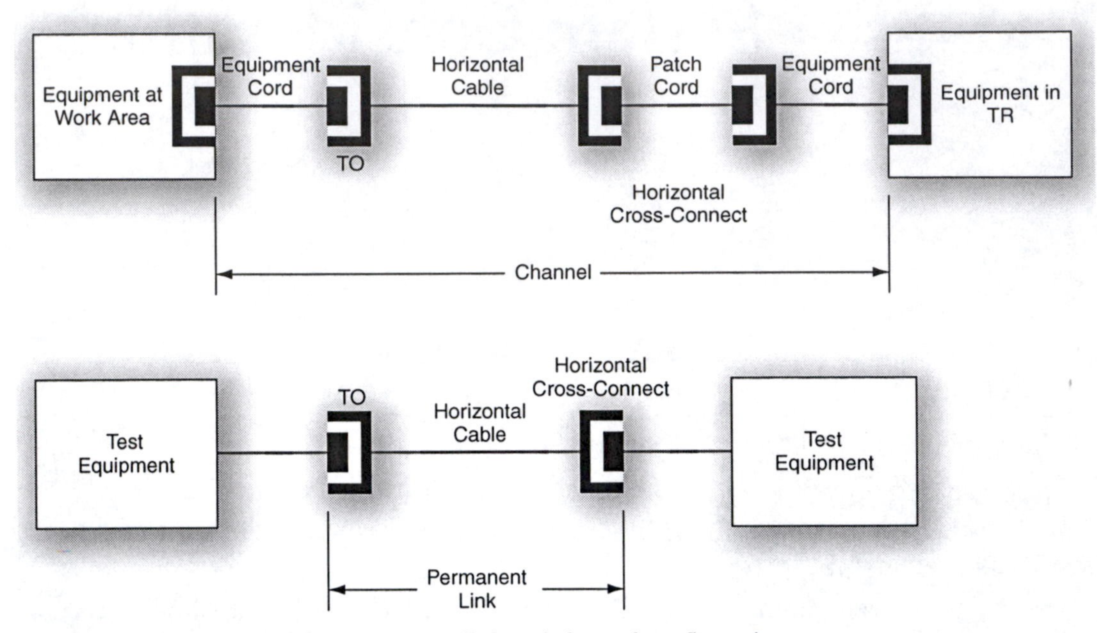

Figure 6–1. TIA and ISO/IEC permanent link and channel configurations

The channel configuration includes all the cable, connecting hardware, and cords which are present when the cabling system is being used. The channel has a maximum length of 100m which consists of up to 90m of installed cable and up to 10m of cords. Channel measurements give the best indication of the overall performance of the cabling system.

The permanent link configuration includes the telecommunications outlet, horizontal cable, and the cross-connect block on which the cable terminates, but does not include any equipment cords or patch cords. The permanent link therefore has a maximum length of 90m. The purpose of the permanent link is to allow the contractor to perform a test when the installation of the cabling system is completed. Patch cords and equipment cords are usually not added to the system until much later, when the telecommunications and data networking equipment are installed. The permanent link test allows cable verification before the equipment is installed. It is usually followed by a channel test after the installation. Because the channel includes more connectors and cordage than the permanent link, specifications for the channel will always be worse (i.e., higher attenuation, lower NEXT loss, etc.) than equivalent specifications for a permanent link.

Note that the modular plugs at the two ends of the channel are ***not*** included in the channel configuration. For laboratory measurements, these modular plugs are typically cut off and the pairs are terminated on a network analyzer test fixture. For measurements made in the field, or for non-destructive measurements in the lab, this is not practical. In such a case, the measurement must be made with the jacks in place, as indicated in Figure 6–2.

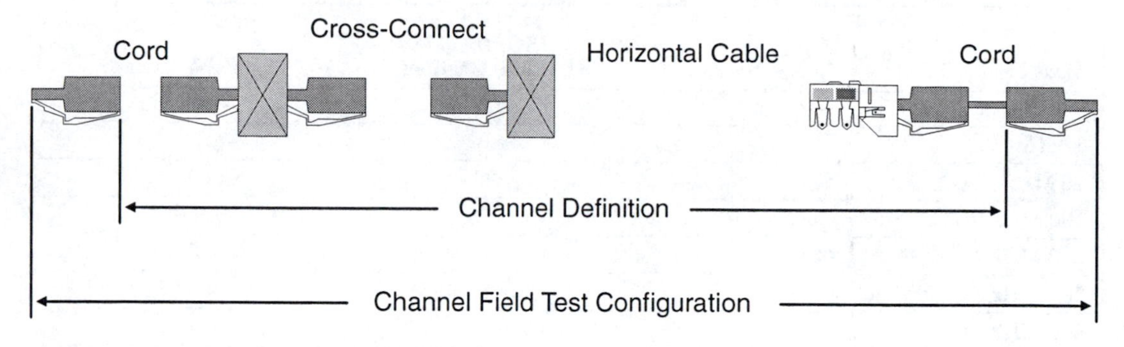

Figure 6–2. Field and laboratory channel measurements

The presence of these "extra" modular jacks results in a slight degradation in NEXT performance for the field test channel configuration. For Category 6 and below, this degradation is not significant. At the higher frequencies used in Category 6A, however, it becomes more important. To compensate for this, the Augmented Category 6 specification includes a small NEXT allowance when channels are measured in the field test configuration.

TIA permanent link and channel performance standards are based on Categories 3, 5e, 6, and 6A UTP and fiber. Coaxial and STP installations should be tested specifically to the requirements of the applications for which they are installed.

ISO/IEC 11801 defines seven classes of applications:

Class A: Voice and low-frequency applications. Cables are rated to 100 kHz.

Class B: Medium-bit-rate applications, with cables rated to 1 MHz.

Class C: High-bit-rate applications, with cables rated to 16 MHz (i.e., Cat 3).

Class D: High-bit-rate applications, with cables rated to 100 MHz (i.e., Cat 5e).

Class E: Very high-bit-rate applications, with cables rated to 200 MHz (i.e., Cat 6).

Class F: Extremely high-bit-rate applications, with cables rated to 600 MHz (i.e., Cat 7).

Optical: Applications whose bandwidth is not a limiting factor in premises cabling.

The distinction between *classes* and *categories* can sometimes be confusing. TIA-568B uses the term *category* to refer to both the link/channel performance and the cable and connecting hardware performance. ISO/IEC 11801 uses the term *class* to refer to link/channel performance while still using the term *category* to refer to cable/connecting hardware performance. So, for example, an ISO/IEC Class D channel would be made from Category 5e components. Notice that Classes C, D, E, and F map directly into Categories 3, 5e, 6, and 7 respectively. This difference in terminology is summarized in Figure 6–3.

Max. Freq.	TIA-568B Link/Channel	ISO/IEC 11801 Link/Channel	Cable Type
100 kHz	—	Class A	—
1 MHz	—	Class B	—
16 MHz	Cat 3	Class C	Cat 3
100 MHz	Cat 5e	Class D	Cat 5e
250 MHz	Cat 6	Class E	Cat 6
500 MHz	Cat 6A	Class E-2006	Cat 6A
600 MHz	Cat 7	Class F	Cat 7
1000 MHz	—	Class F-2006	—

Figure 6–3. Comparison of TIA Categories and ISO/IEC Classes

As indicated in Figure 6–3, the ISO/IEC 11801 classes will be redefined (probably in 2006) so that Class E extends to 500 MHz and Class F extends to 1 GHz. Official names for the new classes have not been defined yet.

In Chapter 1 we talked about the levels of cross connects required in a building. Let's look closer at these and how everything fits into the system. While a cabling system deals primarily with the cable and related components like cross connects, patch panels, and connectors, a functional system must also consider other network hubs and telephone PBX. While a performance-driven cabling system exists independently of specific network and telephony equipment, neither one can work without the other.

Like today's networks, the building cabling system uses a star configuration, consolidating cabling in different telecommunication closets.

A main cross connect is required as the center of every installation. The main cross connect can serve a single building or a campus of buildings. A horizontal cross connect is recommended for each floor of the building to act as a dividing point between the backbone and horizontal cabling. The intermediate cross connect exists to simplify larger installation by providing a cross connect between the main cross connect and the horizontal cross connect. In a campus application, for example, an intermediate cross connect can be used in each building. The main cross connect in one building feeds the intermediate cross connect in other buildings. (The main cross connect can also serve to distribute to a horizontal cross connect in the same building. Or it can feed an intermediate cross connect in the same building, or even in the same room.)

The main cross connect can also serve as a demarcation point between outside telephone cables coming into the building. This provides a clear separation between the telephone company's cable and responsibility, and the building's cabling.

While it's popular to show the main cross connect in the basement of a building, it can be placed anywhere in the building. In a high-rise building, placing the main cross connect in a middle floor can simplify cable routing. Similarly, a telecommunications closet in the middle of a floor, rather than on the outside, shortens the maximum distance any cable must be run.

Telecommunications Spaces

TIA-569B gives requirements for telecommunications pathways and spaces in commercial buildings. Each floor of a TIA-569B-compliant building must contain a telecommunications room (TR) as illustrated in Figures 1–2 and 1–3. The TR typically houses the horizontal cross-connect and other telecommunications equipment (such as LAN hubs, etc.).

In addition to (but not instead of) the TR, TIA-569B also permits the use of one of more *telecommunications enclosures* (TE) on each floor. Like the TR, a TE may contain telecommunications equipment, cable terminations, and associated cross-connect cabling. However, a TE is much smaller, often consisting of a cabinet that is mounted in a wall or ceiling. Figure 6–4 shows an example of a telecommunications enclosure. Because of their small size, TEs are sometimes referred to as "tiny TRs."

A Note on TIA/EIA and ISO/IEC Terminology

While the TIA/EIA and ISO/IEC approach to building cabling is essentially the same, the terms they use to describe the parts of the system differ. Figure 6–5 provides a quick cross reference to these differences.

Figure 6–4. Ceiling-mounted telecommunications enclosure *(Courtesy of American Access Technologies, Inc.)*

TIA/EIA	ISO/IEC
Main cross connect	Campus distributor
Intermediate cross connect	Building distributor
Telecommunications room	Floor distributor
Backbone cable	Campus backbone cable Building backbone cable
Horizontal cable	Horizontal cable
Telecommunications outlet	Telecommunications outlet

Figure 6–5. Differences in TIA and ISO terminology

Cross Connects and Distribution Frames

The term *cross connect* usually refers to a passive system, one that does not include such devices as network hubs or routers. The cross connect is concerned solely with the wiring. When devices are added to the system, the term distribution frame is often used in place of cross connect. Distribution frame includes both the cabling interconnections—cross connects and patch panels—and the hubs, routers, and other devices. Distribution frames have the same hierarchy as cross connects: main distribution frame, intermediate distribution frame, and horizontal cross connect.

System Architectures

The general layout of a premises cabling network is divided between the backbone and horizontal cabling. There are, however, different ways of connecting the cabling and the attached network equipment. These architectures affect the complexity of the system, the efficient use of network equipment, and the ease with which the system can be maintained. For example, you can consolidate all the equipment in one room or you can locate equipment in different closets throughout the building—one on each floor, for example. The three most common architectures are

- Distributed
- Centralized
- Zone

Each is briefly discussed below.

Distributed Network

Figure 6–6 shows a distributed system. Here equipment is located in the telecommunications room of each floor and usually in the main cross connect room. The vertical backbone cable connects the main cross connect room to each telecommunications closet. The backbone cable between floors and the horizontal cable running to offices are clearly separated. The network equipment (hubs and switches) sits between the backbone and horizontal cable. In a typical system, the backbone runs to a patch panel and connects to the equipment through

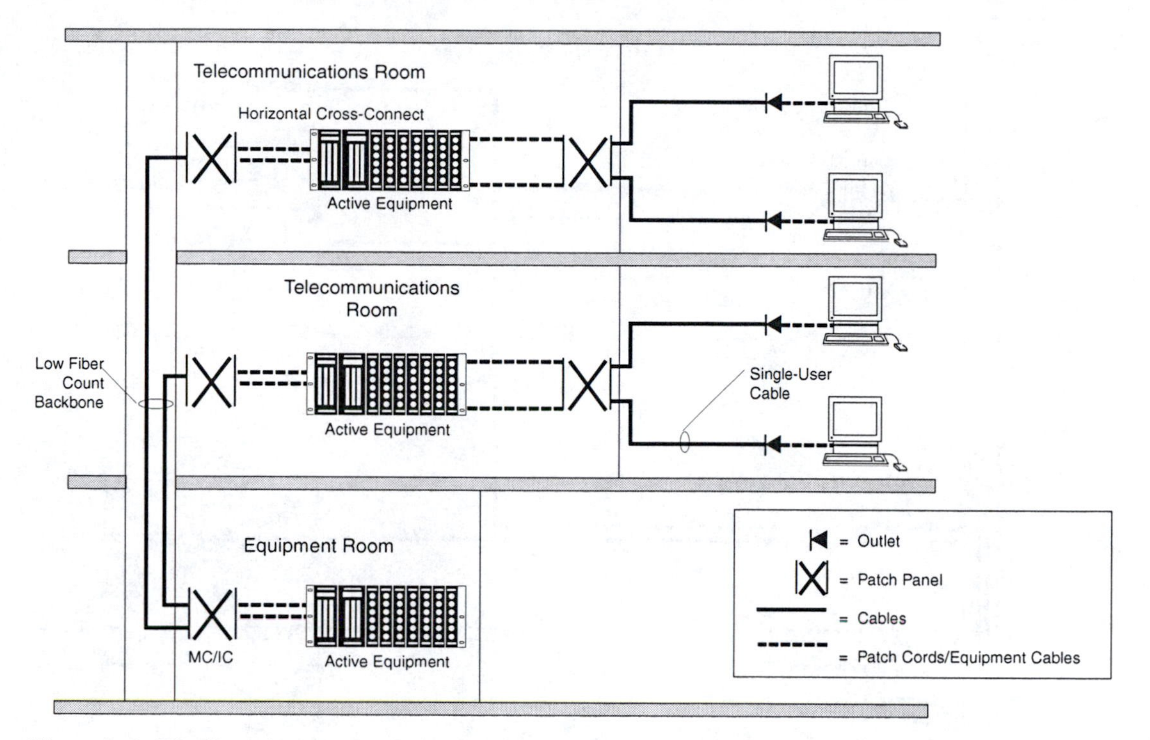

Figure 6–6. Distributed network *(Courtesy of Siecor Corporation)*

a patch cable. The equipment connects through another patch panel to the horizontal cable. The horizontal cable runs are limited to 90 meters plus a total of 10 meters for patch cables in the telecommunications room and office.

The advantage of a distributed network is that it is quite compatible with both copper and fiber cabling. You can think of a distributed system as the "classic" way of connecting a network, and most networks today use a distributed architecture. The drawback to the distributed system, as you will see, is that it typically is less efficient in its use of equipment than a centralized system. In addition, from the standpoint of fiber, the 90 meters allowed for horizontal cable clearly does not take advantage of the transmission characteristics of fiber. The centralized network does take advantage of them.

Centralized Network

Figure 6–7 shows a centralized network. The centralized network takes advantage of the distinct advantages of fiber. The distributed network limits the horizontal cabling to 90 meters. What the centralized approach does is to blur the distinction between the horizontal and vertical cable to allow runs from the main equipment closet to the work area. No equipment is required in the telecommunications closet. The fiber can be spliced to the transition point between vertical and horizontal cable or it can be run directly without splicing. The maximum recommended distance for a multimode fiber is 300 meters.

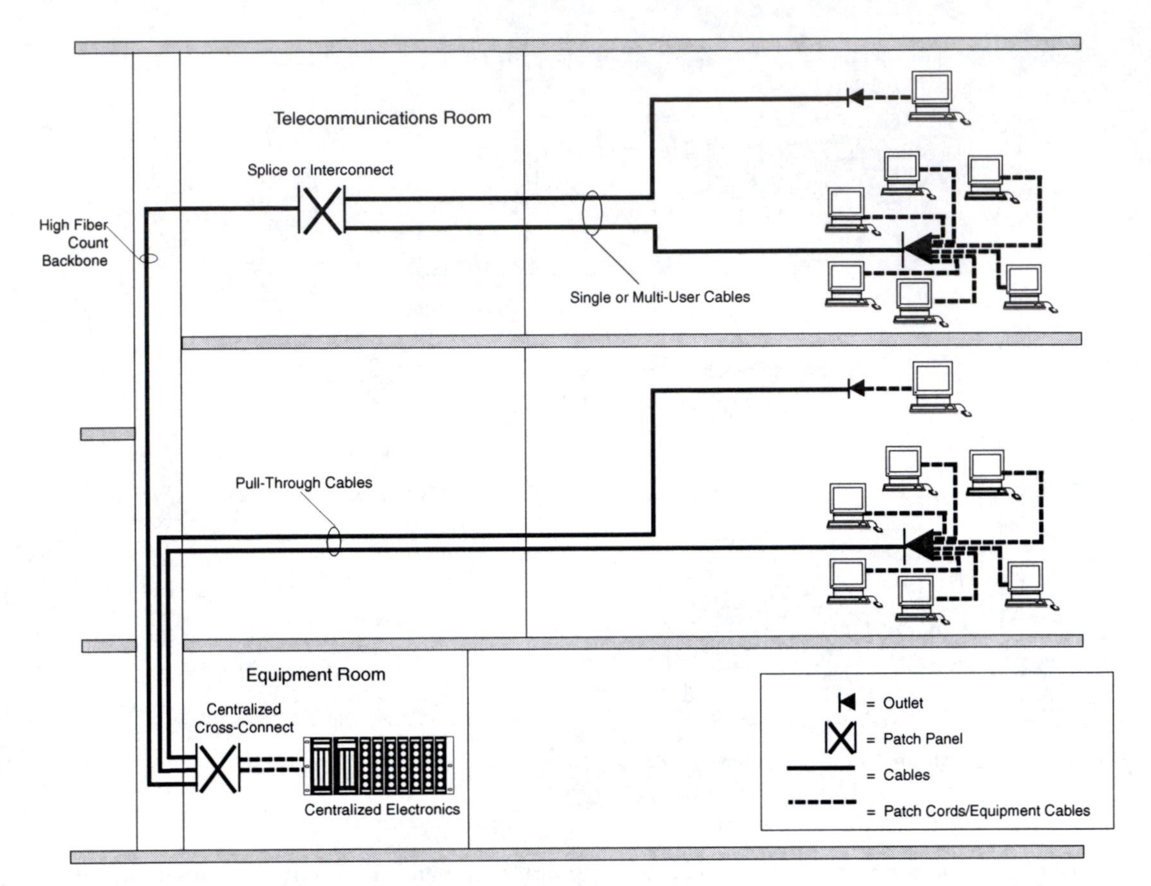

Figure 6–7. Centralized network *(Courtesy of Siecor Corporation)*

Centralized networks offer several advantages: They can cut equipment costs by consolidating equipment in a single location. Consider, for example, an application that requires 17 users on each floor. Most network hubs or workgroup switches are available with 8, 12, 16, or 24 ports. With a distributed network, you would need a 24-port switch on each floor, even though you have need for only 17 ports. For four floors, you need four switches or a total of 96 ports for only 68 users. Seven ports on each floor—or 28 ports total—are "wasted." Because you are buying a 24-port switch, you have an unused switch in terms of capacity and available ports. Plus you might want to connect each floor through a switch in the main closet. With centralized network, you can more closely match the number of ports in your equipment to your actual needs. You need only provide a 72-port switch in the central closet—a savings of 24 ports.

The centralized approach makes more efficient use of equipment, using 94% of the ports. With the distributed approach, only 71% of the ports are used. Figure 6–8 summarizes the advantages the centralized network brings to efficient equipment use.

Architecture	Users per Floor	Total Users (Four Floors)	Total Number of Ports Required*	Unused Ports	Efficiency
Distributed	17	68	96	28	71%
Centralized	17	68	72	4	94%

**Assumes a distributed architecture requires a 24-port switch on each floor.*

Figure 6–8. Example of how centralized network promotes efficient equipment use

Centralized networking also promotes easier expansion. Sure, the unused ports in our example are available for expansion. But what if the expansion is uneven? What if you need eight additional ports on one floor? And three on another? And eight on the third? You need to buy a new switch for two floors, because you only have seven available ports. With a centralized network, you can more easily match capacity to needs. In this case, you only need one new switch (or one new plug-in board for a chassis-based switch).

Beyond the equipment, you also save money by eliminating the need for cross connectors or patch panels in the equipment room on each floor. Hardware requirements are reduced.

Centralized networking can also reduce the time and costs of administering a network. Networks sometimes need to be maintained, by rearranging cables to accommodate users who move offices, by adding new users and removing users who should no longer be connected, and so forth. Because all equipment and cross connects are in one location, you can perform all moves, adds, and changes from a single location. No trudging up and down stairs, no lower productivity from wasted time on those same stairs. Likewise, troubleshooting can be simpler, because there are fewer potential points of failure (every interconnection is a suspect).

Even though centralized networking offers many benefits, it's not suited to all applications. A centralized copper system is constrained by distances. The 90-meter horizontal run still applies, so only smaller installations will benefit. Centralized networks are also not recommended for multitenant buildings. Most tenants will want control of their own network, which makes a distributed system more logical.

Zone Network

A zone system (Figure 6–9) is a specialized form of distributed system. Here the horizontal cable is divided into two sections. A "backbone" zone cable is run to a work area. Individual cables are then broken out and run to the work area.

In a typical copper-based zone system, a 25-pair cable runs from the equipment room to the zone box. The cable then breaks out into individual four-pair connections terminating in modular jacks.

In a fiber-based zone system, an MT-style 12-fiber array connector can be used from the equipment closet to the zone. A breakout box will divide the ribbon fiber into separate interfaces. In other words, the cable on the trunk side is a multifiber cable terminated in an MT-style connector. Internally, the breakout box will have a small section of fiber, with an MT connector at one end and individual duplex interfaces at the other. The box becomes an interface converter between the MT array connector and SC, LC, or MT-RJ connectors.

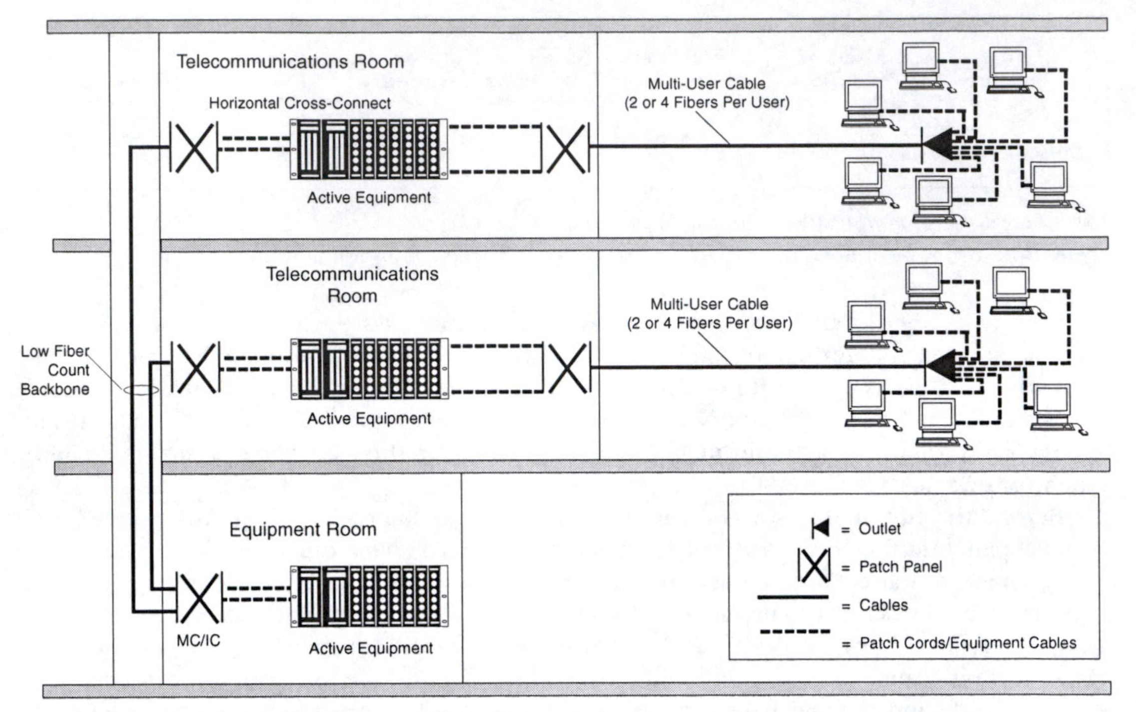

Figure 6–9. Zone network *(Courtesy of Siecor Corporation)*

Multibuilding Systems

A cabling system does not have to be restricted on a single building. Figure 6–10 shows a four-building system, which forms a small campus for business or schools. The main cross connect room is located in one building, with intermediate cross connects in satellite buildings. The cables between the buildings are most often fiber—multimode if distances are short or single-mode if the distances are over 300 meters. Within each building, the cabling system can use a distributed or centralized architecture to each floor and a single-user or zone scheme in the horizontal runs.

The number of cables between buildings is typically low. For example, the cables may simply be a backbone connection between network switches. Only a single fiber pair could accomplish this need. More often, however, several pairs would be run, if only for purposes of fault tolerances.

Customer-owned Outside Plant

In a campus arrangement like the one illustrated in Figure 6–10, the cable connecting the buildings is referred to as outside plant cable. TIA-758, Customer-owned Outside Plant Telecommunications Infrastructure Standard, gives requirements for both the cabling and the pathways and spaces.

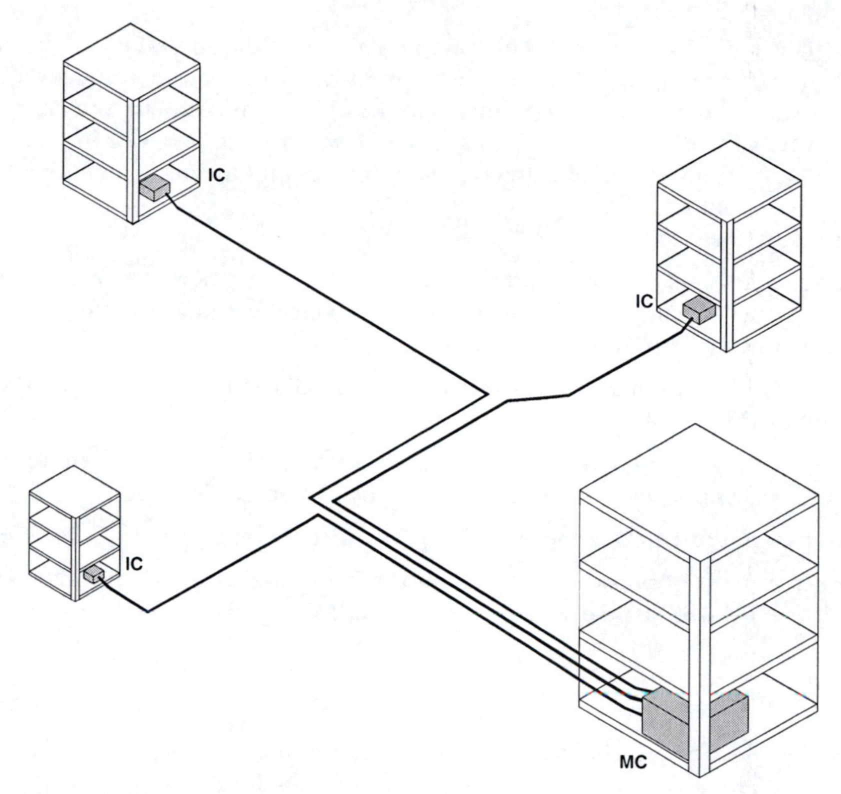

Figure 6–10. Multibuilding application *(Courtesy of Siecor Corporation)*

There are three major types of outside plant cable:

- Aerial cable, which consists of cable strung above ground and supported by poles, buildings, etc.
- Buried cable, which consists of cable buried directly in the ground.
- Underground cable, which consists of cable placed in underground ducts or troughs so that the cable is not in direct contact with the soil.

TIA-758 lists the same recognized cables for outside plant applications as TIA-568B.1 lists for backbone applications, with the addition of coaxial cable.

- 50/125 μm optical fiber cable
- 50/125 μm laser-optimized optical fiber cable
- 62.5/125 μm optical fiber cable
- single-mode optical fiber cable
- 100 Ω twisted-pair cable
- 75 Ω coaxial cable

Fiber is usually the preferred media for campus cabling, although UTP and coaxial cable are sometimes used, primarily for voice telephony and video distribution, respectively.

Much of TIA-758 is devoted to requirements for outside plant pathways and spaces. There are two major types of pathways—subsurface and aerial. Subsurface pathways include conduit, innerduct, and utility tunnels. Aerial pathways include poles and means of attaching the cable to poles and other structures.

Outside plant spaces include:

- Maintenance holes—an underground vault large enough for a person to enter to perform cable pulling and splicing. Maintenance holes are typically entered via a circular grade-level cover (formerly known as a *manhole*).
- Handholes—similar to a maintenance hole, but too small for a person to enter. Handholes are typically used for cable pulling but not splicing.
- Pedestals and cabinets—above-ground enclosures that provide environmental protection, security, and access to cable terminations and transmission equipment.
- Vaults—housing at or below grade-level that contain cable terminations, splices, etc.

A detailed treatment of outside plant cabling is beyond the scope of this book. For more details, consult the TIA-758 standard, or the BICSI TDMM [BICSI 2002].

Fault Tolerance

Fault tolerance refers to the ability of a system to continue to operate if an error occurs in part of the system. The error should not crash the system. Fault tolerance does not mean the system's operation is not diminished; it usually means the problem is isolated and the rest of the system operates. Total fault tolerance is possible so that no system functionality is lost, but this often is too costly to be practical.

In planning for fault tolerance, you must judge how critical each part of the premises cabling system is to your business. Any disruption is irksome and expensive, but some disruptions are much worse than others.

The main approach to fault tolerance is redundancy. Subsystems are duplicated in whole or in part. A backbone cable between buildings can be redundant. But how redundant is redundant? Take a fiber-optic backbone requiring four fibers. You have several choices:

- Run eight fibers in a single cable. If any of the four active ones fails, you can use one of the spares.
- Run two separate 4-fiber cables in the same conduit. You have a spare cable if the first breaks.
- Run two separate 8-fiber cables in the same conduit. This combines the first two approaches.
- Run separate cables in *separate* conduits. Then if an entire conduit fails, you still have a backup.
- Use a mesh structure. Even separate conduits run between the same two buildings can fail. If one building burns down, both links are rendered useless. A mesh structure interconnects

each building to each building. Now the main conduit runs from B to A, the backup from C to A and so forth.

- Combine redundant building-to-building conduits with a mesh structure.

You can see that the fault-tolerant structure becomes increasingly complex—and expensive to implement—as you increase the degree of fault tolerance from spare fibers in a cable to a completely redundant mesh system. Figure 6–11 shows examples of backbone fault tolerance.

Going beyond the building-to-building backbone, fault tolerance in the backbone cable, horizontal cable, and in the interconnections all can be accomplished to varying degrees. Again, this is done through redundancy. As shown in Figure 6–12, fault tolerance is often achieved by having duplicate telecommunications closets feeding the horizontal cable to the work area.

The backbone cable is typically seen as the critical link because its interruption will affect the highest number of users and is probably the most time-consuming and expensive to repair. On the other end, a work area cable is a small blip on the screen of critical concerns. It's easily and quickly replaced.

Many network devices have fault tolerance built in. FDDI, for example, can wrap the ring from the primary to the secondary to avoid a cable break. Some hubs allow redundant backbone links, with automatic switchover from the active to the standby cable if a failure occurs. Other networks offer methods of rerouting traffic to avoid points of failure. The long-distance

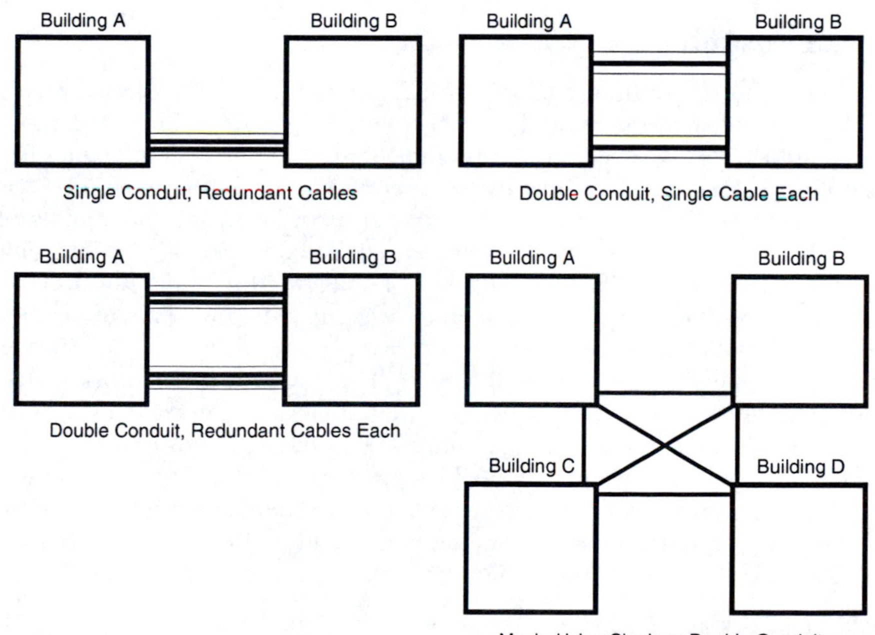

Figure 6–11. Backbone fault tolerance

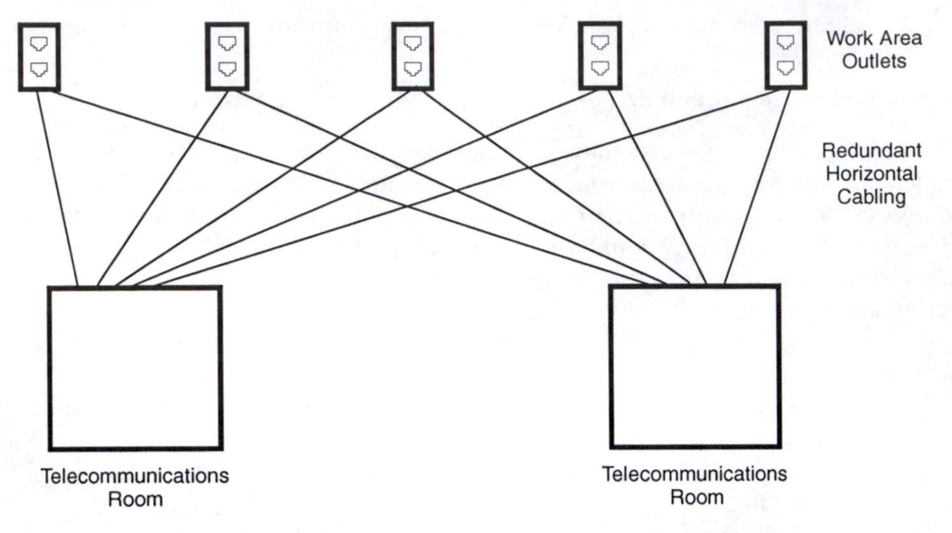

Figure 6–12. Horizontal fault tolerance

telephone service does the same thing by providing multiple routing paths. If one path goes down, many others are available.

Planning an Installation

A fundamental issue in planning a building cabling structure is to first decide what to bring to each work area. How many outlets and of what type are needed? A single data connection? Multiple data connections? How many phone connections? Are phone connections for analog or digital telephony? What level of performance is required today? Tomorrow?

While overspecifying the number of connections in a work area and the number of cables running to it will increase the cost of an installation, it is still cheaper than pulling new cables at a later date. Modular outlets allow the flexibility of changing the outlet interface but don't solve the problem of pulling new cable. Therefore, it is smart to anticipate future needs and pull the cable to the outlet at the very least.

The premises cabling has the longest life cycle of any part of the network or telephone system. Software has the shortest life cycle, with frequent upgrades and changes. Personal computers can become obsolete in increasingly short cycles as the demands of evermore sophisticated software places greater demands on the horsepower of the PC. Network equipment, too, may require upgrading and changes to accommodate newer demands and sophistication. But the premises cabling system should have a minimum useful life of 10 to 15 years. Therefore, take great care in planning the system. Forethought is much more cost-effective than recabling.

- Figure on one work area for every 100 square feet of floor space.
- For each work area, use a minimum of two connections: one for data and one for voice. Two voice and two data connections ensure greater flexibility. Also consider whether you should provide for a direct central-office connection for any voice lines, bypassing the PBX.

- Cable everything. Consider, for example, data/voice outlets for every seat in a conference room and in reception areas. The brave new world of multimedia, video conferencing, portable computers, and anytime/anywhere communications means that you should anticipate the future need to plug into the information infrastructure.
- Be sure to consider the impact of wireless LANs (WLANs). Many areas of the building, such as conference centers, lobby, cafeteria, etc., may be best served by a WLAN. In that case, cabling must be provided for wireless access points (WAPs). WLANs are covered in more detail in the next chapter.
- The telecommunications closet should be located at a common vertical access. Ideally, it should also be located where all work areas are within 100 meters. A single closet can serve an area of about 10,000 square feet. Multiple closets may be required if the area exceeds this. Research by AT&T shows that 99% of the time, workstations are within 93 meters of the closet.
- Carefully evaluate the advantages and disadvantages of cross connects versus patch panels. Cross connects are less expensive, but harder to maintain except by experienced persons. Do you anticipate many changes? Do you have a qualified staff? Patch panels can be easily used by network administrators, who have the in-depth knowledge of the cabling topology but little skill at terminating cable in punch-down blocks. A compromise often made is to use cross connects for voice and patch panels for network data.
- Be generous in anticipating future needs.
- Approximate needs for a telecommunications room are based on the area served and the recommendations of TIA/EIA-569B:

  10,000 square feet: 10 × 11-foot TR
  5,000 square feet: 10 × 7-foot TR

- Equipment can be contained in the telecommunications closet or in a separate equipment closet. The choice depends not only on preference, but on security concerns and the needs of the equipment, including the power and air-conditioning requirements. The equipment room includes the PBX, network hubs, network servers, minicomputers, and so forth.
- Don't try to get double duty out of a cable by combining different signals. While it is possible to use different pairs for voice and low-speed data on a cable (10BASE-T and voice, for example, require only three pairs total), using separate cables is the preferred practice.
- Leave extra cable slack in the walls. While there is a minimum amount recommended for rewiring outlets, a greater amount will provide more flexibility. For example, leaving a 6-meter loop in the ceiling or floor will not only permit an outlet to be rewired, it will allow the outlet to be moved. Remember that offices are rearranged and furniture moved. Don't allow the positions of the outlets to decide the arrangement of offices. In particular, poorly placed floor outlets can be a hazard to flexible office arrangements.
- The premises cabling system should be laid out using scale drawings. Cable distances should be calculated on the actual routing of cables, not on straight-line distances.

- Consider using factory-terminated cable assemblies in the system. This is especially true for patch cables. While the component costs may be higher, installation costs may be lower. What's more, you receive assemblies that are tested and ready to install. With careful installation (avoiding the pitfalls discussed in Chapter 8), the cable plant should be quickly and easily brought up and running. Cable assemblies require careful planning if custom lengths are to be used. You can't make on-the-spot adjustments if the assembly is six inches too short.
- Consider all the telecommunications infrastructure needs of the building, including voice, data, video, and building automation. In the past, there has sometimes been a tendency of "telephone people" and "network people" to maintain their own separate domains. This is becoming increasingly untenable as telecommunications applications converge to a single digital bit stream and applications like VoIP proliferate.

Interpreting Cabling Specifications

When looking at a manufacturer's specifications, there are a number of things to consider. In general, you want to look at worst-case, end-to-end (i.e., either link or channel) specifications. Many times, additional claims will be made that can be confusing. Specific items to watch out for include the following:

- Use of nominal or typical specs. Be wary of any specifications that are given as typical or nominal values. Your idea of what is typical may be dramatically different from the manufacturer's. Typical specifications are often used to conceal the fact that a particular product is good most of the time, but has wide variability among different units.
- Emphasizing the performance of one component. Sometimes an ad will emphasize the performance of a particular component (for example, a patch panel or modular jack) and how much better it is than competitive products. It is important not to be so distracted by this that you lose sight of the link and channel performance.
- Emphasizing unimportant specifications. Another potentially misleading technique is to greatly emphasize a particular specification (e.g., impedance smoothness of a cable), which is not particularly important. It may well be that this claim is true, but has no impact on the link and channel performance.
- Measuring from one end of the channel only. According to the standards, channel measurements must be taken from both ends of the channel, and the worst measurement is the one that counts. Remember from Chapter 4 that when two connectors are placed in close proximity, they give worse performance than when they are spaced further apart. In Figure 6–13, note that on the left (closet) side of the diagram, there are two connectors near the end of the link, while on the right (work area) side, there is only one. Therefore, a NEXT measurement from the closet end will probably be a couple of dB worse than the same measurement taken from the work area. Be cautious of specifications that are taken from only one end of a link or channel.

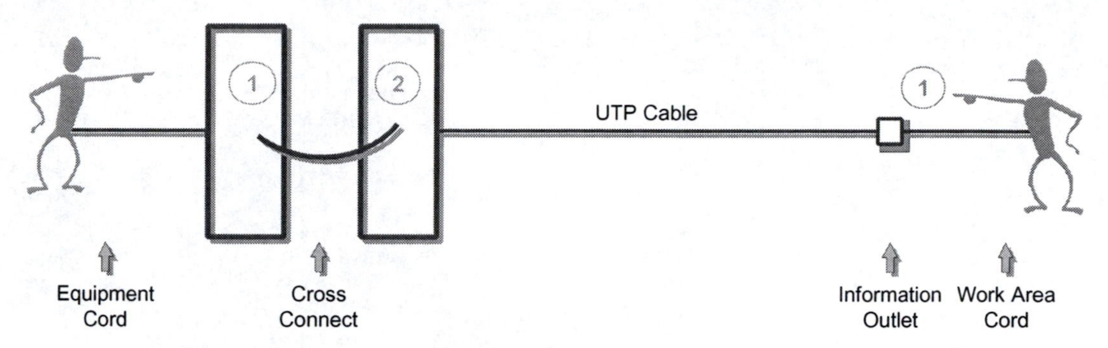

Figure 6–13. The cable plant must meet specifications from both ends of the channel

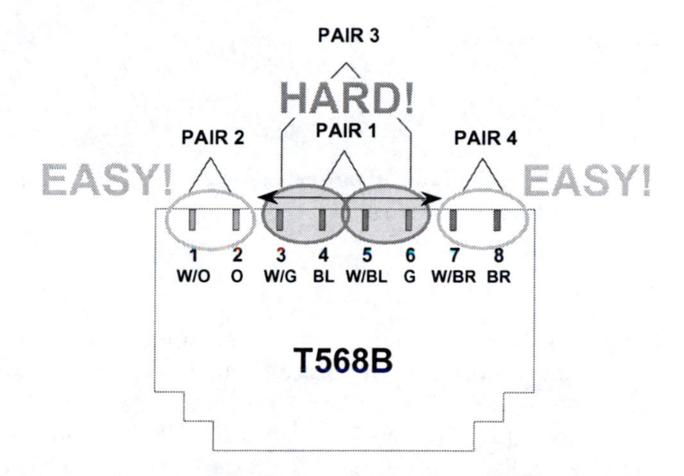

Figure 6–14. The cable assembly must be *all* pair combinations—some are harder to achieve than others

- Particular pair combinations only. For a pair-to-pair crosstalk measurement, the worst pair combination is the one that should be specified. With reference to Figure 6–14, it is relatively easy to get a very high NEXT reading between the two outer pairs (pins 1-2 and 7-8). It is much harder to get a high reading on the inner pairs (pins 3-4 and 5-6). There is nothing wrong with publicizing the crosstalk measurements on the outer pairs, but you must be careful not to compare that rating with the inner pair reading from another product.
- Margins given at a particular frequency. Occasionally, a specification will be given for ACR at a particular frequency, for example, 14 dB ACR at 100 MHz. As shown in Figure 6–15, this can be extremely misleading. The product with 14 dB ACR at 100 MHZ has a much lower rating at about 80 MHz.

To be sure that you are getting a cabling system that meets your expectations, emphasize end-to-end channel (or link) performance, not individual component specifications, and insist

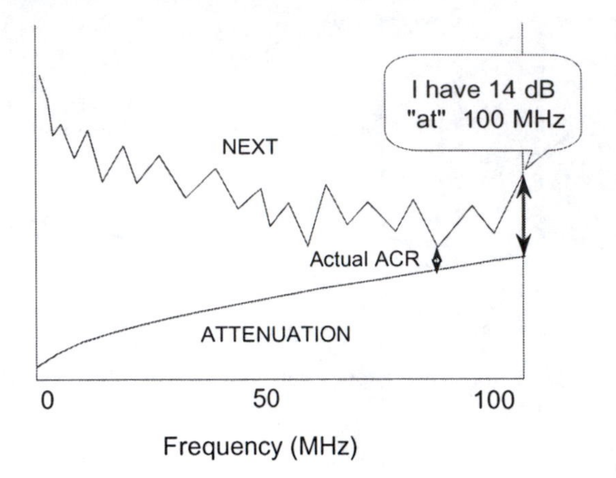

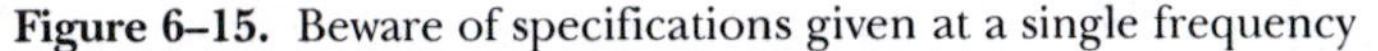

Figure 6–15. Beware of specifications given at a single frequency

on worst-case, guaranteed performance specifications based on swept-frequency measurements on all pairs from both ends of the link/channel.

Cabling System Warranties

Most vendors of premises cabling systems provide an application warranty of anywhere from 15 to 25 years, depending on the vendor. Basically, the warranty says that the system will accommodate any application designed to run over your level of cabling. If you install a Category 5e cabling system, for example, the warranty states that all applications designed to run over Category 5e (including applications designed for lesser systems like Category 3) will run over the cabling system. If it doesn't, the vendor will correct the problem until the system carries the application properly.

Component warranties are different: some components are guaranteed for life. But it is the installed *system* that's important. Proper installation, after all, is critical.

There are three issues you should remember about most warranties.

First, the entire system must use vendor-approved parts. Typically, this means all cable and connecting hardware comes from one company or a group of affiliated companies. Some manufacturers offer both cable and connectors. Others offer one or the other, but work closely with other companies to fill the gap. A connector vendor, for example, may work with two or three cable suppliers from a warranty standpoint. The point is that you can use only certain approved products. Thus the issue of interoperability of connectors and cables from different manufacturers is not a concern. All components have been tested together and no minor interoperability glitches should occur. The single-vendor approach reduces headaches and eliminates fingerpointing among multiple vendors who all feel it's the other guy's fault.

Second, the system must usually be installed by a vendor-approved installer to qualify for the warranty. Most manufacturers have extensive training programs to teach installers the proper use of their components. After installing the system, the installer will test it and certify that it meets specifications. Most installers will provide the owner of the cabling system with extensive test documentation showing how each cable performs.

Third, the system must be recertified after any moves, adds, or changes to make sure that changes do not result in a noncompliant system failing to meet specifications. As we mentioned, patch cables, in particular, can be a source of problems. You simply can't order up a bunch of patch cables and automatically expect them to work, especially with high-speed applications. It's always good to retest any channels or links that are changed.

Some people may think that warranties are confining because they limit the choice of components and lock you into a single solution. But from the vendor's point of view, reluctance to give warranty support to a mixed-vendor system is understandable. Besides, most vendors have a rich enough assortment of cables and hardware to meet most needs. What's more, many cable installations are single-vendor systems anyhow. The peace of mind that comes from performance warranties may outweigh any confinements that are encountered.

SUMMARY

- A permanent link defines the cabling system from equipment cross connect to the telecommunications outlet; it does not include the patch cables at either end.
- A channel defines the end-to-end cabling system, including patch cables.
- ISO/IEC 11801 defines seven classes of cabling system.
- The distributed network architecture is the traditional configuration, with equipment distributed to each floor; it allows 90-meter horizontal cabling runs to the telecommunications outlet.
- The centralized network architecture consolidates all equipment in a single location and allows cabling runs of up to 300 meters over fiber-optic cable.
- The centralized architecture provides efficiencies in equipment use, cabling complexity, and ongoing network and cabling plant maintenance.
- The zone network architecture runs high cable counts to the larger work area and then breaks them out to individual users.
- TIA-758 gives requirements for a customer-owned outside plant, including cabling, pathways, and spaces.
- Fault tolerance is a method of providing redundancy in the cabling system so that a failure of a cable does not result in system failure; it basically provides a backup path.
- To be sure that you are getting a cabling system that meets your expectations, emphasize end-to-end channel (or link) performance, not individual component specifications, and insist on worst-case, guaranteed performance specifications based on swept-frequency measurements on all pairs from both ends of the link/channel.

REVIEW QUESTIONS

1. What is the difference between a permanent link and a channel?
2. Name the three main cabling architectures.
3. Which architecture makes the best use of the capabilities of fiber? Why?
4. What is the purpose of fault tolerance in the cabling system?
5. What is the single basic principle for achieving fault tolerance in a cabling system?
6. Which ISO class is not compatible with TIA/EIA standards?
7. How big an area can a single telecommunications closet service?
8. Can a single cabling system serve multiple buildings?
9. Which cabling architecture is most compatible with modular offices and cubicles?
10. Why is it necessary to measure channel performance from both ends of the channel?
11. Does cabling system performance get monotonically worse with increasing frequency?
12. Name at least three advantages of a centralized cabling architecture.
13. Name two disadvantages of a centralized cabling architecture.
14. Discuss the pros and cons of zone cabling.
15. What is the difference between a telecommunications room (TR) and a telecommunications enclosure (TE)?
16. How many TRs and TEs are required on each floor? Can a TE be used instead of a TR?
17. List three constraints generally imposed by cabling system warranties.
18. Name at least four types of potentially misleading cabling specifications.
19. What are the three main types of outside plant cabling?
20. Name at least three types of telecommunications spaces used in the outside plant.

CHAPTER 7 Application-Oriented Cabling Standards

Objectives

After reading this chapter, you should be able to:

- Define an intelligent building.
- Understand the role of structured cabling systems in building automation.
- Understand the requirements of data center cabling.
- Describe the types of redundancy that may be required in data centers.
- Define an industrial cabling system.
- List the cabling media used in industrial cabling systems.
- Discuss the categories of environmental conditions in industrial facilities.
- Identify the two major types of wireless networks.
- Describe the cabling requirements for wireless access points.

Since TIA-568 was issued in 1991, structured cabling has achieved almost universal acceptance. It would be almost unthinkable to construct a commercial building today without at least considering the requirements of the various structured cabling standards. In recent years, another trend has emerged. Standards are being developed that specify how to apply the basic structured cabling techniques for particular applications. In this chapter, we will consider cabling for four such applications:

- Intelligent buildings, or building automation systems (BAS)
- Data centers
- Industrial cabling systems
- Wireless LANs[1]

[1] *Although "cabling for wireless LANs" may sound like an oxymoron, providing adequate connectivity for wireless access points is actually an important part of cabling system design, as we shall see later in the chapter.*

Building Automation Systems

So far, we have been considering premises cabling from the standpoint of networks and communications. A structured cabling system can also be used for the wiring of functions concerned with operating the building, such as fire alarms, security, lighting control, and heating, ventilation, and air-conditioning (HVAC). All these systems are increasingly becoming more and more sophisticated, employing computer-based controllers to monitor conditions and adjust them accordingly. The goal is threefold:

- Increase safety
- Ensure comfort for occupants and enhance productivity
- Control energy costs and lower the costs of operating the building

Such systems are called building automation systems. An intelligent building is one that achieves a high degree of automation to control the building for increased safety, comfort, and energy savings. Figure 7–1 summarizes the types of typical building automation systems and the functions they perform.

The traditional approach to building automation meant that each individual system was designed and installed individually, without reference to other systems. Just as network and communications systems benefit from a common structured cabling system, so can the building automation system. The building-control systems can share the same vertical and horizontal

System	Typical Functions
Security/Access Control	Surveillance (closed circuit TV) Intrusion detection Access control Alarms
Fire	Monitoring Automatic alarms Manual alarms Reporting
HVAC	Temperature control Humidity control Rate of airflow
Lighting Control	Fixed power reduction Occupancy-based or time-of-day on/off control Daylight compensation (varying light with changing ambient conditions) Lumen depreciation compensation (adjusting light levels to account for reduced output of fluorescent tubes over time)
Energy Management	Scheduling Control of peak demand Energy consumption and demand monitoring Optimal starting and stopping of equipment

Figure 7–1. Typical building automation systems and their functions

cabling pathways, equipment rooms, and interconnection hardware. The advantage is that all systems can be installed at one time and use the same pathways and facilities.

Cabling for building automation systems is addressed in TIA-862, *Building Automation Cabling Standard for Commercial Buildings.* TIA-862 describes its contents as follows:

> This Standard specifies minimum requirements for building automation system (BAS) cabling within a commercial building and between buildings in a campus environment. It specifies cabling requirements for cabling topology, architecture, design, and installation practices, test procedures and requirements for components that comprise the cabling system.[2]

Before discussing BAS cabling issues, we will review the fundamentals of building automation systems.

Parts of a Building Automation System

A building automation system consists of three basic parts: sensors, actuators, and controllers (Figure 7–2). The active and passive components are connected, configured, and controlled through a cabling system. Consider the HVAC system. Sensors monitor temperatures and humidity throughout the building. Actuators are used to control the HVAC equipment, such as opening or closing dampers, varying the speed of fans, and so forth so that the temperature and humidity are maintained at the required level. Many actuators also have sensors built in so that the controller can read the state of the device. The controller sits in the middle, reading the information from the sensors and operating the actuators. Most systems, of course, are more complex. The controller must poll many sensors, evaluate the conditions, and then operate the appropriate devices. The controller's program contains the instructions that relate sensor reading to actions.

Building automation and intelligent buildings take things a step further by adding smart control to the systems. For example, temperature levels can be adjusted by time of day, lowering temperatures at night when the building is unoccupied. The system can also control the system to achieve maximum energy savings. For example, bringing temperatures up slowly in the morning can use less energy than bringing them up quickly.

Beyond simple control is *adaptive* operation, so that the system can continuously adapt the building to changing conditions. Adaptive systems can "learn." This means that the system can use past conditions to calculate and predict what will be needed in the future. For example, it will learn the best speed to bring up the heating or cooling in the morning so that the proper temperature is reached with the least use of energy. The system must adapt to seasonal changes and to changes in the load. For example, the population and distribution of population within

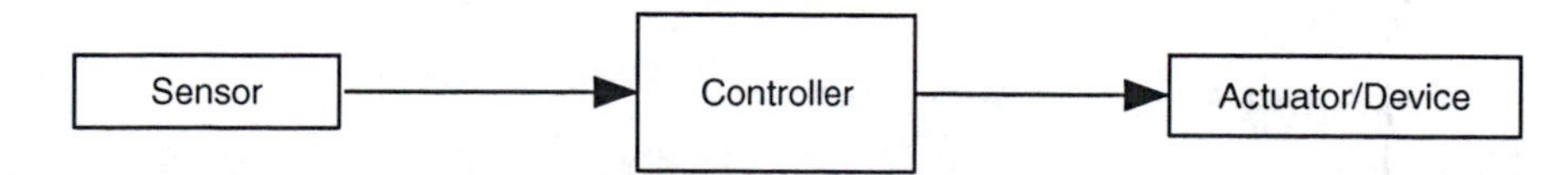

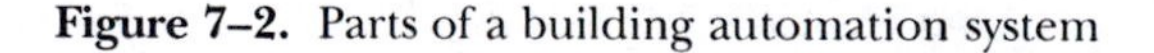

Figure 7–2. Parts of a building automation system

[2]*Excerpted from clause 2.1 (Scope, General) of TIA-862.*

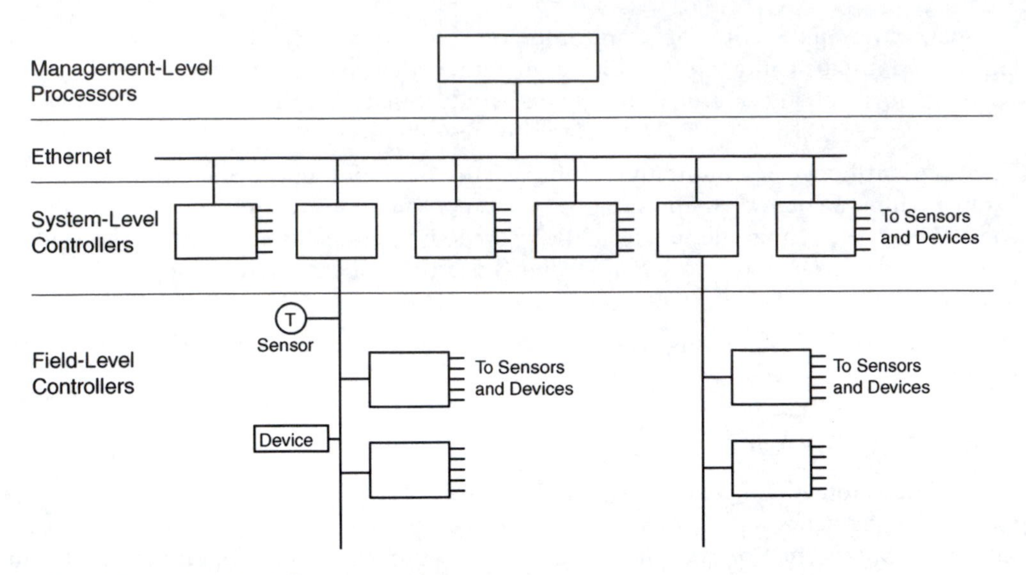

Figure 7–3. Hierarchical building automation system

the building will change over time and so will the heating or cooling requirements. The intelligent system will use adaptive strategies to optimize heating and cooling for any condition.

One key to successful building automation strategies is the low-cost computer power and sophisticated software that allow advanced control strategies. The intelligence is often distributed hierarchically in several levels. Microprocessors, costing only a few dollars but providing the powerful computing capabilities, allow intelligence to be moved outward from a central location. These nodes communicate with a central controller over some form of bus or network. Coupled with this processor power is the control software that permits greater levels of control and the ability to monitor and control building conditions with more precision.

Figure 7–3 shows a hierarchical system. A central PC provides overall management control and is connected over an Ethernet LAN to system-level controllers. These controllers can be connected directly to sensors and actuator devices or they can be connected over a bus to yet another level of controller, termed *field level,* that are used in different zones in the building. Field-level controllers can be either standalone units or they can be built into equipment. These lower-level buses are typically application-specific (and sometime proprietary) devices operating at low speeds. Communication moves up and down the hierarchy, as well as to and from the sensors and actuators.

Cabling the Building Automation System

Building automation systems use the same backbone and horizontal cabling structure as telecommunications and network systems, with the same advantages of application independence and flexibility. There are also some notable differences that must be accommodated.

Data network and communications systems typically have a single connection at the work area or in the equipment closet. A computer or a telephone plugs into the outlet. This is called a single-point connection. Single-point connections may also be used with building automation systems. In this case, each device (sensor or actuator) connects directly to the controller.

In addition to single point, building automation systems use three other types of connections—daisy-chained, fault tolerant, and tapped. This is illustrated in Figure 7–4. In the daisy-chained configuration, the devices are connected serially, one after another. This is often used for simple sensors, such as normally closed relays. If any of the relays in a daisy chain opens, then current flow is interrupted, and the controller is notified that some type of event has occurred (e.g., a fire, if the relays are part of a fire alarm system).

The fault tolerant configuration is a serial connection like the daisy chain, except that the last station in the chain is connected back to another port on the controller. In this configuration, the controller typically communicates with the devices using some type of message protocol. In this case, if the connection between two devices is broken, the controller can still communicate with all of the devices through one of the two controller ports.

In the tapped configuration, the devices are all connected in parallel to a central bus. Typically, the devices are no more than 3 feet (1 meter) from the central bus.

Another difference between building automation cabling and data cabling is that building automation systems often send power for sensors and actuators over the same cable used for

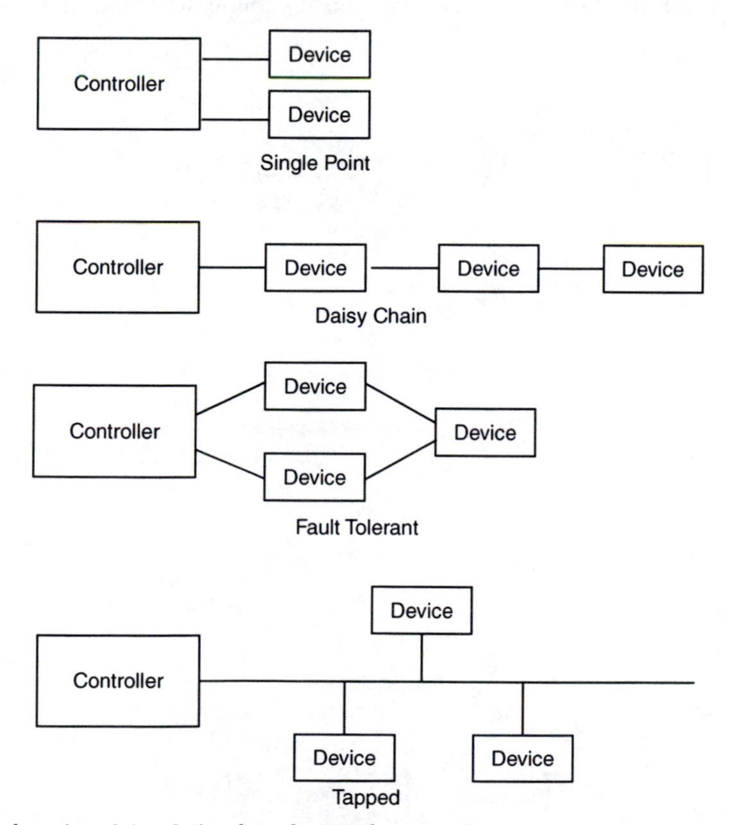

Figure 7–4. Single-point, daisy-chained, and tapped connections

signaling. Low-voltage DC or AC power is typically used. In some cases, such as a fire alarm bell, the devices can use a substantial amount of power. It is therefore necessary to make sure that the cable has enough ampacity to supply the device without an excessive voltage drop.

Network and communications systems typically have a single connection at the work area or in the equipment closet. A computer or a telephone plugs into the outlet. Building automation systems can have several devices or sensors connected through a daisy-chaining or taped connections. In a single-point system, each device connects directly to the controller. Daisy-chaining means that devices are connected serially, one after another. The tapped connection uses a central bus off of which the devices connect. Figure 7–4 shows the difference between single-point, daisy-chained, and tapped connections. The use of daisy-chained and tapped connections allows several BAS devices to be connected over a single cable. The area served by a single device attached to a single BAS cable is called a *coverage area.*

TIA-862 Cabling Architecture

BAS cabling uses an architecture very similar to ordinary structured cabling. Backbone and horizontal cables follow the same cabling rules we have covered in earlier chapters. The horizontal run is limited to 90 meters to the last device in the chain. The device cabling connects through an outlet to the system or it can connect directly to the cross connect in the telecommunications closet. Figure 7–5 shows a typical cabling system. You'll see that it looks

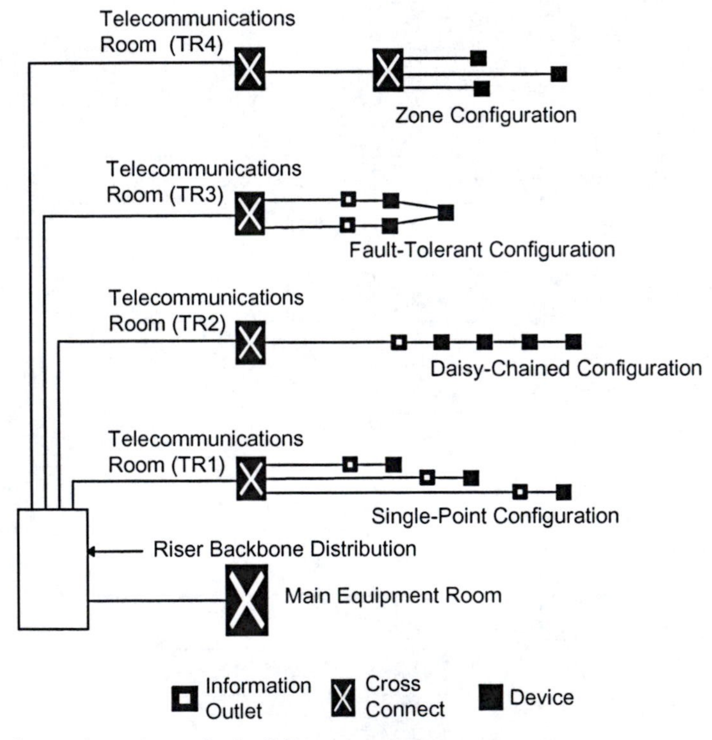

Figure 7–5. Typical premises system for building automation

similar to the cabling system for telephones and computers. The main difference is that the devices connected in the "work areas" are BAS devices, rather than telecommunications equipment (such as telephones and computers).

The key differences between BAS cabling and TIA-568 are in the horizontal and "work area." TIA-862 defines several new concepts:

- The horizontal cable may terminate either on a BAS outlet or on a *horizontal connection point* (HCP). The HCP is intended to make it easier to rearrange or modify the connections to the BAS devices.
- The area served by all the BAS devices attached to a horizontal cable is called a *coverage area,* as previously noted.
- The cable from the BAS outlet or HCP to the BAS device is called a *coverage area cable.*

Figure 7–6 illustrates these concepts. Coverage area 1 shows three BAS devices (which might be smoke detectors, for example) that are daisy-chained to a single BAS outlet. Coverage area 2 shows a single BAS device (a thermostat, for example) that is connected via a coverage area cable to an HCP.

TIA-862 contains guidelines for the size of coverage areas in various situations, as shown in Figure 7–7.

Because BAS devices are typically powered over the horizontal cable, the current-carrying capacity of the cable is very important. TIA-862 permits the use of either 22 or 24 AWG UTP cable. The nominal current capacity of this cable is summarized in Figure 7–8.

One purpose of a generic premises cabling system is to accommodate any application. With a network, this idea is straightforward. You plug your switches or hubs into the system in the

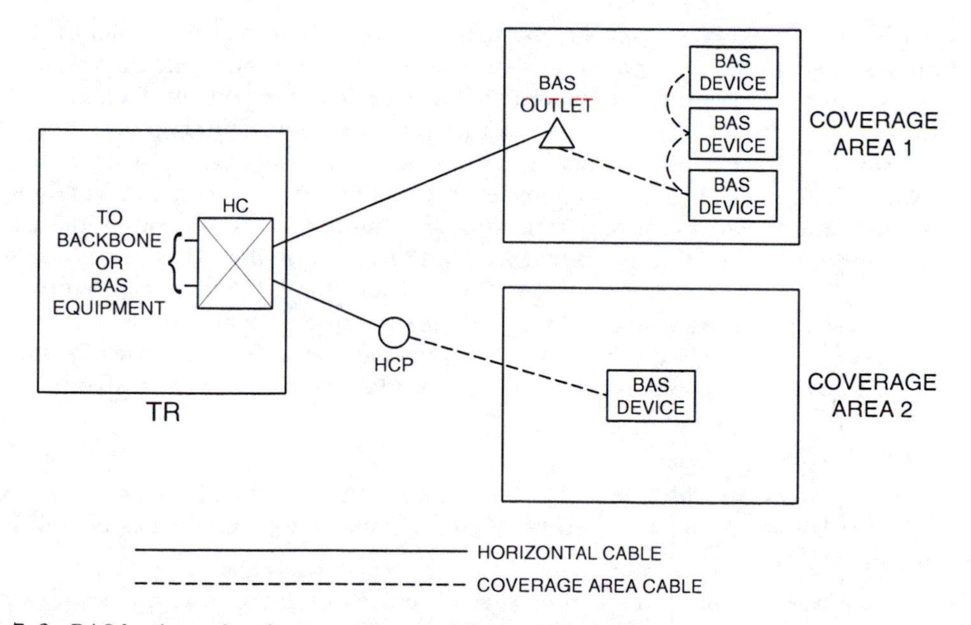

Figure 7–6. BAS horizontal and coverage area cabling

Type of Space	Typical Coverage Area (square meters)	Approx. Size (feet)
Indoor parking	50	23 × 23
Factory		
Office	25	16 × 17
Retail		
Hotel		
Classroom		
Hospital		
Mechanical room	5	7 × 8

Figure 7–7. Typical size of coverage areas

Cable Gauge	Maximum Temp. (°C)	Max. Current (A)	
		Single conductor	4-pair cable
24 AWG	25	1.5	3.36
	55	0.75	1.68
22 AWG	25	1.5	6.00
	55	0.75	3.00

Figure 7–8. UTP current handling capacities

equipment room or main cross connect. You plug your computer in at the outlet in the work area. With building automation systems, this same idea holds true—with one important exception. A system can be more proprietary outside of the generic cabling system. That is, the HVAC system wiring and controls can differ from manufacturer to manufacturer so that the "outside the system" wiring is more vendor specific. The types of connections at equipment, wiring rules (such as whether daisy-chained connections are recommended), the communication speeds and protocols between devices can vary from vendor to vendor. Thus it becomes important to consult the vendor about cabling specifics. Even with this caution, the premises cabling system can greatly simplify the system by providing the interconnection cabling throughout the building. The building automation system can plug into the premises cabling system.

What's more, the situation is improving as vendors move away from proprietary systems to open solutions that allow equipment from different vendors to communicate with one another.

Some Cabling Guidelines

- Horizontal cables should follow TIA-568B and TIA-862 guidelines and not exceed 100 meters. Distances may be extended if the system vendor's guidelines permit longer distances.
- Use a different cable for each function, such as security or energy management. Don't try to get double-duty out of a cable.

- Although TIA-862 does not specify any particular category of UTP, it is recommended that Category 5e be used at a minimum.
- Check for shared-sheath restriction on certain applications when considering using high-pair-count cables (such as 25-pair cable). Some applications, such as audio or analog input/output (I/O) should not be used in the same cable as digital signals because of the potential for interference.
- Carefully evaluate how many devices should be connected on a single cable run. While one device per cable is recommended, this increases the amount of cabling. But it also presents the greatest flexibility in meeting future needs. Daisy-chaining several devices reduces cabling expenses, but at the cost of flexibility.
- On backbone cables, do not mix network/telecommunication and building automation signals in the same 25-pair binder group. A binder group is a group of 25 pairs; a 50-pair cable has two binder groups. Keeping the two separate helps decrease interference.

Building Automation Architectures

Building automation systems are compatible with the distributed, centralized, and zone architectures discussed in Chapter 6.

The traditional building automation implementation is to distribute controllers in the mechanical equipment rooms—the rooms containing heating and cooling equipment and so forth. This is the approach used when the different systems were designed and installed separately. The contractor positions the controllers in the same area as the equipment they served. The controllers are often underused and have extra capacity.

A more efficient approach is to centralize the controllers as much as possible to make full use of their capacity. It may still be necessary to distribute some controllers because of the size and layout of the building. Controllers that are distributed may often be placed in the telecommunications closets. This has the advantage of requiring fewer secure areas.

The centralized approach allows one or more large controllers to replace a number of smaller controllers. For example, consider a four-story building with 10 devices on each floor. The distributed approach would have one 10-device system-level controller on each floor. With the centralized approach, a single 40-device system-level controller could operate the system from a single location. It also becomes more economical to have spare capacity to accommodate changes or upgrades. A single, centralized controller with additional capacity can be less expensive than four smaller controllers on each floor. Field-level controllers are typically still distributed, because they often are built into equipment. You'll notice that this is essentially the same argument made for centralized network equipment (see Figure 6–5).

While the use of fiber-optic cabling is less common with building automation systems, it is still available, especially for connections between controllers. Media converters or equipment-specific transceivers are also available to allow conversion between copper and fiber. In considering or designing a centralized system, you must carefully gauge allowable distances. Consult with the automation equipment vendor about allowable distances and about the availability of fiber-optic interfaces.

Zone cabling also allows flexibility in the cabling system, particularly in modular offices where walls are prone to be relocated.

Data Center Cabling

A data center is defined as a building or portion thereof whose primary function is to house a computer room and its support areas. Data centers include a lot of different kinds of equipment, including computers, networking equipment (LAN and WAN), storage area networks (SANs), multiplexers, transmission equipment, and so on. TIA-942, which is entitled *Telecommunications Infrastructure Standard for Data Centers,* actually includes a good deal more than the title implies. In addition to cabling and pathways and spaces in the data center, TIA-942 deals with many data center design issues, such as:

- Facility planning
- Data center sizing
- Site selection
- Architectural and structural issues
- Security and fire protection
- Electrical and mechanical systems
- Coordination with access providers
- Data center redundancy

As a widely cited example of this type of data center issue, TIA-942 recommends arranging telecommunications and computing equipment in aisles with the air intakes and exhausts facing each other to create "hot and cold" aisles, as shown in Figure 7–9. The air conditioning system would then be designed to supply cold air to the cold aisles, and exhaust hot air from the hot aisles.

For the most part, the items in the previous list are beyond the scope of this book. We will concentrate on the requirements relevant to the data center cabling and pathways.

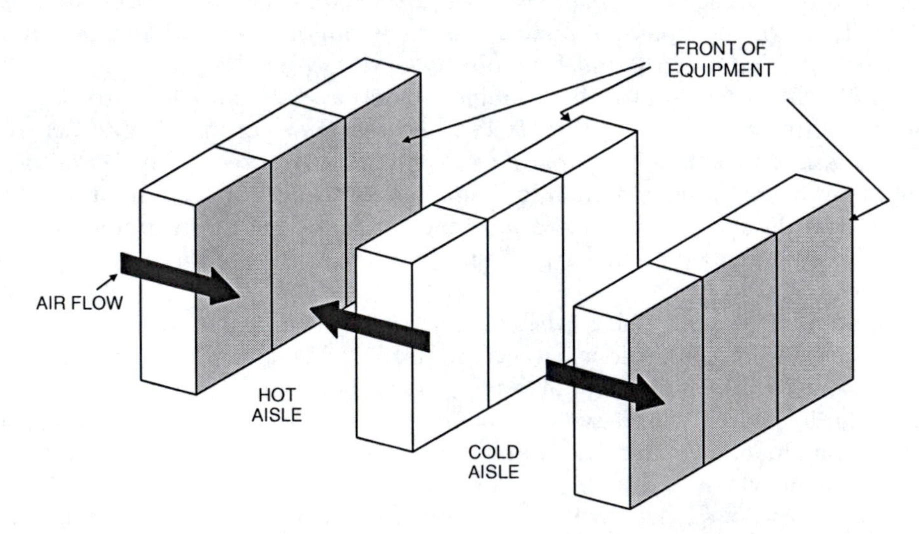

Figure 7–9. Arrangement of a data center into hot and cold aisles

Data Center Architecture

A data center consists of the following major components:

- Active equipment (usually lots of equipment) such as computers, networking equipment, storage networks, etc.
- One or more entrance facilities (EF), where transmission facilities and services enter the building.
- A main cross connect (MC).
- Interconnects that allow flexible connections to the equipment.
- Optional horizontal crossconnects (HC).
- Horizontal (H) and backbone (BB) cabling to connect the MC, HCs, and interconnects together.

A simplified diagram of the topology of a data center is shown in Figure 7–10. There are three ways that equipment can be connected to the MC. It can be directly connected by a horizontal cable, as shown in the left side of the figure. It can be connected through a HC as indicated in the center of the figure. Or, the interconnect can be separated from the equipment in a zone distribution configuration as shown in the right side of the figure.

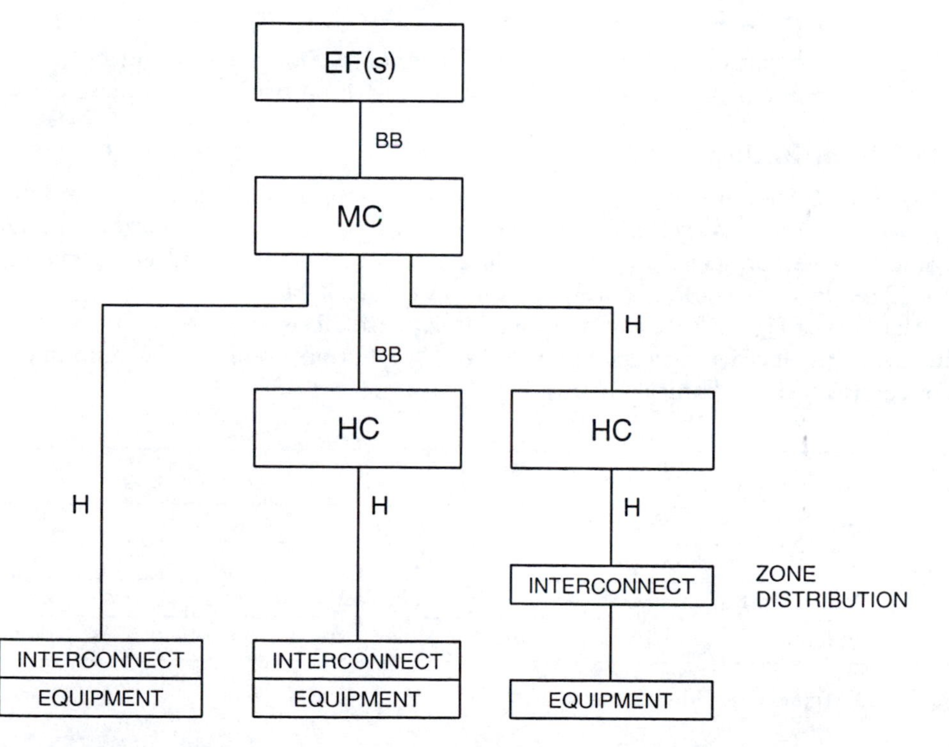

Figure 7–10. Data center topology

Data Center Cabling

Data center cabling uses a star architecture. The recognized media are essentially the same as those in TIA-758, except that the higher-performance options are recommended:

- 100Ω UTP cable—Category 6 is recommended[3].
- Multimode optical fiber—50 μm laser-optimized fiber is recommended.
- Single mode optical fiber.
- 75Ω coaxial cable (Type 734 and 735) to support T-3, E-1, and E-3 circuits.

Channels constructed from any of these media are required to meet the performance specifications of the appropriate sections of TIA-568B.

Data center cabling is designed to support a very broad range of applications, including but not limited to:

- Analog telecommunications services, such as voice, modem, and fax.
- Keyboard/video/mouse (KVM) connections.
- Various types of data communications used for management and control functions.
- Wide area networks (WAN).
- Local area networks (LAN).
- Storage area networks (SAN).
- Building automation systems (BAS) such as HVAC, security, and so on.

The primary pathways used in data centers are raised floors and overhead cable trays. TIA-942 makes a number of recommendations regarding the provisioning of these pathways.

Data Center Redundancy

As we discussed in Chapter 6, there are many different subsystems where redundancy can be provided to increase reliability and reduce downtime, including: entrance facilities, access providers, cross connects, backbone cabling, horizontal cabling, and electronic equipment. TIA-942 defines 4 tiers of reliability, as shown in Figure 7–11.

The tiers of Figure 7–11 can be applied independently to many different facets of the data center, such as electrical power generation and distribution, cooling, telecommunications, and architectural and mechanical systems.

Tier	Name	Description
1	Basic	No redundancy
2	Redundant components	Only one path
3	Concurrently maintainable	Multiple paths and components, but only one active at a time
4	Fault tolerant	Multiple active paths and components

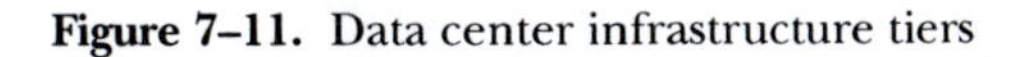

Figure 7–11. Data center infrastructure tiers

[3] *Once TIA-568B.1, Addendum 10 is issued, the recommendation will probably be increased to Category 6A UTP.*

Industrial Cabling Systems

Industrial cabling refers to systems installed in factories and similar environments which are used for manufacturing, processing, refining, and so on. The industrial cabling environment differs from a normal commercial cabling environment in several ways. First, industrial buildings are often very large and can have long cable runs. Second, industrial equipment is often hooked up in a bus or ring configuration. Finally, the environmental requirements are much more demanding in industrial applications, which may be characterized by:

- Extreme temperatures
- Mechanical vibration
- Moisture
- Corrosive liquids or vapors
- Dust and other particulate matter in the air
- EMI from large motors, arc welders, and other equipment

Industrial networks often use a variation of Ethernet known as Ethernet/IP (Industrial Protocol). In this section, we will only be concerned with physical layer issues. For a discussion of higher-layer protocol issues with industrial networks, see Sterling and Wissler 2003 or the Open DeviceNet Vendors Association (ODVA) Web site (www.odva.org).

The TIA is working (in mid-2005) on a document entitled *Industrial Telecommunications Infrastructure Standard for Manufacturing, Process, & Refining.* When completed, this document will be issued as TIA-1005[4]. It will address telecommunications cabling, pathways, and spaces in industrial buildings.

Industrial Cabling

The architecture of industrial cabling is similar to TIA-568-B, main, intermediate, and horizontal cross connects connected by backbone and horizontal cabling. Seven types of cable are recognized:

a. four-pair 100-ohm unshielded twisted-pair (UTP)
b. four-pair 100-ohm screened twisted-pair (ScTP) cables
c. optical fiber multimode cable 62.5/125 μm
d. optical fiber multimode cable 50/125 μm
e. optical fiber single-mode cable
f. two pair, 100-ohm unshielded twisted-pair (UTP)
g. two pair, 100-ohm screened twisted-pair (ScTP)

Note that ScTP and two-pair cables (items b, f, and g in the list) are not recognized cabling media for commercial buildings. Because of the long distances and harsh EMI environment, fiber has traditionally been the preferred cabling medium for industrial applications. However, there is a growing trend towards using UTP in industrial buildings.

[4]*Keep in mind that the industrial cabling standard is in rough draft form as this chapter is being written, so changes are likely.*

Environmental Categories

For industrial applications, environmental conditions are categorized into *MICE Levels.* MICE is an acronym for Mechanical, Ingress, Climatic, and Electromagnetic. For each of these four areas, three MICE levels are defined, as indicated in Figure 7–12.

For each of the parameters, level 1 is the least stringent and level 3 is the harshest. For example, for the vibration displacement amplitude specification, C_1 = 1.5 mm, C_2 = 7.0 mm, and C_3 = 15.0 mm. The four ratings can be applied independently—a given location in the building might be rated at level 3 for mechanical (M_3) and level 1 for the other parameters (I_1, C_1, and E_1).

Due to the importance of environmental considerations, TIA-1005 will include specifications for sealed connectors for both copper and fiber. These sealed connectors will have a MICE rating to indicate what environmental conditions they are designed to handle. Figure 7–13 shows an example of a sealed RJ-45 connector.

Type	Levels	Parameters
Mechanical	M_1, M_2, M_3	Shock—peak acceleration Vibration—displacement amplitude Acceleration amplitude Tensile force Crush Impact Bending and flexing
Ingress	I_1, I_2, I_3	Particulate ingress (min. diameter) Immersion
Climatic	C_1, C_2, C_3	Ambient temperature Rate of temp change Solar radiation Humidity Liquid pollution (6 parameters) Gaseous pollution (10 parameters)
Electromagnetic	E_1, E_2, E_3	Electromagnetic discharge (contact) Electromagnetic discharge (air) Radiated RF Conducted RF Electronic fast transient Surge Magnetic Field (2 parameters)

Figure 7–12. MICE Levels

Figure 7–13. An environmentally sealed RJ-45 connector *(Photo Courtesy of the Siemon Company)*

Cabling for Wireless LANs

Wireless networks have become very popular in the last few years. There are two main types—wireless local area networks (WLANs) and wireless personal area networks (WPANS). Figure 7–14 summarizes some of the key characteristics of WLANs and WPANs.

WPANs are typically highly mobile and are set up on an ad-hoc basis without any impact on the premises cabling. We will not consider WPANs further. WLANs, however, often require cabling to support the wireless infrastructure. The remainder of this section will address WLANs in more detail.

WLAN Architecture

The simplest WLAN configuration is called an Independent Basic Service Set (IBSS), although it is sometimes referred to as an ad-hoc network or peer-to-peer network. In an IBSS, all of the stations communicate directly with each other without any central point of control or administration, as illustrated in Figure 7–15.

A more common configuration is called the Basic Service Set (BSS), as shown in Figure 7–16. In a BSS, there is a central node called the Wireless Access Point (WAP) that provides control,

Type	Data Rate	Distance	Specifications	Name	Typical Application
WLAN	2–100 Mb/s	10s of meters or more	IEEE 802.11a IEEE 802.11b IEEE 802.11g IEEE 802.11n	Wi-Fi	Replacement for wired LAN
WPAN	Tens of kb/s to hundreds of Mb/s	< 10m	IEEE 802.15.x	Bluetooth, UWB, Zigbee	Cordless connection of peripherals to PC or PDA

Figure 7–14. Characteristics of WLANs and WPANs

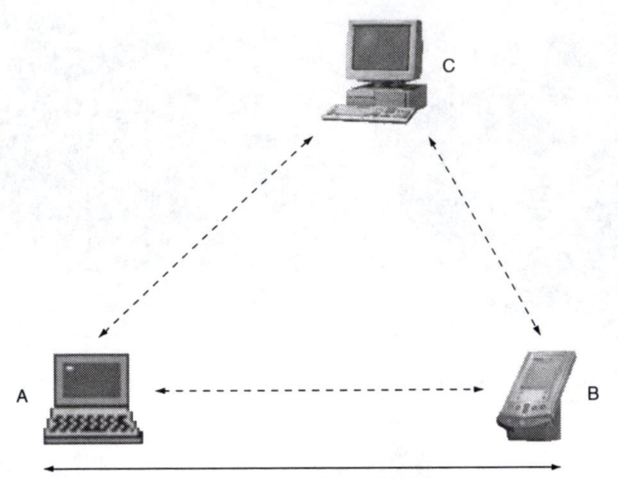

Figure 7–15. Independent Basic Service Set (IBSS)

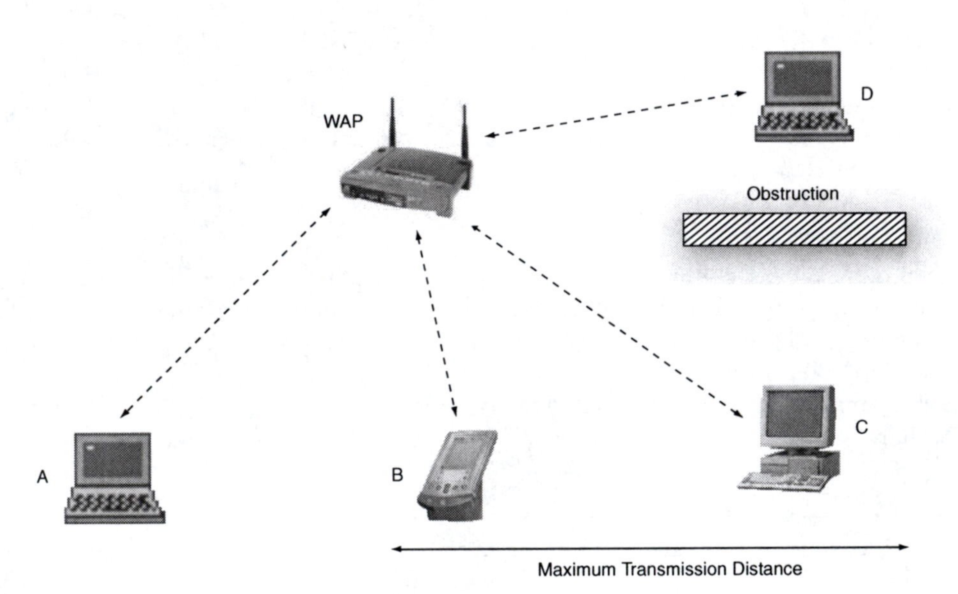

Figure 7–16. Basic Service Set (BSS)

buffering, and synchronization. Each node in a BSS communicates only with the WAP which broadcasts received packets to all the stations. This allows communication among stations which have an obstruction between them, such as stations C and D in Figure 7–16.

The BSS configuration is often adequate in a residence, but for a commercial building, coverage over a wider area is usually required. This requires an Extended Service Set (ESS), as shown in Figure 7–17.

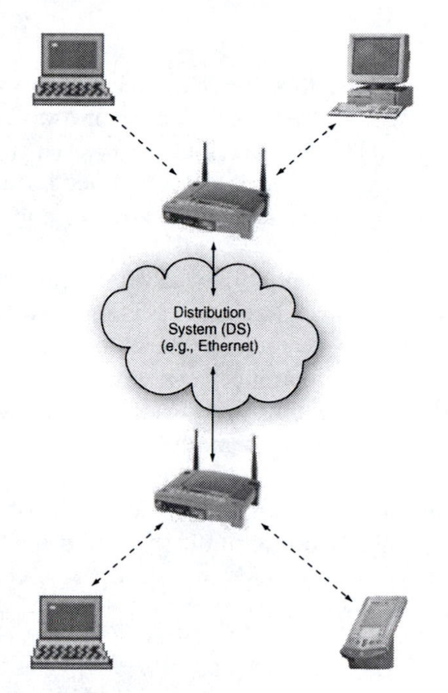

Figure 7–17. Extended Service Set (ESS)

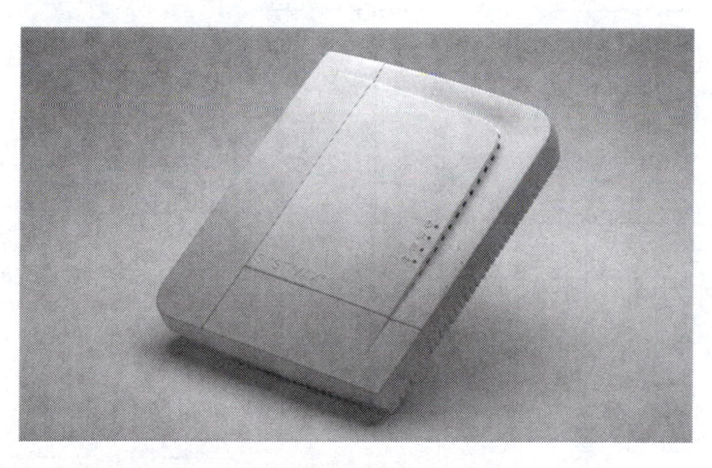

Figure 7–18. Typical wireless access point (WAP) *(SYSTIMAX Solutions graphics are courtesy of SYSTIMAX Solutions™, a CommScope company)*

In an ESS, the WAPs are connected together with another type of network, typically an Ethernet. This requires that cabling be provided from the site of each WAP back to a telecommunications room.

A typical WAP (Figure 7–18) is a small unit that may sit on a shelf, or be mounted on a wall or the ceiling. To achieve the greatest range, ceiling mounting is often the preferred alternative.

WLAN Cabling Considerations

To facilitate the extension of structured cabling systems to include the infrastructure to support wireless networks, the TIA[5] has issued a Telecommunications Systems Bulletin (TSB-162) entitled "Telecommunications Cabling Guidelines for Wireless Access Points." This TSB includes information on horizontal cabling for WAPs, telecommunications outlet locations, remote powering of WAPs over the horizontal cabling (using PoE), mounting and installation guidelines, and testing guidelines.

To precable a building for WAPs, the building is divided into *cells* and a WAP is placed in the center of each cell. The actual distance that wireless transmissions can support depends on many factors, including the type of network, the type of antenna, and the construction of the building. The TIA TSB recommends a default cell size of 60' square, as indicated in Figure 7–19. One or more TOs is placed in each cell and wired back to the nearest TR.

Category 5e or better UTP cable is normally used, although multimode fiber is also permitted.

Cabling for WAPs deviates slightly from the normal horizontal cabling allocation of 90 meters plus 10 meters of cords. WAP cabling limits the horizontal cable to 80 meters, with up to 13 meters of cord from the TO to the WAP and 6 meters of cord in the TR. The TSB also includes provisions for defining alternative cell sizes if desired.

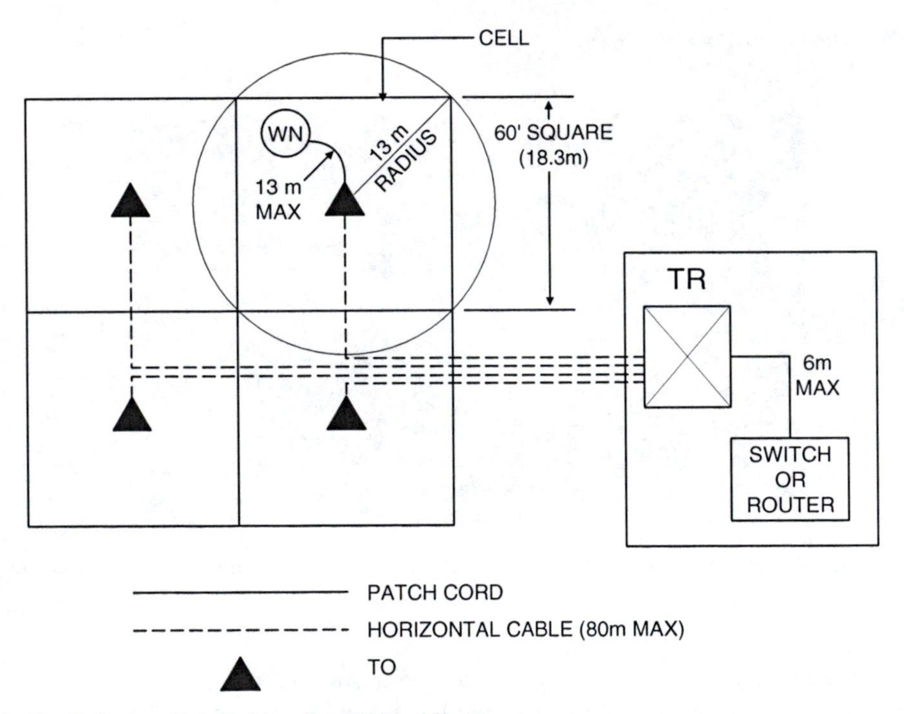

Figure 7–19. Default cell definition for WAP cabling

[5] *Both ISO/IEC and BICSI have also issued similar documents.*

The WAPs may be mounted either on the ceiling or on the wall, either above or below the dropped ceiling. To facilitate field testing, insertion loss specifications are given for the 80m permanent link.

Note that, because the TOs for the wireless network are installed in the ceiling, this cabling and TOs are in addition to the cabling infrastructure that is normally required for voice, data, and video networks.

SUMMARY

- Building automation systems increase safety, ensure comfort, enhance productivity, and lower costs.
- Typical building automation systems include security/access control, fire, HVAC, lighting control, and energy management.
- A building automation system consists of sensors, controllers, and actuators/devices. Devices are connected to controllers by single-point, daisy-chain, fault-tolerant, and tapped connections.
- TIA-862 provides requirements for building automation cabling.
- A data center is a building or area whose primary function is to house a computer room and its support areas.
- TIA-942 gives requirements for telecommunications infrastructure in data centers, including pathways, spaces, and cabling.
- Redundancy is a key element of data center design. TIA-942 defines four tiers of redundancy.
- Industrial cabling refers to systems installed in factories and other facilities engaged in manufacturing, processing, and refining.
- TIA-1005 defines requirements for telecommunications pathways, spaces, and cabling in industrial buildings.
- Environmental considerations are a key factor in industrial buildings. TIA-1005 defines four levels of environmental conditions, which are referred to as MICE levels.
- Wireless local area networks (WLANs) and wireless personal area networks (WPANs) are frequently used in commercial buildings.
- TIA TSB-162 provides guidelines for cabling to support WLAN access points.

REVIEW QUESTIONS

1. What TIA standard gives requirements for building automation cabling?
2. What are the three goals of building automation systems?
3. What three components make up a typical building automation system?

4. Explain the difference between the methods of connecting devices to controllers (point-to-point, daisy-chained, and tapped).
5. What is a coverage area?
6. What TIA standard gives telecommunications infrastructure requirements for data centers?
7. Name at least five applications that are typically supported by data center cabling.
8. What four types of media are recommended for data centers?
9. Name the four tiers of redundancy for data centers.
10. List at least five non-cabling-related issues that the data center standard covers.
11. What TIA standard gives telecommunications infrastructure requirements for industrial cabling systems?
12. What seven types of media are recognized for use in industrial cabling systems?
13. What are the three main differences between industrial and commercial cabling?
14. List at least six environmental conditions that are encountered in industrial applications.
15. What does MICE stand for? Give an example of a parameter for each of the four areas: M, I, C, and E.
16. What TIA document gives cabling requirements for wireless LANs?
17. What are the two main types of wireless networks? Which of them needs to be considered when designing the premises cabling infrastructure?
18. Where are wireless access points (WAPs) most commonly mounted?
19. What is the most common type of cabling used to support WAPs?
20. What is the default cell size recommended for WLANs?

CHAPTER 8

Installing Cabling Systems

Objectives

After reading this chapter, you should be able to:

- List and define the basic guidelines for installing cabling systems.
- Describe the basic termination procedures for modular plugs and jacks.
- List guidelines for termination fiber-optic connectors.
- Define polarization requirements for fiber-optic connectors.

The most common mistakes in installing a premises cabling system involve either improper cable handling or improper terminations.

In general, proper cabling handling becomes increasingly important with better grades of cable. For instance, Categories 5e and 6 UTP and optical fibers are more sensitive to improper installation than Category 3 cable. To avoid problems, installation practices should be based on the requirements of Category 6 UTP and fiber. Doing so avoids the need to have multiple procedures based on the grade of cable. Even with this recommendation, there are still differences that must be considered in dealing with copper and fiber.

Cables can be run in raceways and conduits installed in open spaces. A raceway is an enclosed path through which the cable is run. A conduit is a circular raceway. They can be run horizontally in the floor and ceiling as well as vertically. If a cable is run through an air-handling space, it must either be enclosed in a raceway or be plenum rated.

A pathway is the path between two spaces in a premises cabling structure. A space can be a wiring closet or a work area; the pathway connects them. The pathway itself can be structured—a raceway running under the floor or through a modular office wall. In some cases, cables are simply laid on the supports of a suspended ceiling in an unstructured manner, although this practice runs counter to the requirements of TIA/EIA-569B.

A number of conduits and raceways are available for cable. If the run is long enough, the raceway will have intermediate access points used to pull the cable during installation. Such pull boxes are necessary to keep tensile loading on the cable within specified limits.

General Cable Installation Guidelines

Neatness Counts

The premises cabling plant should never resemble a rat's nest of disorganized wiring. Cables should be organized and dressed with cable ties to keep them neatly bundled. If the need arises to troubleshoot or revise the cabling, a neat system is much easier to work with. Even so, neatness shouldn't be carried to the extreme where it stresses the cable in unacceptable ways (as described below).

Neatness counts. Cables must be routed, interconnected, and organized in a neat and manageable way. Once there are more than a few cables in an installation, they need to be managed neatly and efficiently. Otherwise, you will end up with a tangled mess of cabling and a hopeless mass of confusion. The purpose of management hardware is to allow fiber to be organized so they can be installed correctly and maintained efficiently. It is quite common for cables to be rearranged as needs change. If you cannot identify the cables and where they are routed, you cannot properly maintain an installation.

Figure 8–1 shows an example of an installation in both an unmanaged and managed state. In the unmanaged state, cables are run directly between the input and output points, with no thought given to keeping things neat and tidy. Cables cannot be easily traced for identification. Changing a cable might involve untangling of cables. The likelihood of error is high. In short, pity the poor technician responsible for maintaining the system.

The managed version is obviously more coherent, organized, and installed with forethought and care. The payoff is an installation that is easy to manage and maintain.

While all installations have management hardware to some degree or other, the successful system is one in which the hardware is properly installed and used.

Figure 8–1. Neatness counts: Management hardware can be the difference between a mess and a tidy, manageable installation *(Courtesy of Siecor Corporation)*

Cabling Installation Standards

TIA-568B.1 contains a clause on installation requirements. In 2001, a separate standard, ANSI/NECA/BICSI 568-2001 (Standard for Installing Commercial Building Telecommunications Cabling) was also issued jointly by BICSI and the National Electrical Contractors Association. This standard is, in general, compatible with the installation requirements in TIA-568, but provides a little more detail. It covers cable support structures, pulling and terminating cable, firestopping, and field testing.

The BICSI *Information Transport Systems Installation Manual* also contains a great deal of information on installation practices. The remainder of this chapter will cover the main aspects of telecommunications cable installation in commercial buildings. Testing and certified will be covered in Chapter 9.

Never Exceed the Minimum Bend Radius

All cables have a minimum bend radius. If this radius is exceeded, performance degrades. Even straightening the bend may not return the cable to its original performance level.

Cable should be installed in swept curves, avoiding sharp bends. Be careful when pulling cable that the conduit does not cause excessive bends or kinks. If necessary with conduits, use a fitting that increases the radius of the turn (Figure 8–2).

Too much attention to a neat appearance in a wiring closet can also cause the bend radius to be exceeded. Don't bend cables sharply simply to keep them close to equipment.

In the work area, cable is often stuffed into the outlet box. Be careful here to maintain the radius and avoid sharp bends and kinks. Some outlets have built-in features to allow a generous radius to be maintained.

Hardware such as cross connects, patch panels, and fiber-optic distribution boxes often offer cable management features. Figure 8–3 shows a fiber-optic distribution box. The box uses a tray to organize the fibers neatly and orderly, maintaining a generous bend radius on each fiber. Such boxes typically allow fibers to be connected either by splices or connectors.

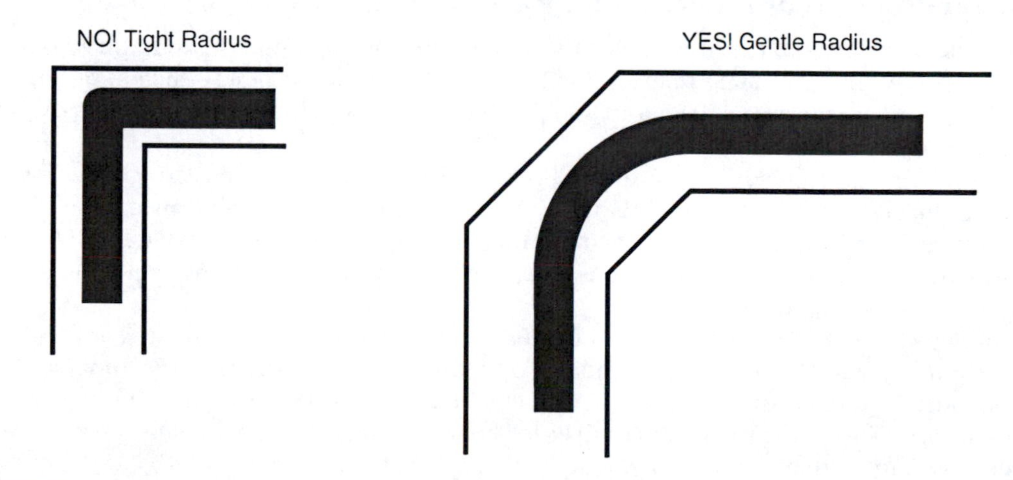

Figure 8–2. Use proper fittings to maintain swept bend radius

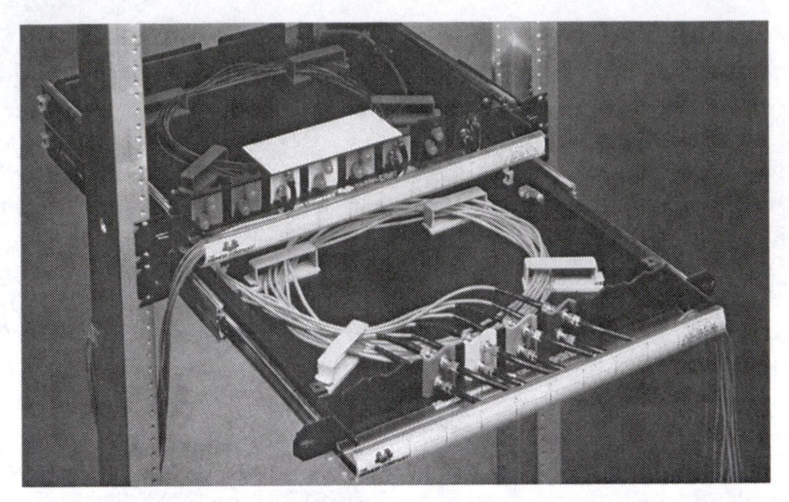

Figure 8–3. Fiber-optic distribution box demonstrates the cable management features of hardware *(Courtesy of The Siemon Company)*

Optical fiber cable has two minimum bend radii: one to be observed during the installation procedure and one for long-term use. During installation, the tensile loads of pulling necessitate a greater radius. The installed cable can be wrapped in a tighter radius.

Here are general rules for the minimum bend radius:

UTP: 4 times the cable diameter for 4-pair horizontal cable; TIA/EIA-568B recommends a 1-inch minimum radius.

10 times the cable diameter for multipair cable

Fiber: 4 times the cable diameter (about 1.8 inch for a zipcord construction)

Maintain Proper Tensile Loads During and After Installation

Tensile load refers to the stresses exerted on the cable during pulling or after installation. A cable's tensile strength rating refers not to the failure point of the cable—such as pulling so hard the conductors break—but to a reasonable limit above which performance degrades or the long-term reliability is impaired.

Too much tensile loading on a twisted-pair cable, for example, can stretch the twists, changing the geometry of the cable. With Category 5e cable, whose performance depends on maintaining the proper geometry, exceeding the tensile load can ruin the cable for 100-MHz performance. Too much tensile loading is a quick way to turn a Category 5e cable into a Category 3 cable—permanently.

The tensile load on an optical fiber involves its long-term reliability. Overstressing a fiber can cause tiny flaws to grow, possibly leading to ultimate failure of the fiber. Fortunately, the cable's construction, especially the strength members, significantly decouples the fiber from pulling forces. Despite the reputation of fiber for being fragile, fiber-optic cables can withstand higher tensile loads than copper cable.

Conductor Size (AWG)	Type	Maximum Loading (lb)
26	Solid	2.0
	Stranded	2.2
24	Solid	3.2
	Stranded	3.6
22	Solid	5.0
	Stranded	5.6

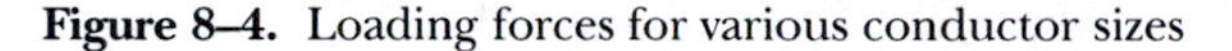

Figure 8–4. Loading forces for various conductor sizes

Tensile loads should be monitored during installation. Problems occur during pulling from several sources:

- Pulling too long a cable
- Pulling around corners
- Pulling over obstacles
- Pulling through a conduit too small for the pull being performed
- Not observing the recommended loading limit for the type of cable being pulled

Cables should be well supported, especially in hanging applications, to ensure that after-installation tensile loadings are not exceeded.

Common UTP cable (24 AWG, four-pair cable) has a tensile loading limit of 25 pounds. For other conductor sizes, use the information in Figure 8–4, which shows the maximum tensile loading for a given conductor size. Multiply the maximum tensile strength by the number of conductors in the cable. Notice that solid and stranded conductors of the same wire gauge have different load limits.

A typical two-fiber optical cable for horizontal application has a tensile load of 100 pounds during installation and 14 pounds after installation; other two-fiber cables can offer pulling loads ranging up to 210 pounds for installation and 65 pounds long term.

Always follow the manufacturer's recommendation on loads.

Cables can be hung from J or T hooks or bridle rings above suspended ceilings. Make sure the hooks are spaced closely enough to keep the tensile load of drooping cable within specified limits. TIA/EIA-569B specifies a maximum spacing of five feet, but shorter spacing may be required. If cables are run in trays or conduits, be careful that heavier cables don't load lighter cables by lying atop them.

Provide Support for Vertical Cables

Vertical cables must be well supported. Because a cable rising vertically must support its own weight, it must be supported at regular intervals. Starting at the top of a building and pulling downward allows the weight of the cable to help the pull.

Avoid Deforming the Cable When Supporting It

Cables should not be deformed by fastening hardware. Cable ties, straps, clamps, and staples can compress the jacket and deform conductors. Such compression can degrade performance, often irreparably. This is particularly important in installations carrying high-frequency data, because tight bundling of cables can significantly increase alien crosstalk. Always use care to prevent overtight placement of support hardware. Be sure cable ties aren't overcinched and that Velcro straps and nail-on clamps are snug without being over-tight. Staples, for both hand and gun application, are available that bridge the cable without crushing it. Always leave staples snug enough to support the cable but loose enough not to stress it.

For vertical applications, split wire mesh grips support cable without crushing it. The mesh spreads the holding force over a wider area to prevent overstressing a single point on the cable.

Keep Cables Away from EMI Sources

Because power cables can interfere with signals on copper cables, they should be kept separated. If it is necessary to run power and signal cables in the same conduit, separate inner ducts must be used for each.

Fluorescent lights, radio frequency sources, large motors, induction heaters, and arc welders also present serious sources of EMI that must be considered in routing cable. Closed metal conduits generally provide sufficient shielding of most sources. If the noise source is inductive (large changes in current), a special ferrous induction-suppressing conduit may be required. Fluorescent light fixtures represent the most common noise source in offices. Keep cables at least five inches away from any fluorescent fixtures.

Open or nonmetallic pathways offer no shielding from noise sources, therefore cables must be separated from them. Figure 8–5 shows ANSI/NECA/BICSI-568 recommendations for separation distances for power wiring.

Separation distances are not a concern with all-dielectric fiber-optic cables. These are immune to electromagnetic interference. For hybrid optical cables that also contain copper conductors or metallic strength members, the recommended separation distances should be observed.

Condition		Minimum Separation Distance (inches)		
Power Line Pathway or Electrical Equipment	**Signal Line Pathway**	**Less than 2 kVA**	**2–5 kVA**	**Over 5 kVA**
Unshielded	Unshielded	5	12	24
Unshielded	Shielded	2.5	6	12
Shielded	Shielded	—	3	6

Shielded: Metallic, grounded conduit
Unshielded: Open or nonmetallic conduit

Figure 8–5. Recommended separation from high-voltage power sources

Observe Fill Ratios for Conduits

The cross section of a conduit should be only 50% filled with cable. This ratio applies to each inner duct of a conduit. Keeping a low cable-to-conduit area ratio helps reduce pulling loads.

Reduce Loads by Pulling from the Center

Most often cable is pulled from one end to the other, that is, from the telecommunications closet to the work area. If you think tensile loading will be exceeded because of any reason—long pulls, obstructions, bends in the path—you can start at a midpoint and pull half the cable in one direction and half the cable in the other direction. This effectively reduces the loading over that of a single pull.

Avoid Splices

Cable should, whenever possible, be run in a continuous length, without splices. Splices are prohibited in copper horizontal cable (UTP and coax). With some backbone applications, splicing may be necessary. Always locate splices in an accessible point, never behind walls or above inaccessible ceilings. Any mechanical terminations—connector or splice—is a potential point of problems that must be left accessible.

One possible exception to this splice-free recommendation is the 300-meter fiber homerun system. Splices are sometimes used in place of a disconnectable horizontal cross connect. All rearrangements of cables can be done in the equipment room where hubs are located. Because rearrangements are not made at the horizontal cross connect, splices can be used without degrading the flexibility of the system.

Interbuilding Cables

Cables run between buildings can be directly buried, run through underground conduits, or hung aerially on poles. Underground conduits are the favored method in most cases and are used in around 80% of all premises cabling interbuilding applications. For long runs over a mile, where the cost of conduit becomes prohibitive, direct-buried cable is a reasonable alternative. Direct-buried cable requires armor that can withstand environmental extremes of temperature and moisture and the hazards of ravenous rodents. Because fiber is the preferred medium for interbuilding runs, we concentrate on it.

While fiber and copper share common pulling practices, fibers can also be blown through conduits. First, a *tube cable* is laid. This consists of an outer jacket surrounding up to 19 inner tubes. Each inner tube accepts a single fiber bundle of up to 18 fibers. The blowing device bleeds compressed air into the inner tube to float the fiber, which is paid off a motor-driven reel at a rate of about 100 feet per minute.

Air-blown fiber depends on continuous, splice-free runs, which typically is not a problem with premises cabling applications. One advantage of air-blown fiber is that it is easy to blow new fibers through empty tubes if it becomes necessary to expand capacity. Damaged fibers can also be easily blown out of the tube and new ones installed.

Most outside plant fiber uses a loose-tube design in which the fibers "float" within the tube. Gel fills the tube to prevent moisture from forming on the fibers. Burial can be either by trenching or plowing. In trenching, a trench is dug, the cable is placed in it, and the trench is filled. Plowing uses special cable-laying equipment to dig a ditch, lay the cable, and cover it in one operation. It's advisable to place the cable below the frost line in your area.

Conduit provides greater flexibility than direct burial. First, it makes repairs and installation of new cable easier because no additional digging is required. Using inner ducts to segregate cables makes pulling new cables easier by eliminating snags, frictions, and blockages that can occur when new cable is installed over old. Inner ducts can also be color coded by function. Some installers place pull ropes or tapes in empty ducts to make future installations even easier.

Terminating Cables

Misapplied connectors can be a source of problems in premises cabling. Particularly with Categories 5e and 6 UTP and fiber, careless application can degrade performance considerably. Again, a reminder of an earlier statement: most network problems occur at the physical layer, with the cables and connectors.

Terminating Modular Connectors

Figure 8–6 shows the basic procedure for applying a modular plug to a connector. Remember, however, that the examples here are illustrative. Always follow the manufacturer's installation instructions for a particular product. The general termination procedure is simple:

- Strip away the cable jacket to expose the individual pairs.
- Untwist the pairs just enough to insert them into the plug—no more than ½ inch for Category 5 cable or higher.
- Insert the cables into the plug, being careful to use the correct order for the wiring pattern being used.
- Cycle the crimping tool.
- Inspect the termination. The clear plastic of the plug allows you to inspect the termination area and check that the pairs are in the right order.

> ***The importance of carefully inspecting each termination cannot be overemphasized. Careless or inadvertent terminations are a major source of problems and errors in premises cabling.***

Figure 8–7 (pages 208–210) shows a typical termination procedure for terminating a modular jack with 110-style punch-down blocks. The wires are carefully positioned in the contacts (taking care to limit untwisting of pairs) and then pushed into place by squeezing the cover with pliers.

Terminating a cable in a punch-down block is fairly straightforward. The wire is positioned in the IDC contact and the punch-down tool is applied to seat it in the contact. The tool also cuts off the end of the wire.

Here are some things to pay attention to:

- Remove as little of the outer jacket as possible.
- Maintain the twist on the cables as close to the connector as possible. *With Category 5e UTP or higher, do not untwist more than 0.5 inch.* For Category 3 cable, the limit is one inch.

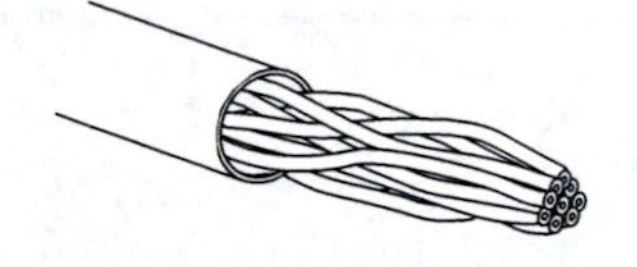

1. Strip the cable jacket

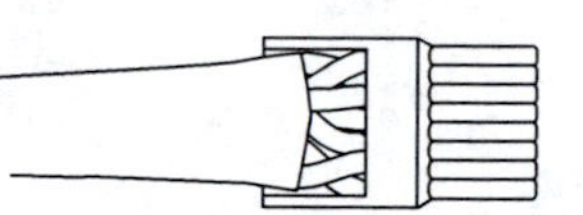

2. Assemble the load bar on the conductors

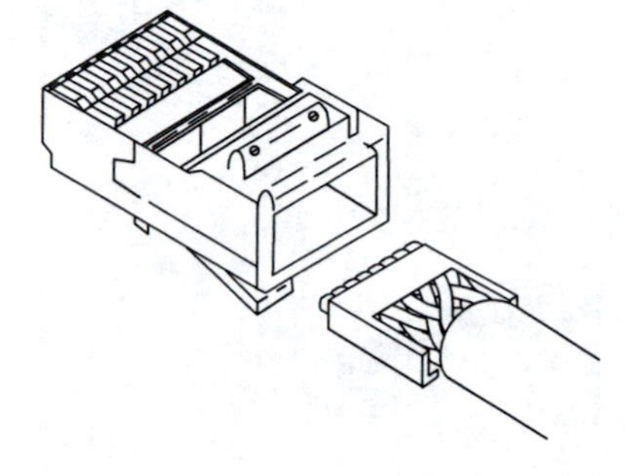

3. Insert cable in connector housing

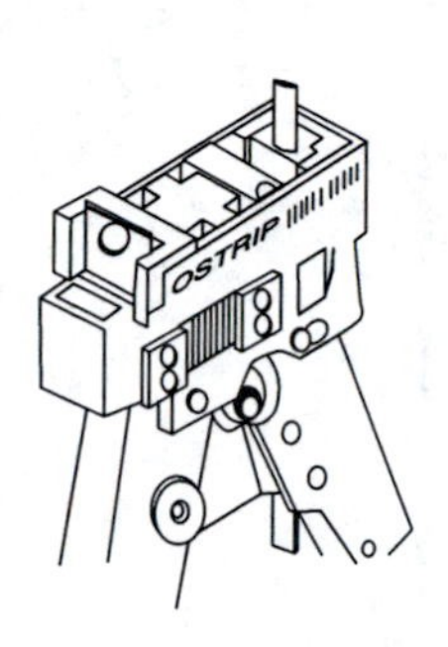

4. Crimp

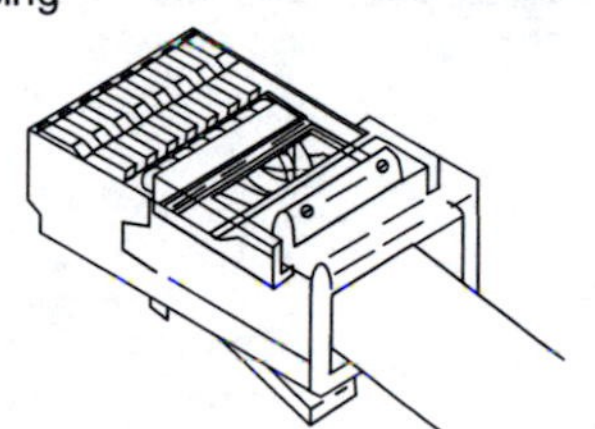

5. Inspect the finished termination

Figure 8–6. Terminating a modular plug

- Make sure all components are rated to the category being installed.
- Use the same cabling scheme throughout the installation. That is, terminate each plug in the same wiring pattern. The preferred pattern is T568A or T568B. Don't, for example, mix T568A and T568B patterns. Using a mixed wiring pattern will only lead to future confusion and incompatibilities.
- Crossover cables and connections, such as those required to connect two hubs (see Figure 5–12) should not be part of the premises cabling system. These should be treated as special cases outside of the system. Trying to incorporate them into the cabling plant leads to long-term confusion and incompatibilities.
- Pay close attention to the position of each conductor within the connector. The pattern should be the same at both ends of the cable.

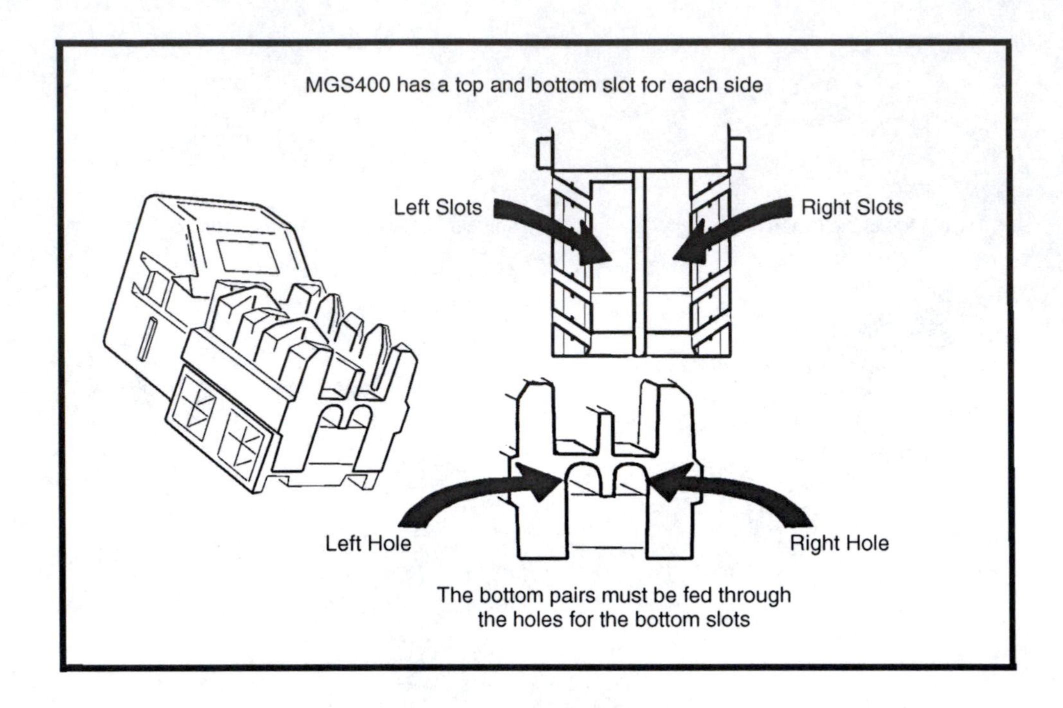

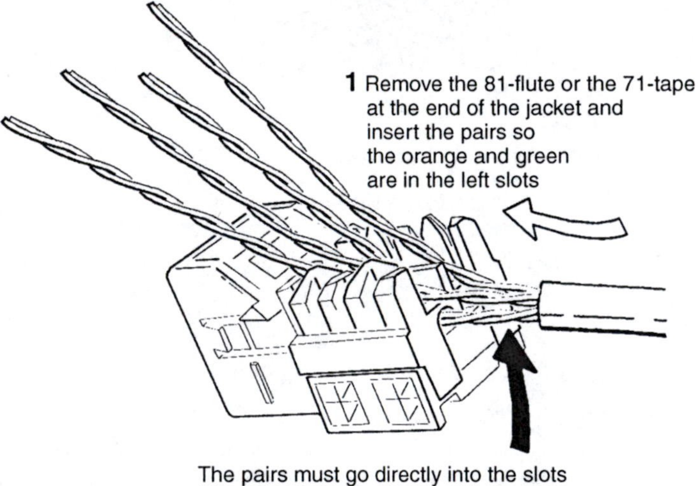

Figure 8–7. Installation procedures for a modular jack with insulation displacement connectors *(SYSTIMAX Solutions graphics are courtesy of SYSTIMAX Solutions™, a CommScope company)*

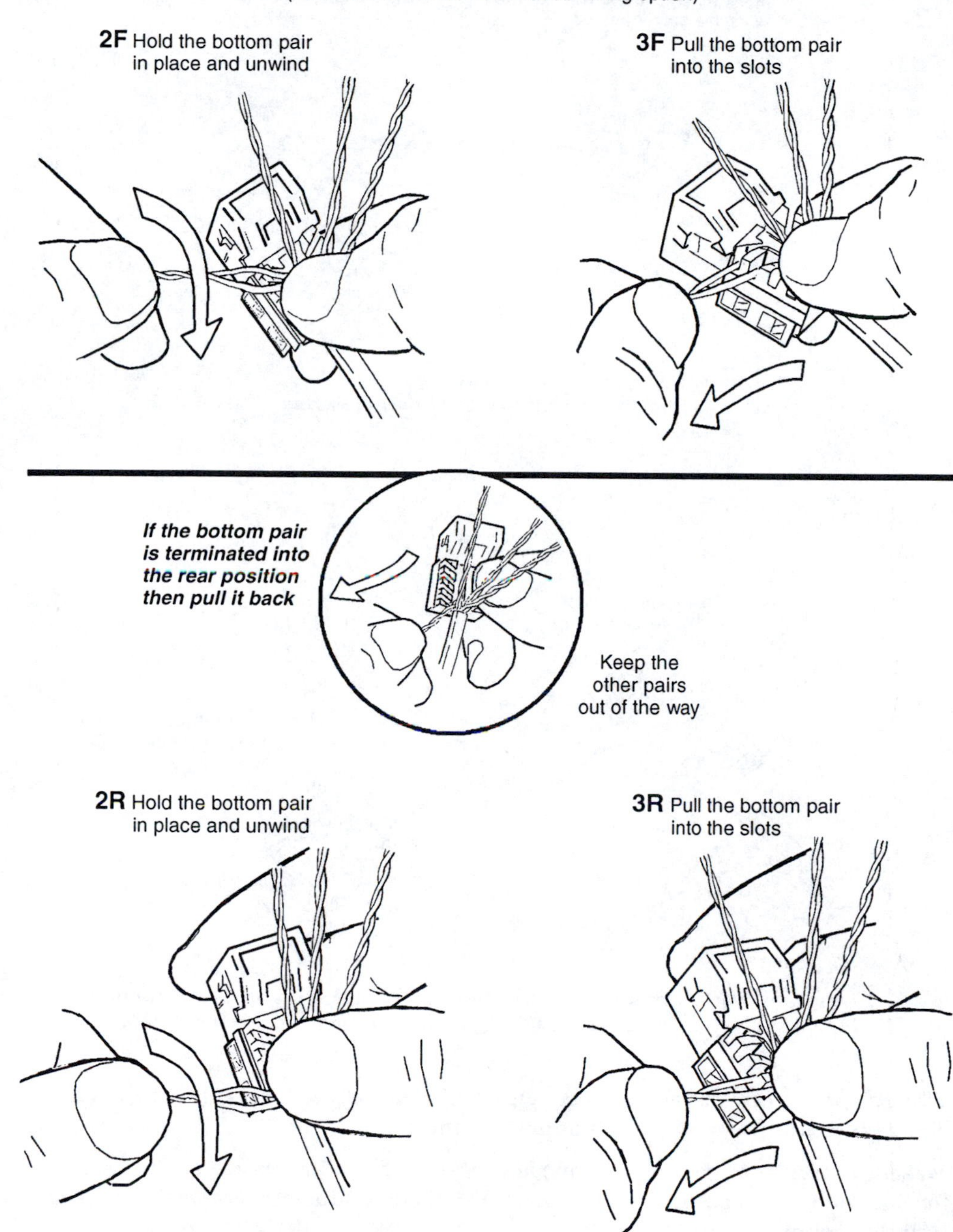

Figure 8–7. Installation procedures for a modular jack with insulation displacement connectors *(SYSTIMAX Solutions graphics are courtesy of SYSTIMAX Solutions™, a CommScope company)*

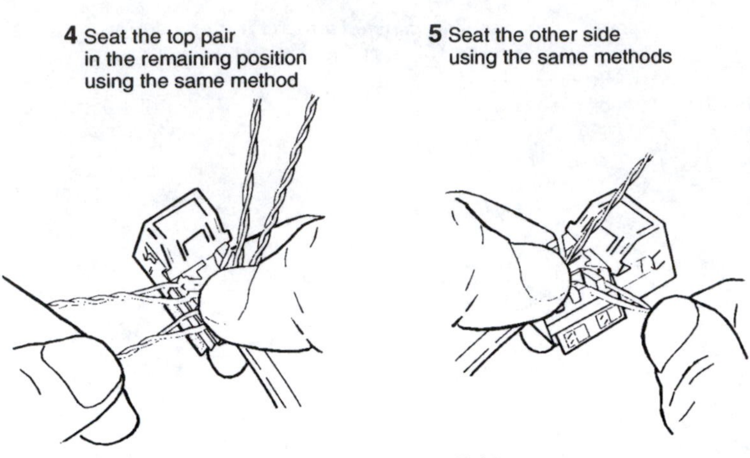

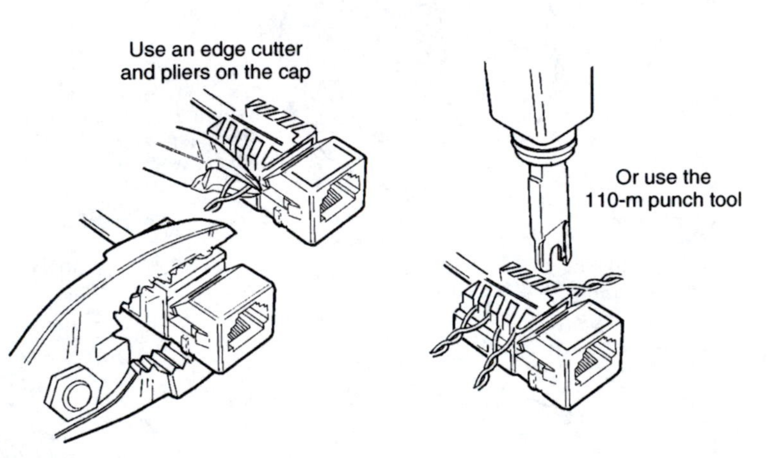

Figure 8–7. Installation procedures for a modular jack with insulation displacement connectors *(SYSTIMAX Solutions graphics are courtesy of SYSTIMAX Solutions™, a CommScope company)*

- While a connector can sometimes be reused, it's generally faster and more cost-effective to cut off and discard an incorrectly applied connector.
- Make sure you have the proper connector for the cable. Some modular plugs are designed for stranded conductors, while others accept solid conductors. Some plugs accept both stranded or solid conductors.
- Maintain proper cable-management practices with regard to tensile loads and bend radii. Neatness counts, but not at the expense of degraded cable.

Factory-terminated patch cables are preferred in the work area and at cross connects. This is especially true for Category 5e and higher UTP cables, which are more prone to misapplication. Single-ended cables are available terminated at one end with a modular plug and unterminated at the other end. The unterminated end can be punched down at the cross connect.

Terminating Fiber-Optic Connectors

Earlier, in Chapter 4, we showed a typical installation procedure for a fiber-optic connector. Here are some useful things to remember about working with fiber:

- Use epoxyless connectors to speed application time. Eliminating epoxy cuts the number of consumables, cuts the potential for mess, and slashes installation time. The lack of epoxy also speeds polishing time because epoxy is harder to remove than glass. The connectors have proven themselves reliable and easy to use.
- Be careful with the fiber ends removed from the connectors. They can puncture the skin like a splinter and are hard to remove. Some installers use a piece of tape to collect them.
- Cleanliness is essential. Keep the fiber clean and dry when working with it. Dirt, films, and moisture are the enemies of fiber performance and reliability. Likewise, keep tools and supplies clean so they don't contaminate fibers.
- Limit the number of passes with the stripping blade. One pass is recommended. Then clean the fiber with a lint-free pad soaked in pure (99%) isopropyl alcohol. Isopropyl alcohol of less than 99% purity can leave a film on the fiber.
- Maintain the proper bend radius. Most outlets and organizers have built-in fiber management features that will allow generous radii to be achieved.
- Leave slack at each interconnection, such as inside wall and floor outlets and organizers. At least one meter of fiber is recommended. This allows for future repairs or rearrangement of cabling.

Polarization of Fiber-Optic Cables

TIA/EIA-568B recommends duplex connectors as the standard interface. The connectors are labeled A and B. This standard requires a crossover at every coupling adapter. At a wall outlet, for example, the in-wall A-connector mates with the work area B cable, as shown in Figure 8–8. The crossover simply means that one end of a specific fiber is labeled A and the other end is labeled B.

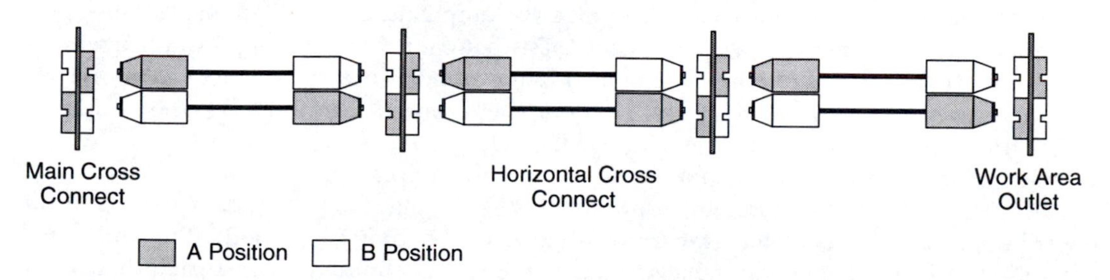

Figure 8–8. Polarization of duplex fiber-optic cables

The example in Figure 8-8 shows SC connectors, which are polarized by a key on the connector and keyway on the adapter. It becomes important to maintain polarization throughout the installation so that transceivers on equipment are always properly connected. Adapters at each end of a cable run should be installed in the opposite manner from each other. In other words, if the adapter keyway is up at one end, it should be down at the other end.

For multifiber cables, pair even-numbered fibers with odd-numbered fibers. Fibers 1 and 2 form a pair, 3 and 4 form a second pair, and so forth.

This may seem overly complex, but it has a purpose. Consider two electronic devices, for example a hub and an NIC. For devices, the receiver port is A and the transmitter port is B. If the NIC connects directly to the hub, the transmitter port (B) of the NIC connects to the receiver port (A) of the hub. In other words, port A at one end connects to port B at the other end. What the 568 rules do is to maintain this A-to-B and B-to-A orientation at every adapter throughout the system. By thoughtfully ensuring polarization throughout the system, you save users from having to think about it later.

Grounding

Grounding and bonding are important to the cabling system. Grounding involves establishing a low-impedance path to earth. Grounding has two purposes. Most important is safety to prevent the buildup of voltages that can result in hazards to persons or equipment. The second purpose of grounding is to help preserve the signal transmission quality.

Structural steel in a building can be used for ground, or a dedicated ground riser can be installed. The riser is dedicated to the telecommunications and network equipment and is separate from the grounds in electrical circuits. As such it is not intended to carry fault currents or perform safety functions associated with the building's power system. A second conductor, the coupled bonding conductor, can be installed in parallel to the telecommunications riser grounding bar to provide surge protection. While the telecommunications grounding system is separate from the electrical grounds, they are bonded together at the service entrance.

In the work area, equipment is grounded to the power grounding conductor. Any surge protection equipment is likewise grounded to the local power ground.

Earth ground is important with some electronic equipment to prevent ground loops. A ground loop occurs when two separate pieces of equipment are grounded without reference to earth. They are not connected to a common ground. Although grounded, their ground potentials may be different. Problems arise when shielded cables are run between the two pieces of equipment. The shields are grounded to equipment at each end, but because their potentials are different, a voltage difference exists between them allowing noise currents to flow in the shield. Do not rely on equipment chassis ground to provide an adequate, noise-free ground termination for shielded cables. The equipment should be grounded to the equipment rack, and the rack should be grounded to the telecommunications earth ground system. Unshielded interconnections do not require grounded equipment.

Bonding is the method used to ensure that the safety path to earth has electrical continuity and the capacity to conduct any currents that occur. Bonding is the method by which electrical contact is made between metallic parts of the system. Bonding is important because it determines the quality of the ground system. Poor bonding results in joints that exhibit high impedance and discontinuities to the flow of current. It also allows the buildup of differences in potential in the ground system.

Bonded joints should be joined by welding, brazing, swaging, soldering, screws, and so forth. Attention must be paid to the surfaces of the metal to be joined. The surfaces should be bare metal, cleaned of any nonconductive coating such as paint or grease. The joint should also be protected against corrosion, which decreases conductivity. It's a good idea to seal the bond.

There are three standards widely used for grounding and bonding:

- *National Electrical Code®* has grounding requirements that are the basis of local building codes. The *NEC®* is concerned with safety issues and requires the earth ground.
- J-STD-607-A. *Commercial Building Grounding (Earthing) and Bonding Requirements for Telecommunications*—presents a general overview of a grounding system for buildings.
- IEEE P1100, known as the Emerald Book, covers powering and grounding for sensitive electronic equipment.

Metallic cable sheaths must be grounded at the building entrance. The Emerald Book recommends grounding telecommunications cables to structural steel; J-STD-607-A requires bonding them to a grounding bar in the entrance space. The grounding bar, in turn, is bonded to the power service equipment. The advantage of 607 is that it is compatible with buildings that don't have structural steel. It is also important to note that telecommunications cables should not be bonded to any structural steel members that are used for lightning protection.

In short, the grounding system must provide a low-impedance (less than 25 ohms) path to earth. In addition, the grounding conductors must be well bonded and sufficiently sized to conduct any anticipated currents.

Surge Protection

Surge protection isolates the cabling system from damage from surges in voltage and current from lightning, power faults, and electrostatic discharges. Buildings have primary protection built in by codes to protect people and property. But the primary protection may not be fast enough to protect sensitive electronic circuits. Nor do they often provide sufficient clamping to prevent damaging voltages or current from reaching the electronics. Clamping refers to how quickly and to what level the protective device shunts surges to ground. With electronic circuits, surge energy must be dissipated quickly so that the total power passing through the protection device is limited.

Secondary protection can be designed into the premises cabling system. Surge protectors can be included at cross connects and outlets. It's a good idea to install protection as close as possible to the equipment being protected.

National Electrical Code®

The *NEC®* establishes flammability ratings for both copper and fiber cables, as was previously discussed in Chapters 2 and 3 (refer to Figures 2–37 and 3–21.) During installation, it is important to ensure that properly rated cables are used in all parts of the cabling system.

A fairly recent change (in the 2002 edition of the *NEC*®) requires that abandoned cable be removed from plenums and risers. Abandoned cable is defined as cable that is not terminated on both ends or identified for future use with a tag. This requirement was instituted because abandoned cable provides additional fuel in the event of a fire and can also affect air flow in plenums.

Under this definition, dark fiber or other cables that have been installed for future use are not considered abandoned as long as they are properly labeled. However, bundles of old cable with the ends cut off would be considered abandoned.

Note that the removal of abandoned cable can be a significant expense, especially if there are issues with hazardous materials such as asbestos. If possible, the amount of old cable to be removed (and who will pay for it) should be agreed on before the start of the installation. It may also be useful to check with the authority having jurisdiction (AHJ) to determine how aggressively they are interpreting and enforcing this requirement.[1]

SUMMARY

- Proper installation of cable and connectors is essential to obtaining the desired levels of performance.
- Neatness counts.
- Cables should not be stressed by tight bend radii, high tensile loads, or forces that deform them.
- Copper cables should be kept separate from EMI sources, including inductive loads and high-voltage sources.
- Avoid overfilling conduits: keep the fill ratio under 50%.
- Follow the manufacturer's instruction when applying connectors.
- Always comply with the grounding and bonding requirements in the *NEC*® and J-STD-607-A.

REVIEW QUESTIONS

1. What is the bend radius for UTP and fiber cable?
2. Name three conditions that can result in too high a tensile loading during cable installation.
3. Should you run UTP cable near a large motor? Why?
4. Should you run fiber near a large motor? Why?
5. What is the general fill ratio for conduits?

[1] *It is not clear from NEC® Article 800, for example, whether all abandoned cable must be removed, or only cable which has been abandoned since the requirement took effect in 2002. The AHJ in a particular area could interpret this requirement either way, or they could decide not to enforce it at all.*

6. What is the maximum untwist allowed for Categories 5e and 6 UTP at the point of termination?
7. What is the easiest way to correct a misapplied 110 jack?
8. What is the recommended amount of extra fiber to leave at a termination point?
9. How can you tell if a cable tie is installed too tightly?
10. What system parameter will be most adversely affected if a UTP cable is untwisted too far during termination?
11. Which has the higher tensile strength, UTP or fiber cable?
12. What is the maximum tensile loading of 4-pair UTP cable?
13. How many 4-pair UTP cables (¼-inch diameter) can be pulled through a 2-inch (inside diameter) conduit?
14. What is an abandoned cable?
15. What parameter will probably be most negatively affected by cinching a cable bundle too tight?

CHAPTER 9

Certifying the Cabling System

Objectives

After reading this chapter, you should be able to:

- List tests required for UTP.
- Distinguish between time-domain and frequency-domain tests.
- Describe basic setup for testing the link and the channel.
- Interpret testing results.
- Troubleshoot a link based on test results.
- Describe basic test procedures for testing fiber links.
- Discuss the advantages of OTDR.

After the cabling system is installed, it should be tested and certified so that it meets the performance specifications. Testing and certification have different implications in your building cabling. Testing implies that certain values are measured. Certification, on the other hand, compares these measured values to standards-derived values to see if the values are within the limits specified. If you want a Category 6 UTP cabling plant according to TIA/EIA specifications, the certification process ensures that all relevant characteristics meet the TIA/EIA Category 6 requirements. Thus testing is a quantitative procedure, while certification is both quantitative and qualitative.

Certification is especially important for high-speed applications. Simply installing Category 6 UTP cabling does not guarantee 250-MHz performance. The pitfalls of an incorrect installation are many, but testing will easily identify most. An understanding of the tests and their significance will speed troubleshooting if problems arise.

Component, Link, and Channel Testing

We've emphasized throughout this book the need to use components with the proper rating throughout the cabling system. Failure of even part of the system to meet Category 5e performance, for example, could result in the entire link failing to allow 100-MHz performance.

Likewise, having all components rated to Category 5e does not guarantee performance. Installation errors—too much untwisting of pairs in a connector, the wrong modular jack or plug for the wire being used, overstressed cables, cable ties too tight, and so forth—can degrade performance. It is important, therefore, to test the assembled system.

Certification Tools for Copper Cable

Of the many instruments available for testing twisted-pair and coaxial cable, the cable tester is the most versatile for testing and certifying a cable plant. A cable tester is a handheld unit that can perform the variety of tests required for certifying a cabling installation.

Because cable must be tested from both ends, a tester is required at each end of the permanent link or channel. Figure 9–1 shows a typical cable tester. A similar unit, typically without the keypad and display, would be used at the far end.

A typical tester can perform the following tests:

- Wire map (required by TIA/EIA-568B)
- Length (required by TIA/EIA-568B)
- Attenuation (required by TIA/EIA-568B)
- NEXT loss (required by TIA/EIA-568B)
- Power-sum NEXT (required by TIA/EIA-568B)

Figure 9–1. Cable tester *(Photo courtesy of Fluke Networks)*

- ELFEXT (required by TIA/EIA-568B)
- Power-sum ELFEXT (required by TIA/EIA-568B)
- Return loss (required by TIA/EIA-568B)
- Propagation delay (required by TIA/EIA-568B)
- Delay skew (required by TIA/EIA-568B)
- Attenuation-to-crosstalk ratio
- Impedance
- Capacitance
- Loop resistance
- Noise

The sophisticated capabilities of testers give you great flexibility in the tests performed. The tester will, for example, allow you to perform all tests automatically—called *autotesting*—or to perform specific tests. And to make testing even easier, high-end testers have a built-in database of values to be tested. Even further, you can specify the performance criteria to which you want to test. The tester shown in Figure 9–1, for example, will automatically test:

- TIA/EIA and ISO links and channels
- Ethernet, 10, 100, and 1000 Mbps
- ATM
- FDDI
- Vendor specific cabling systems

You can test to either application- or performance-driven criteria. The tester maintains the database of specifications for different applications in a type of memory called flash memory. You update the database by downloading data from a PC into the tester's flash memory.

You can also define custom tests so that special cases or new standards can be included.

Using the tester is easy in most cases. You simply connect it to the cabling system, select the type of testing you want done, and tell it to start testing. The high level of automation means the tester will perform all necessary tests and report the results.

Time Domain: Where Is the Fault?

Some newer testers use time-domain techniques to help you locate the source of a problem. Time domain refers to measuring results versus time. The tester, by measuring the time it takes a signal to reach it, can calculate where along the cable an event occurred. For example, suppose a cable has unacceptable NEXT results. Normally, you would be left finding the source of the problem. With time-domain techniques, a tester can show you the measured values versus distance. You can find that the unacceptable NEXT occurred, for example, at about 6 meters out from the tester. If you find that there's an interconnection at 6 meters, you have quickly located the source of the problem.

Figure 9–2 shows general setups for permanent link and channel testing. The basic difference between the two tests is whether patch cables to equipment are included.

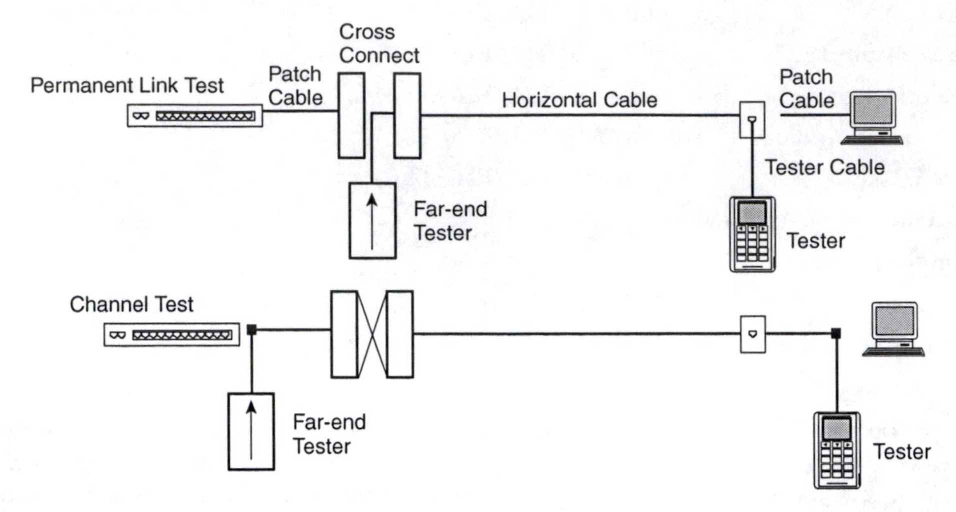

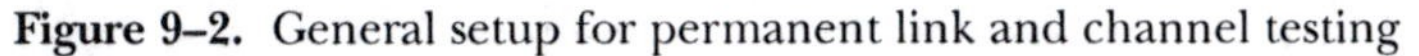

Figure 9–2. General setup for permanent link and channel testing

Wire Maps

A continuity test answers the fundamental question: Is there a circuit from end to end on each wire? Absence of continuity usually indicates a broken wire or faulty connector termination.

A wire map test checks whether each conductor (and pair) is properly terminated in the correct connector position at each end of the wiring under test (which may be a patch cord, permanent link, or channel).

When performing a wire map test, the tester verifies the integrity of each pair in addition to checking each wire for opens and shorts. The most common wiring errors, illustrated in Figure 9–3, are reversed, crossed, and split pairs.

- A reversed pair occurs when the two conductors within a pair are interchanged (for example, interchanging the blue and white-blue conductors.)
- A crossed pair occurs when two complete pairs are interchanged. Note that crossing pairs 2 and 3 will convert a T568A jack to a T568B and vice-versa.

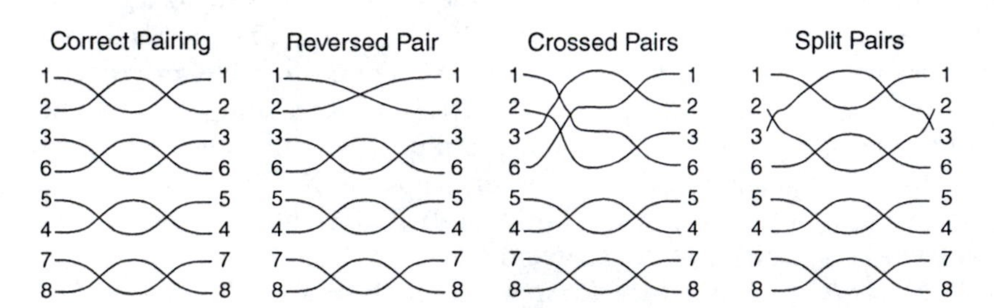

Figure 9–3. Continuity test for crossed, reversed, and split pairs

- Split pairs occur when conductors from two different pairs are intermixed. For example, if the blue and orange conductors are paired together, the white-blue and white-orange conductors remain to be paired together also, thus creating two split pairs.

Cable Length

A cable tester can measure the length of the cable by time-domain reflectometer techniques. It sends a pulse down the cable. The pulse travels to the end and reflects at the impedance mismatch. The tester measures the time it takes to travel down the cable and back. The cable's velocity of propagation allows travel time to be equated to distance. The tester can calculate the distance based on the pulse's travel time and the cable's velocity of propagation.

One cable specification you might want to custom enter for a test is the cable's velocity of propagation. High-end testers offer a database of specific cables from major vendors. An error in the velocity of propagation will translate into an error in the calculated length. This is the same as calculating driving distance by knowing driving time. Velocity of propagation is the speed you drive. If you drive 60 miles an hour and a trip takes two hours, we can calculate you travelled 120 miles. If you drive 55 miles an hour for two hours, you have travelled 110 miles. If our database says velocity of propagation is 60 miles an hour, but your true speed is 55, then our calculations will be wrong by 10 miles. A tester will base velocity of propagation either on those recommended by the standard or by having specific cables in the database. If you have a different cable with a different velocity of propagation, you should enter it so that measurements will be more accurate. Remember that specifications for UTP cable, for example, are worst-case specifications. Many cables will have performance that will exceed the minimum requirements of the standard. In most cases, this simply gives you extra margin in meeting specifications. With velocity of propagation, the differences can lead to miscalculated lengths and distances.

A tester can calculate the nominal velocity of propagation on a known length of cable. If you are installing a type of cable not found in the database, you can determine the velocity of propagation on a cable 15 meters or longer.

Length measurements can also be used to locate breaks in the cable. A break will cause an end-of-cable reflection; the length reading will be obviously short, showing the distance to the break.

NEXT Loss

To measure NEXT, the tester injects a signal onto the transmit pair and records the energy on the receive pair.

A tester has multiple RF generators to generate radio-frequency signals over the bandwidth of interest, typically from 700 kHz to 100, 250, or 500 MHz. The instrument sweeps the frequencies from low to high. Actually, it doesn't generate every frequency; rather, it works in user-selectable steps of 100, 200, or 500 kHz to generate over 1000 readings if readings are taken to 100 MHz. If you test Category 3 cable, the tester only generates frequencies to 16 MHz. For NEXT loss measurements, the standard requires a maximum step size of 150 kHz for frequencies to 31.25 MHz and a step size of 250 kHz for frequencies above 31.25 MHz.

The tester also contains a tunable narrow-band receiver. Narrow band means that it only responds to a very small range of frequencies at a given time. A television or radio is a tunable narrow-band receiver. They tune to a station being received on a specific frequency range. They reject all other frequencies. The receiver tunes itself to the same frequency as the input signal. Extraneous noise from the surrounding environment—including strong radio signals, fluorescent lights, motors, and even noise from other cables—can be coupled onto the receive cable and added to the NEXT energy coupled from the transmitting or active line. A wide-band receiver would detect such noise and give invalid results, while a narrow-band receiver rejects it. The narrow-band receiver filters out any extraneous noise.

Attenuation

The attenuation measurement is straightforward. The signal injector launches a signal of known strength into one end of the cable, the tester measures the strength of the attenuated signal at the other end, and the attenuation is calculated in decibels. Results can also include a simple pass/fail indication. As with NEXT measurements, attenuation measurements are made over the entire frequency range.

Attenuation-to-Crosstalk Ratio

ACR is calculated from the NEXT and attenuation measurements. If problems occur with the ACR reading, it is best to review the NEXT and attenuation measurements and look for problems related to either of these parameters.

Capacitance

Capacitance is a measure of the energy stored between the two conductors. For twisted pairs, capacitance is measured between the two wires of a pair. For coaxial cable, it is measured between the center conductor and shield. A given capacitance has an effective charging and discharging time constant. By measuring the time constant, capacitance can be calculated. Capacitance must fall into a medium range between two failing extremes.

DC Loop Resistance

DC loop resistance is the total resistance of both conductors of a pair or the center conductor and shield of a coaxial cable. A cable fails the loop resistance if its resistance exceeds the value allowed for the maximum cable length (most often, 100 meters). This means, for example, that a 50-meter cable fails only when its loop resistance exceeds that allowed for a 100-meter cable.

Impedance

A well-controlled impedance in the cable is important to reduce reflections and promote signal transmission without undue loss. The tester uses time-domain reflectometry to measure impedance at a point in the cable, typically between 25 and 40 meters away. The tester measures the reflected energy from that point. It's important to measure impedance at enough distance to go beyond any patch cables and cross connects. In a short cable, the impedance measurement will be the termination impedance.

While the impedance of a cable will vary along its length, this single measurement will tell whether the cable falls within the tolerances allowed. UTP cables have a nominal impedance of 100 ohms and a tolerance of ±15 ohms. This means the allowed impedance can be anywhere from 85 ohms to 115 ohms. Most testers will display the actual measurement from each pair, as well as a pass/fail indication.

Pass/Fail and Accuracy

In making a pass/fail evaluation of the cable, the tester evaluates all pairs of the cable. If a single pair fails, the entire cable fails on that test. In other words, the tester evaluates on worst-case conditions. The tester will highlight the failure, indicating the pair that failed.

For each category of cabling, the TIA has defined a corresponding accuracy level of field tester, as shown in Figure 9–4.

Figure 9–5 gives an example of the performance that can be expected from an Accuracy Level II-E tester at 100 MHz.

Accuracy Level	Cabling Category	Defined in
I	3	TSB-67
II	5	TSB-67
II-E	5e	TIA 568-B.2, Annex I
III	6	TIA 568-B.1-1, Annex B
III-E	6a	TSB-155, Annex B

Figure 9–4. Field tester accuracy levels

Test Parameter	Accuracy @ 100 MHz	
	Permanent Link	**Channel**
Attenuation	1.7 dB	1.9 dB
NEXT Loss	2.4 dB	3.6 dB
Power-Sum NEXT	2.5 dB	3.9 dB
ELFEXT	3.1 dB	4.4 dB
Power-Sum ELFEXT	3.2 dB	4.8 dB
Return Loss	2.6 dB	2.4 dB
Propagation Delay	25 ns	25 ns
Delay Skew	10 ns	10 ns
Length	5 m	5 m

Figure 9–5. Accuracy requirements for Level II-E testers

The TIA accuracy level specifications are very technical and are intended primarily for manufacturers of test equipment. Users of test equipment must simply make sure they use the proper type of tester for the cabling under test. For example, testing Category 5e cabling requires an Accuracy Level II-E tester or better, while testing Category 6 cabling requires an Accuracy Level III or better tester.

These accuracy requirements mean that a test result close to the specification limit is inconclusive—you can't be sure that the link or channel is actually within spec or not. Consider a test of pair-to-pair NEXT loss. TIA/EIA-568B allows 30.1 dB for a Category 5e channel at 100 MHz. Any value higher than 30.1 dB means the cable should pass. But what if the value is 32 dB? The measurement accuracy for channel NEXT is 4.5 dB, which means the measured value could be wrong by as much as 4.5 dB. The actual NEXT loss could be a failing 27.5 dB.

Any measurement that falls within the uncertainty region of the tester's accuracy should be flagged. Instead of reporting *Passed,* the tester will report **Passed,* with the asterisk indicating that the reading is close to the specification limit and within the uncertainty region.

Reporting the Results

Each cable, link, or channel tested should be given an ID. While testers can store hundreds of tests, many also allow you to download the results to a PC. Newer testers provide a wealth of detailed results, including graphical displays, data displays, summaries, and so forth. You receive a complete record of each test, which, stored on the computer, provides a comprehensive collection of data on the cabling plant (assuming, of course, that comprehensive testing was done!). These records are an invaluable aid to administering the cabling system. Figure 9–6 shows an example of computer-based results. Important to these advanced testers is the ability to provide such information as:

- Pass/fail for each pair
- Pass/fail for the cable, link, or channel
- Worst-case frequency and value (either passing or failing) on a cable or pair basis
- Graphical display of results (with comparison to spec)

While is seems that the testers provide you with more information than you'll ever want to know, it also answers almost any question anyone might think to ask. Knowledge is power.

Troubleshooting UTP Cabling

Faults detected during testing and certification must be corrected. Some errors can easily be determined, others not so easily. The following gives some tips to help find the source of failure.

Testing can be done on the channel, link, and component level. If certification is done on a link level, a failure can occur on any cable or at interconnection in the path. To help target the point of failure, test individual cables in the link. This will help eliminate good cables and interconnections and highlight the bad one.

Wire Map Troubleshooting

Crossed pairs found by the wire map test are easily corrected by rewiring the connection. Be sure the tester is set for the wiring pattern being used. While all four pairs should be

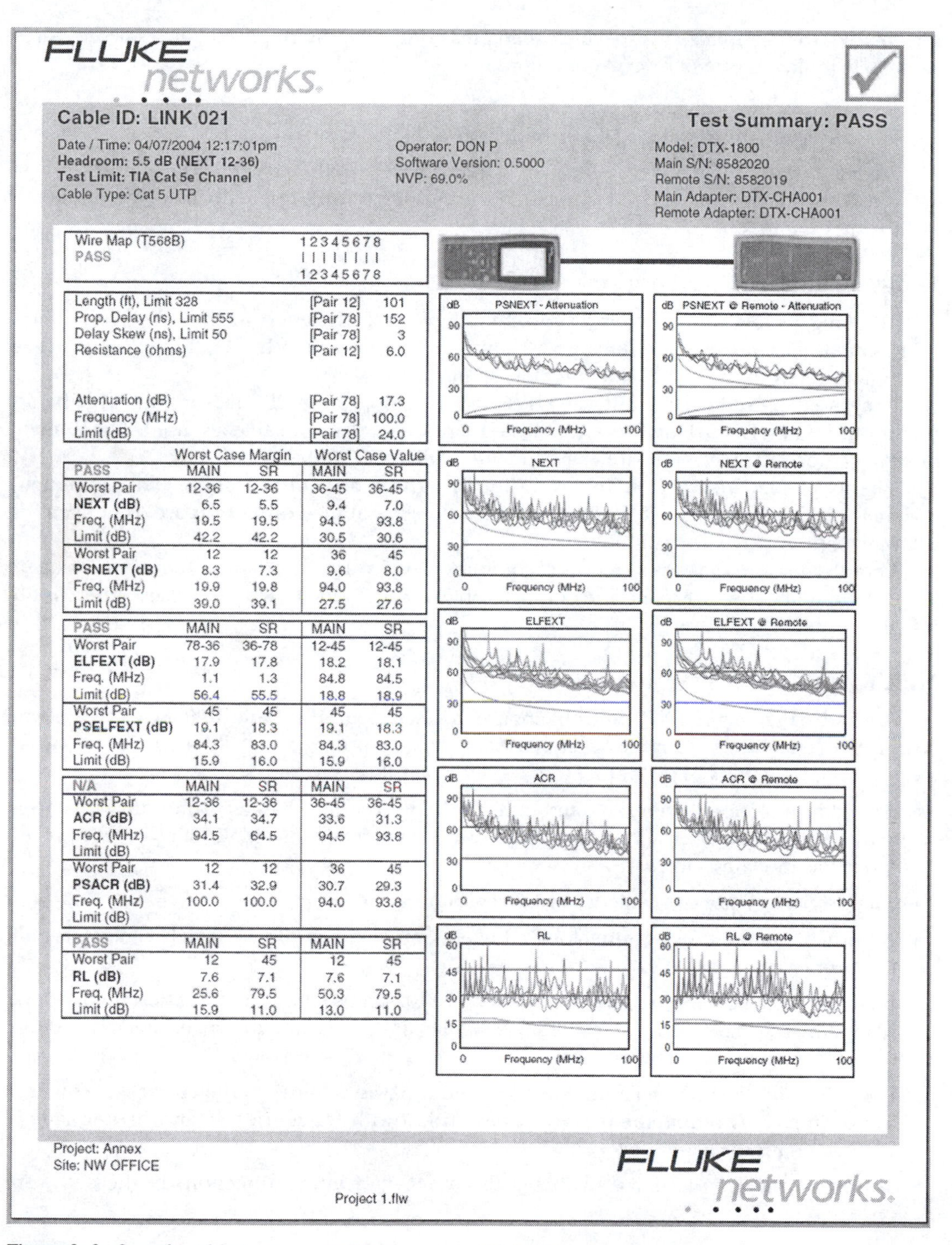

FLUKE networks.

Cable ID: LINK 021 — **Test Summary: PASS**

Date / Time: 04/07/2004 12:17:01pm
Headroom: 5.5 dB (NEXT 12-36)
Test Limit: TIA Cat 5e Channel
Cable Type: Cat 5 UTP

Operator: DON P
Software Version: 0.5000
NVP: 69.0%

Model: DTX-1800
Main S/N: 8582020
Remote S/N: 8582019
Main Adapter: DTX-CHA001
Remote Adapter: DTX-CHA001

Wire Map (T568B) — PASS
12345678
||||||||
12345678

Length (ft), Limit 328	[Pair 12]	101
Prop. Delay (ns), Limit 555	[Pair 78]	152
Delay Skew (ns), Limit 50	[Pair 78]	3
Resistance (ohms)	[Pair 12]	6.0
Attenuation (dB)	[Pair 78]	17.3
Frequency (MHz)	[Pair 78]	100.0
Limit (dB)	[Pair 78]	24.0

	Worst Case Margin		Worst Case Value	
PASS	MAIN	SR	MAIN	SR
Worst Pair	12-36	12-36	36-45	36-45
NEXT (dB)	6.5	5.5	9.4	7.0
Freq. (MHz)	19.5	19.5	94.5	93.8
Limit (dB)	42.2	42.2	30.5	30.6
Worst Pair	12	12	36	45
PSNEXT (dB)	8.3	7.3	9.6	8.0
Freq. (MHz)	19.9	19.8	94.0	93.8
Limit (dB)	39.0	39.1	27.5	27.6

PASS	MAIN	SR	MAIN	SR
Worst Pair	78-36	36-78	12-45	12-45
ELFEXT (dB)	17.9	17.8	18.2	18.1
Freq. (MHz)	1.1	1.3	84.8	84.5
Limit (dB)	56.4	55.5	18.8	18.9
Worst Pair	45	45	45	45
PSELFEXT (dB)	19.1	18.3	19.1	18.3
Freq. (MHz)	84.3	83.0	84.3	83.0
Limit (dB)	15.9	16.0	15.9	16.0

N/A	MAIN	SR	MAIN	SR
Worst Pair	12-36	12-36	36-45	36-45
ACR (dB)	34.1	34.7	33.6	31.3
Freq. (MHz)	94.5	64.5	94.5	93.8
Limit (dB)				
Worst Pair	12	12	36	45
PSACR (dB)	31.4	32.9	30.7	29.3
Freq. (MHz)	100.0	100.0	94.0	93.8
Limit (dB)				

PASS	MAIN	SR	MAIN	SR
Worst Pair	12	45	12	45
RL (dB)	7.6	7.1	7.6	7.1
Freq. (MHz)	25.6	79.5	50.3	79.5
Limit (dB)	15.9	11.0	13.0	11.0

Project: Annex
Site: NW OFFICE

Project 1.flw

FLUKE networks.

Figure 9–6. Sample cable test report *(Photo courtesy of Fluke Networks)*

terminated, some applications-specific cabling may only wire the required pairs. Missing wires in the test indicates one of several conditions:

- Broken conductor
- Missed termination in the RJ-45 or punch-down block
- Wrong test: for example, testing for a 568A wiring pattern when the connection is wired specifically for 10BASE-T or Token Ring. Or testing for 568B patterns when the wiring is 568A.

Length Fault Troubleshooting

Too long a cable is not usually a problem in a properly designed system. Check for coiled cable somewhere in the run or improper routing by an indirect path. In addition, length may not be a problem if the NEXT and attenuation limits are okay.

Length errors will occur if the cable's velocity of propagation differs from the value being used by the tester. Be careful if several different brands or types of cable are used. The velocity of propagation may have to be adjusted to handle different cables.

The tester may inaccurately report a short length because of reflections that are punch-down blocks or patch panels. Normally, this is not a concern if all other tests are okay. Even so, you may want to check the connections.

Length measurements using TDR techniques can also locate opens and shorts. An open will cause a reflection indicating the end of a cable. A short will absorb the pulse so there is no reflection.

NEXT and ELFEXT Troubleshooting

After wire map errors, NEXT problems are the main source of faults in the cabling system. NEXT errors derive from three sources: substandard components, installation errors, and too many interconnections in the path. Here are the main things to check:

- An old or improperly calibrated tester. If your tester is more than a few months old, check with the manufacturer about an upgrade to the latest version. Make sure even a recent tester has the latest firmware.
- The proper category of cable is used throughout.
- All other components—connectors, couplers, punch-down blocks, patch cables—are of the proper category.
- Incorrect patch cables. Double-check patch cables because they are the easiest to misapply. Make sure the category is correct. For any application, including low-speed voice, avoid the use of untwisted (silver satin) cables.
- Excessive untwist at the termination. Maintain the twist as close to the connector contacts as possible. This requirement is especially critical with Categories 5e and higher cables, where the untwist cannot exceed 0.5 inch.
- Too many interconnections. Reducing the number of interconnections in the path can reduce NEXT.

- Nearby noise sources. Even though the tester uses a narrow-band receiver to reduce its effects, ambient noise can affect readings. Power cables, fluorescent light, and so forth, can couple energy onto the cable and be read as crosstalk. If necessary, reroute the cable away from noise sources.

Be sure to test NEXT at both ends of the cable. A cable or link can pass at one end and fail at the other. ELFEXT troubleshooting is similar to NEXT troubleshooting.

Attenuation Troubleshooting

Attenuation errors almost always result from either using the wrong category of components or having a bad termination. Any point of termination increases resistance and hence attenuation. While a correctly terminated connection maintains an acceptably low resistance, the poor connection does not.

Attenuation-to-Crosstalk Ratio Troubleshooting

Because ACR measurements are based on NEXT and attenuation, the same troubleshooting tips apply. In particular, check categories of components and the quality of terminations.

Capacitance Troubleshooting

Capacitance errors most often are caused by using a lower category of cable. Also check for broken conductors, shorts, opens, and excessive noise.

Loop Resistance Troubleshooting

Too high a loop resistance typically has two causes:

- Excessive cable length. Check the cable length.
- Poor terminations. An improper termination can add significant resistance.

Impedance and Return Loss Troubleshooting

Impedance errors should not occur if the proper cable has been installed. Impedance errors most often occur if a substandard cable is installed. Notice that all UTP cables share the same impedance value, so impedance measurements will not indicate the wrong category. Measurements showing improper impedance will also detect other problems that should be investigated.

Low return loss (meaning high reflected energy) can result from impedance mismatches. If return loss values are unacceptable and the proper type of cable is being used, check all interconnections. A poorly terminated connector or a loose interconnection can also increase reflections.

Recurring Problems

It's quite common to find a few errors to correct during certification. The most common are wiring errors at interconnections, either crossed and split pairs, or poor workmanship. But what do you do if large parts of the installation fail?

First, make sure the tester is properly calibrated. This includes setting the NVP for the type of cable being used and making sure that the tester has the latest firmware.

Next, make sure you are using the proper suite of tests. For example, running a permanent link test on a channel, or a Category 6 test on Category 5e cabling, will cause most of the tests to fail.

This type of confusion often results from misunderstandings about what is being tested. Many installations are to be tested to TIA-568, but without specifying exactly what is meant. For some, this means link testing; for others, it means cable testing. There are lots of numbers flying around—specifications for cables, connectors, and links—so that it's easy to compare one test to another set of results. A link test will always fail if only the horizontal cable specifications are used. Patch cable specifications are about 20% looser than horizontal cable specifications. Always make sure the tester is actually set for the test being performed.

Standards address these problems by providing clear definitions of link and channels, performance specifications for each, testing requirements, and accuracy levels. These results should allow much less confusion about what's being tested.

Fiber Testing

Fiber testing is simpler than UTP testing in that fewer characteristics are measured. In fact, testing can be done simply by measuring power with an optical power meter. The key to any fiber-optic link is the power budget: loss of optical power through the link cannot exceed specified limits. Power loss is the main measurement made.

Certification is done on link level. Begin with a basic link test. If the link passes, component-level testing is not necessary. If the link fails, test individual components to isolate the fault.

Continuity Testing

Simple continuity testing can be achieved by a flashlight: does the light come through the fiber? Fancy flashlights—called visual continuity testers—are available specifically for fiber-optic testing. A red light is preferred because it is easiest to see.

Warning: Never Look into an Energized Fiber

Never look into a fiber that might be energized. Fiber-optic systems use infrared light that cannot be seen. Laser and some LED sources emit light powerful enough to damage the eye. Before looking into a fiber, either make sure the system is powered down or that the fiber is disconnected from any transmitters.

Link Attenuation

Link attenuation is simply the sum of the individual losses from the cable, connectors, and splices. Cable loss can be assumed to be linear with distance. For the 90-meter horizontal cable run, a 62.5/125-µm cable has an attenuation of 0.34 dB at 850 nm and 0.14 dB at 1300 nm. Maximum insertion loss values are 0.75 dB for connectors and 0.3 dB for splices. To estimate the total loss in the link, simply add up the individual losses. Consider a horizontal run with two connectors. The total loss is 1.84 dB:

Cable loss: 3.75 dB/km × 90 m = 0.34 dB

Connectors: 0.75 dB × 2 = 1.5 dB

Total loss: = 1.84 dB

Link attenuation, then, is simply the sum of the losses contributed by individual components:

Total Loss = Cable Losses + Connector Losses + Splice Losses

For horizontal links, the following are the maximum attenuation values allowed:

Distributed link (90 meters): 2.0 dB

Centralized link (300 meters): 3.3 dB

Centralized/zone (300 meters): 4.1 dB

Zone link (90 meters): 2.0 dB (closet to multiuser outlet)
2.75 dB (closet to outlet)

These values assume there are two connections for distributed and zone links (one in the telecommunications closet and one at the outlet). The centralized value assumes three connections—one at the equipment room cross connect, one at the telecommunications closet, and one at the outlet. You can have more connections in a link as long as you don't exceed the allowed loss values. Additional connectors, however, require low-loss terminations. The 0.75 dB allowed by the standard quickly chews up link margin. The best connectors will achieve losses as low as 0.2 dB, allowing more connections in the link path.

The 3.3 dB value for centralized cable deserves comment, because it exceeds at least one variation of Gigabit Ethernet. Figure 9–6 summarizes the margins allowed by representative network standards. Notice that short-wavelength Gigabit Ethernet has a maximum channel attenuation of 3.2 dB. The recommended distance for this variation is 220 meters. Realistically, most installations will fall under 3.2 dB, but you can't count on it. The important thing to see from Figure 9–7 is that most slower networks provide a great deal of margin. Higher speed systems, such as Gigabit Ethernet or 622-Mbps ATM, are less forgiving in the loss allowed.

Network	**Wavelength**	**Distance (m)**		**Channel Margin**	
	(nm)	**62.5μm**	**50μm**	**62.5μm**	**50μm**
10BASE-FL	850	2000		12.5	7.8
100BASE-FX	1300	2000		11.0	6.3
ATM, 155 Mbps	850	1000		7.2	7.2
	1300	2000		10.0	5.3
ATM, 622 Mbps	850	300		4.0	4.0
	1300	500		6.0	1.3
1000BASE-SX	850	220	550	3.2	3.9
1000BASE-LX	1300	550	550	4.0	3.5

Figure 9–7. Supportable distances and channel attenuation for various networks

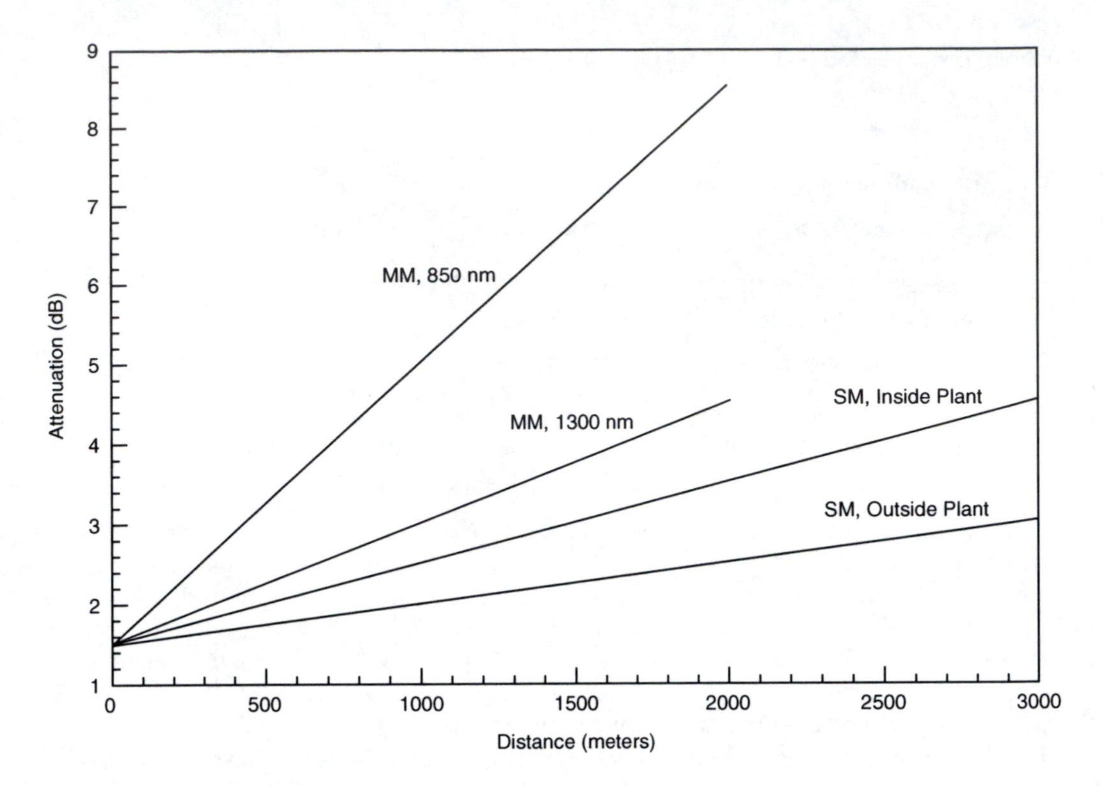

Figure 9–8. Backbone cabling distances versus attenuation

Backbone cable distances depend on attenuation and the number of connections, with a maximum recommended distance of 3000 meters for single-mode cable and 2000 meters for multimode cable. The acceptable link attenuation is based on distance. Figure 9–8 shows attenuation values for various distances, based on two connectors in the link. The graph is derived from the attenuation values for connectors and cable given earlier and scaled for different distances. The lowest loss is thus 1.5 dB (for two connectors and a very short length of cable). If additional splices or connectors are used, then these values should be factored in. The result is then used for testing and certifying the cable plant. In some cases, you must ensure that the backbone attenuation does not exceed that of the margin for the application. For example, Gigabit Ethernet will not run 1000 meters over multimode cable. Single-mode cable is typically the preferred solution for backbone cables running distances over a few hundred meters.

Notice, too, that TIA/EIA-568B link performance does not include passive elements like couplers, switches, and other optical devices. These can add appreciable loss in the link and must be accounted for if present. Links using such devices will not meet 568 requirements and must be evaluated on an individual basis. For example, 10BASE-FP networks use a passive star coupler to divide light among many fibers. This system would have to be evaluated in terms of IEEE 802.3 10BASE-F specifications rather than EIA-TIA-568B specifications.

In estimating cables losses, use the following values:

62.5/125-µm or 50/125-µm cable at 850 nm:	3.5 dB
62.5/125-µm cable at 1300 nm:	1.5 dB
Single-mode outside plant cable at 1310 nm:	0.5 dB
Single-mode outside plant cable at 1550 nm:	0.5 dB
Single-mode inside plant cable at 1310 nm:	1.0 dB
Single-mode inside plant cable at 1550 nm:	1.0 dB

Measured loss must be less than those calculated by worst-case values. However, some care in testing is necessary because the test procedure can significantly affect results.

Understanding Fiber-Optic Testing

A fiber carries light in modes (possible paths for the light). The modal conditions of light propagation can vary, depending on many factors. We generally speak of three general conditions: overfilled fiber, underfilled fiber, and equilibrium mode distribution.

- **Overfilled fiber.** When light is first injected into a fiber, it can be carried in the cladding and in high-order modes. Over distance, these modes will lose energy. The cladding, for example, attenuates the energy quickly. High-order modes can be converted into lower order modes. Imperfections in the fiber can change the angle of reflection or the refraction profile of high-order modes.
- **Underfilled fiber.** In some cases, light injected into the fiber fills only the lowest order modes. This most often happens when a low NA laser is used to couple light into a multimode fiber. Over distance, some of this energy shifts into higher modes.
- **Equilibrium mode distribution.** For both overfilled and underfilled fibers, optical energy can shift between modes until a steady state is achieved over distance. This state is termed *equilibrium mode distribution* or *EMD.* Once a fiber has reached EMD, very little additional transference of power between modes occurs.

EMD has several important consequences in light propagation in a fiber and its effects on testing. Both the active light-carrying diameter of the fiber and the NA are reduced. For example, the fiber recommended for premises cabling has a core diameter of 62.5µm and an NA of 0.275. These are based on physical properties of the fiber. With EMD, we are interested in the *effective* diameter and NA. In a 62.5-µm fiber at EMD, the effective (light-carrying) core diameter is 50µm and the NA is less than 0.275.

So what's the point in being concerned with EMD? The modal conditions of the fiber affect the loss measured at a fiber-to-fiber interconnection. It is easier to couple light from a smaller diameter and NA into a larger diameter and NA. Even so, the light coupling into the second fiber will not be at EMD. It will overfill the fiber to some degree so that EMD will not be achieved until the light has travelled some distance.

In testing, it is important to specify the modal conditions. Testing a connector at EMD can yield significantly different results than testing it under overfilled conditions. A connector can yield an insertion loss of 0.3 dB under EMD conditions and 0.6 dB under fully filled conditions.

In testing, we speak of launch and receive conditions:

Long launch:	Fiber is at EMD at the end of the launch fiber.
Short launch:	Fiber is overfilled at the end of the launch fiber.
Long receive:	Fiber is at EMD at the end of the receive fiber.
Short receive:	Fiber is overfilled at the end of the receive fiber.

EMD conditions can be easily simulated without a long length of fiber. The standard method of achieving EMD in a short length of fiber is by wrapping the fiber five times around a mandrel. This mixes the modes to simulate EMD.

The launch and modal conditions used have a practical consequence in fiber bandwidth. The test conditions should reflect the type of source to be used. In multimode fiber, this is the difference between an LED and a VCSEL. You saw in Chapter 3 (see Figure 3–9) that the bandwidth of the fiber is greatly determined by the source. The next-generation 50/125-μm fiber offers a bandwidth of 500 MHz-km when used with an LED and 2200MHz-km when used with a VCSEL. The test for LED-related bandwidth is based on an overfilled launch. The test for VCSEL-related bandwidth is based on an underfilled launch that matches the small spot size and narrow beam of the source. This difference is extremely critical for testing and certifying fiber. The test equipment must provide the appropriate launch conditions because the difference in bandwidth can be over 1700 MHz-km!

For years, LEDs were the main source used with multimode cables, because lasers were not a cost-effective alternative. Testing was based on overfilled launches because this mirrored the reality of LED-based systems. However, these conditions do not reflect the higher bandwidth that can be achieved with a VCSEL. To address this issue, the TIA issued TSB-140 (Additional Guideline for Field-Testing Length, Loss, and Polarity of Optical Fiber Cabling Systems) in February 2004. This document provides guidance for field testing fiber systems, using optical loss test sets (OLTS), optical time domain reflectometers (OTDR), and visual fault locators (VFL).

One of the key elements of TSB-140 is the use of a mandrel to remove higher-order modes from overfilled fibers when testing with an LED-based optical source. Figure 9–9 illustrates this concept.

When light is provided by an LED source, the fiber is in an overfilled state, as shown on the left side of the figure. The mandrel (sometimes called a mode filter) attenuates the higher-order modes so output approximates an equilibrium mode distribution (EMD).

TSB-140 also defines two tiers of testing. Tier 1 testing requires measuring the attenuation of each fiber with an OLTS and verifying the length and polarity[1] of the fibers. Tier 2 testing also requires an OTDR trace in addition to the Tier 1 tests.

[1]*Polarity testing for fiber is similar to wiremap testing for UTP. A fiber polarity test checks that a fiber is hooked to the proper connector at each end, and in particular, that the two fibers in a pair are not reversed.*

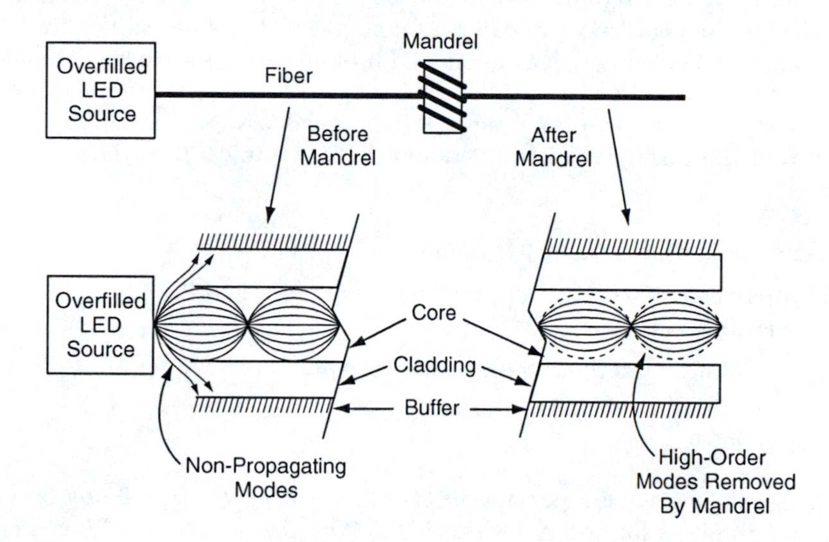

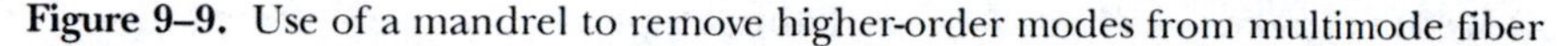

Figure 9–9. Use of a mandrel to remove higher-order modes from multimode fiber

Standard Tests

There are three standard tests applicable to premises cabling:

OFSTP-14:	link certification with power meters
FOTP-171:	patch cable certification
FOTP-61:	link certification with an optical time-domain reflectometer

Let's begin by looking at power meters.

Power Meters

An optical power meter does just that: measures optical power. Light is injected into one end and measured at the other. Some meters incorporate interchangeable light sources into the meter so that it serves as either a light source or a meter (or both simultaneously). Most meters can display power levels in user selectable units, most often dB or dBm. The dB reading is most appropriate for certification. Because fiber performance depends significantly on the wavelengths propagated, the light source is important. TIA/EIA-568A makes the following recommendations for wavelengths and testing:

Horizontal multimode cable:	either 850 or 1300 nm
Backbone multimode cable:	both 850 and 1300 nm
Backbone single-mode cable:	both 1310 and 1550 nm

A power meter typically requires two readings. One is used for calibration, to zero the meter. The other is the actual power reading. Of course, calibration isn't required for each reading; once the unit is calibrated, it can be used for numerous tests. Check the meter's user's guide for direction on how often to zero the unit.

An optical tester similar to a power meter is the optical loss test set (OLTS). The OLTS is distinguished from the power meter by including the signal injector in the same unit as the power meter.

Link Certification with a Power Meter (OFSTP-14)

OFSTP-14 offers two measurement procedures, Method A and B.

Both methods are similar:

1. A calibration reading is taken using one or two test jumpers.
2. The jumpers are then connected to the link.
3. The link loss is displayed.

Method A uses the two test jumpers for the calibration, while Method B uses only one. Both methods use two jumpers for the link evaluation. Why the difference? The single-jumper approach of Method B eliminates the influence of the connectors used to interconnect the two cables during the zeroing procedure. The connectors in the middle of the two jumpers essentially mirror the presence of the connectors in the actual link. Test Method A cancels the influence of these connectors during the test. Method B includes these connectors in the test.

Of the two methods, TIA/EIA-568B recommends Method B. Figure 9–10 shows the general setup and sequence for the test. Keep in mind that the figure is general; it does not show patch panels and other intervening couplings that could be included in the test.

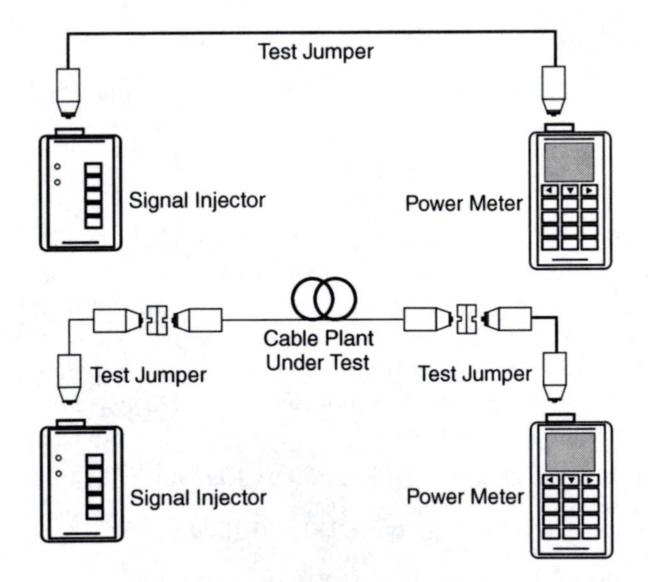

Figure 9–10. Link certification with a power meter

Testing Patch Cables (FOTP-171)

This procedure includes four methods, with three optional procedures for each method. We will restrict our discussion to Method B, which is the best suited to premises cabling and is the method typically used by cable assemblers.

This procedure (Figure 9–11) uses the substitution method to evaluate attenuation in a length of fiber. Power through two jumper cables is measured to zero the meter. The cable assembly to be tested is then inserted between the two jumper fibers. The new reading is the fiber attenuation. Notice that the test requires a mode filter on the launch end to simulate EMD in the short length of jumper cable. This can be achieved by wrapping the cable five turns around a mandrel to mix the modes, (as was illustrated in Figure 9–9).

Optical Time-Domain Reflectometry

An optical time-domain reflectometer (Figure 9–12) is a useful tool for troubleshooting. We discussed the uses of time-domain reflectometry in testing UTP links. Certification tools use TDRs to measure cable lengths. OTDR—optical TDR—is more widely used than metallic TDR. The principles are the same: a pulse is sent down the fiber. Energy reflecting from imperfections and changes in refractive index of the path are detected and displayed. Changes in the index of refraction of a fiber are analogous to changes in impedance in a copper cable: both cause reflections that can be used to get a clear picture of the link.

Most TDRs are highly automated and sophisticated, allowing you to zoom into specific areas for a close inspection. Among capabilities are these:

- Zoom in to specific events like connectors, splices, and faults (all of which will show pronounced bumps in the trace) for a closeup look.

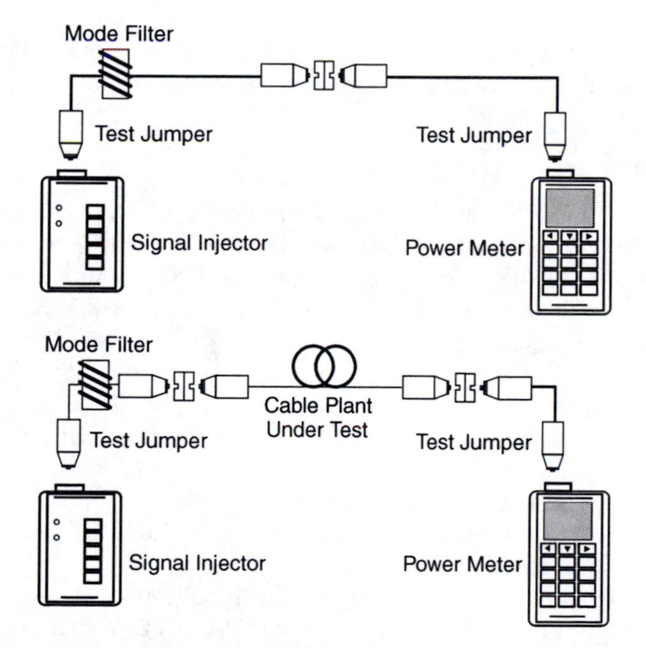

Figure 9–11. Measuring cable attenuation with a power meter

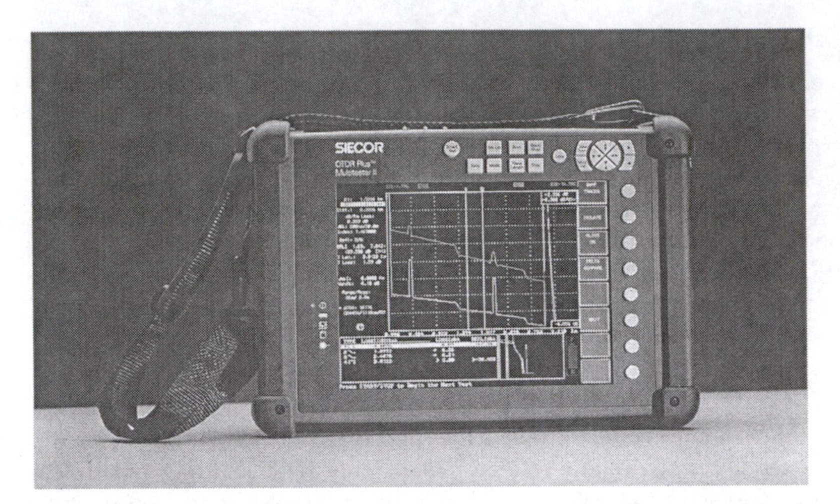

Figure 9–12. Optical time-domain reflectometer *(Courtesy of Siecor Corporation)*

- Zoom out to gain an overall picture of the link.
- Measure the total link length.
- Measure distances to specific spots.
- Measure attenuation for the entire link or for specific areas.
- Compare two sections of a link. You can, for example, compare two different interconnections.
- Compare two different links. Many OTDRs have storage capability that will allow you to store a test. This test can then be called up and compared to the current test.
- Save the test results to a computer database.
- Print the test results.

Figure 9–13 shows an OTDR plot for a link of about 2250 meters. The slow downward slope of the graph indicates the attenuation of the fiber, while the two bumps in the middle are reflections from splices or connectors. (The bumps at each end represent the start and end of the link.) The two vertical dashed lines are markers at 680.5 meters and 763.8 meters—a distance of 83.3 meters—from the beginning of the link. They show the loss at this point (see the horizontal lines of the markers). The right-hand lower box shows the marker information. The most important information is the loss between the two markers: 0.18 dB, which is mainly due to the loss from the splice or connector, because the 83.3 meters of fiber contribute very little loss.

The right-hand box shows fiber and test information.

You can see that the OTDR plot contains much information showing you exactly what is going on in the link. You can easily pinpoint the sources of loss and their distance along the path. OTDR plots are valuable for this reason.

Do you need an OTDR? While link certification can be done with a power meter, nothing beats an OTDR for in-depth troubleshooting. The OTDR has two main drawbacks. First, they are relatively expensive. Second, they require a higher level of operator skill. Even so, OTDRs

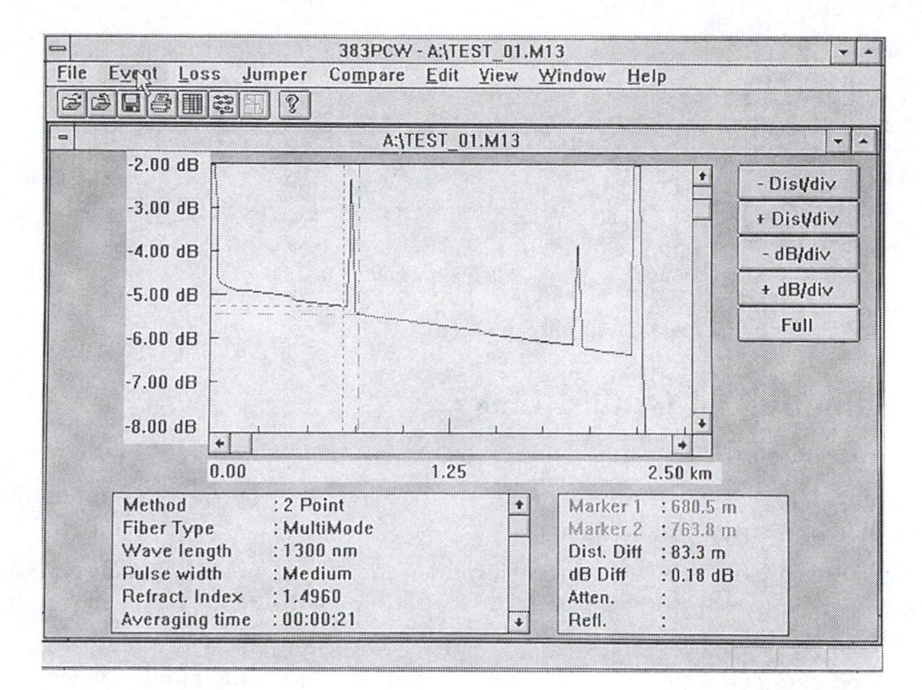

Figure 9–13. OTDR plot *(Courtesy of Siecor Corporation)*

in recent years have become lower in cost and easier to use. Newer handheld models replace the bulkier "lugarounds" that usually required a cart.

A useful strategy is to perform certification using power meters, but have one or two OTDRs available for troubleshooting. This can additionally save on training, because only one or two technicians need to specialize in OTDR use.

Some building owners or network administrators, however, require an OTDR plot for every fiber-optic link. This is a peace-of-mind issue, not a performance issue. Compared to a UTP link report—detailing NEXT, attenuation, ACR, loop resistance, capacitance, and wire mapping—an optical link report seems short and insubstantial. An OTDR plot gives a more rigorous assessment of a link's performance.

Measuring Link Attenuation with OTDR (FOTP-61)

Figure 9–14 shows the general setup for measuring link attenuation with an OTDR. A dead-zone fiber is required at each end of the link. An OTDR cannot "see" the fiber close to it. In order to overcome this deficiency, a short length of fiber is inserted between the instrument and the cable being tested. How long the dead-zone fiber must be depends on the specific OTDR being used. OTDRs designed for long-distance telecommunications testing typically have a longer deadzone than units more appropriate for premises cabling. The second dead-zone fiber is required to allow you to evaluate the final connector at the far end. When the OTDR pulse reaches the end of the fiber, it reflects a large amount of energy. If you end on the cable under test, then the reflected pulse will include no information about the quality of that connection.

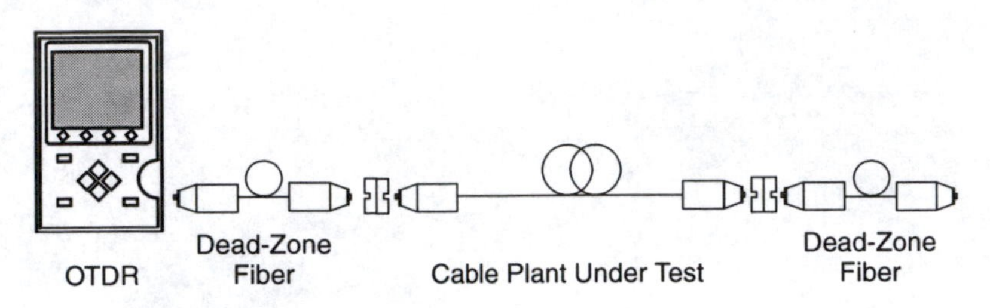

Figure 9–14. Measuring link attenuation with an OTDR

Troubleshooting an Optical Link

This section discusses some of the things to check if the link fails.

- Continuity. Use a visual continuity tester—a flashlight—to perform a continuity check. This will help identify broken fibers.
- Polarity. Were the fibers crossed during termination? Check end-to-end continuity for each section of the link to discover crossed pairs. Reterminate the cable, or reconnect the cable properly.
- Misapplied connector. Check the connector carefully. Check the end finish with a microscope for unacceptable scratches, pits, or other flaws. As appropriate, repolish the connector or reterminate the fiber.
- Dirt. Check the end of the fiber for dirt or films. Clean the connector end with isopropyl alcohol or canned air. Check the core of the coupling adapter for obstructions.
- Make sure the light source and power meter are correctly set and calibrated. For example, using an 850-nm source to test against 1300-nm test conditions will significantly distort the results.
- Check the cable plant for tight bends, kinks in the cable, and other unhealthy conditions. These can either increase attenuation or snap fibers in two.
- Make a habit of checking the cable for continuity before installing it. Test it while it is still on the reel. Considering the time it takes to perform this test versus the cost of installing the cable, it is an inexpensive first step in ensuring a quality installation.

A Note on ISO/IEC Testing Requirements

ISO/IEC 11801-2002 distinguishes three types of testing and specifies the types of measurements required for each. The three types of testing are[2]:

- *Acceptance testing* is used to validate installed cabling "which is known to comply with the implementation requirements of this standard and which is made up of elements complying

[2]*Note that this is a change from the first edition of 11801 (and the 2nd edition of this book) where the three types of testing were: acceptance, troubleshooting, and compliance.*

with the performance requirements for components for the relevant categories." Acceptance testing consists of only a wiremap and continuity test.

- *Compliance testing* is used to validate installed cabling "comprising known or unknown components." This is the type of testing most commonly performed because it verifies the installation as well as the components.
- *Reference testing* is used for laboratory testing of cabling models and for verification of properties which cannot be tested in the field. This type of testing would normally be done by research and development engineers during the product development process.

ISO/IEC 11801-2002 specifies which types of measurement are needed for each type of testing, as shown in Figure 9–15. Each measurement is classified as either normative (N), informative (I), or computed (C). Normative measurements are required, while informative measurements are useful but optional. Computed measurements are typically power-sum parameters that are calculated from the pair-to-pair measurements.

Type	Parameter	Testing for		
		Acceptance	Compliance	Reference
Copper	Wiremap	N	N	N
	Continuity	N	N	N
	Return loss	I	N	N
	Insertion loss	I	N	N
	NEXT	I	N	N
	ACR	I	N	N
	ELFEXT	I	N	N
	DC loop resistance	I	N	N
	Propagation delay	I	N	N
	Delay skew	I	N	N
	TCL			N
	Length	I	I	N
	PSNEXT, PSACR, PSELFEXT	C	C	C
Fiber	Attenuation	N	N	N
	Modal BW			N
	Propagation delay	I	N	N
	Length	C	C	C
	Polarity	N	N	N

Figure 9–15. ISO/IEC 11801-2002 testing recommendations

SUMMARY

- Testing and certifying a cabling system are essential parts of the installation.
- Most copper testers automatically perform all required tests, give pass/fail indication, and numerical results.
- Time-domain measurements help locate the point of failure.
- TSB-140 gives field testing guidelines for optical fiber cabling.
- Fiber testing can be done with a power meter or OTDR.
- Fiber power meters give basic link loss measurements.
- ODTRs give informative and graphical displays of the power levels throughout the link.
- The accuracy of the test equipment must always be considered when evaluating test results, and especially when comparing one measurement to another.

REVIEW QUESTIONS

1. Why is channel-level testing important (as opposed to link testing)?
2. Which tests for UTP links and channels are required by TIA/EIA-568B?
3. Which tests for fiber links are required by 568B?
4. Why must NEXT loss tests be performed at both ends of the link or channel?
5. For NEXT and ELFEXT measurements, can you average readings for different pairs to find if the cable passes? Why?
6. When terminating UTP cable on an RJ-45 jack, if the green and white-brown conductors are reversed, what type of wiremap fault does this cause?
7. What accuracy level of field test equipment is required for Category 5e and Category 6 UTP cabling?
8. What are two advantages of time-domain measurements?
9. What is the difference between testing and certification?
10. Why are launch and receive conditions important in fiber testing?
11. What is the maximum allowed attenuation for a backbone link using single-mode outside plant cable, a distance of 2 km, and two connectors and one splice?
12. What is the difference between continuity in a UTP cable and in a fiber-optic cable?
13. What is the purpose of wrapping the fiber around a mandrel before testing a multimode fiber link?
14. What is actually being checked when fiber polarity is verified?
15. What is the difference between Tier 1 and Tier 2 testing of an optical link?

CHAPTER

10 Documenting and Administering the Installation

Objectives

After reading this chapter, you should be able to:

- List reasons for documenting and administering a cabling system.
- Define the four types of information in a cabling system record.
- Describe the role of cabling management software in documenting and maintaining a cabling system.
- Describe the color-coding conventions for cross-connect fields.
- Understand the advantages of smart patching systems.

The cabling structure is not static. It is dynamic, growing and evolving with the changing needs of users. Maintaining up-to-date, reliable documentation of the cabling system is essential. Without such documentation, the wiring system descends into chaos until nobody knows how the system is arranged.

Careful, accurate administration of the cabling system has several benefits:

- To allow fast, accurate moves, adds, and changes.
- To speed and simplify troubleshooting.
- To increase network reliability and uptime.
- To allow better asset management.
- To increase both management and MIS people's confidence in structured cabling systems. Some MIS people, in particular, may question whether a structured cabling system offers the same reliability as a proprietary, fixed system. As businesses "downsize" from mainframe and minicomputer-based systems to LAN-based systems, this question is significant to those responsible for the computer infrastructure.
- To allow disaster recovery plans for the cable plant to be developed.
- To generate management reports.

- To allow reliable capacity planning to manage the needs for growth and migration in both the cabling plant and in executed applications. In other words, careful administration will help prove (or disprove) the ability of the plant to accommodate emerging applications like ATM.
- To allow new tenants to receive accurate, up-to-date information on the cabling plant. Such information can enhance the value of a building.

TIA/EIA-606-A is a standard entitled *Administration Standard for Commercial Telecommunications Infrastructure.* It provides guidance on documenting a cabling system to make its administration efficient and effective.

Classes of Administration

TIA-606-A defines administration as "The method for labeling, identification, documentation and usage needed to implement moves, additions and changes of the telecommunications infrastructure." Because the administration needs of a structure vary with its size and complexity, TIA-606-A defines four classes of administration.

- Class 1 is the simplest type of administration system. It is used for premises that are served by a single telecommunications space (TS). Because all the telecommunications equipment is located in a single space, there is no backbone or outside plant cable.
- Class 2 administration is for installations with multiple telecommunications spaces within a single building. It includes administration of backbone cable, grounding and bonding systems, and firestopping in addition to the capabilities of a Class 1 system.
- Class 3 administration addresses the needs of a campus installation that consists of multiple buildings with outside plant cabling.
- Class 4 administration is for multi-site systems that consist of multiple campuses with wide area network connections between them.

Class 1 administration systems are normally implemented with either paper records or general-purpose spreadsheet software. Class 4 administration systems are very complex and are usually implemented with custom cable administration software. Classes 2 and 3 may be implemented with any of the three methods (paper, spreadsheet, or custom software).

What's in the Documentation?

Documentation includes labels, records, drawings, work orders, and reports.

Labels

Every cable, piece of termination hardware, cross connect, patch panel, conduit, closet, and so forth, should be labeled with a unique identifier. While the label can simply be a number, it's wiser to create a code system that will provide meaningful information. For example, you could label the cable "C12345." No other cabling component will have this number, but the label tells you nothing about the cable. A better approach would be to label the cable "TR3A-18-WA16." This encoding convention indicates cable 18, which runs from telecommunications room 3A to work area 16.

Recall also the NEC® requirement that all unterminated cables be labeled for future use (as mentioned in Chapter 8).

TIA-606-A places a number of requirements on labels, including:

- Labels must be of a size and color that is easily visible and readable.
- Labels must be resistant to environmental conditions such as moisture, heat, UV, etc.
- Labels must have a useful life at least as long as the component being labeled (often 15 years or more).
- Labels must be mechanically printed—they may not be written by hand.

Ordinary, office-grade labels (such as those commonly used on file folders and similar items) do not meet these requirements and should not be used for labeling cables and connecting hardware. There are a number of vendors who supply labels and printers that meet the requirements of TIA-606-A.

Records

No matter how good your coding scheme for components, you cannot put all the information required for decent documentation on the label. The label also serves as a link to a record that contains complete information on the component. The record can exist either on paper or in a computer file. If you search the administrative database for "TR3A-18-WA16," you will be able to access the complete record for that cable.

Drawings

Drawings are usually based on the architectural drawings for the building. They help locate components within the building. It's of little help to know that a cable terminates in Telecommunications Room 3A, if you don't know where that room is. Drawings also show the locations of conduits, pull boxes, and other components hidden from view behind walls, above ceilings, and under floors.

TIA-606-A provides an informative annex that lists six types of drawings as indicated below.

T0—Campus or Site Plans—Exterior Pathways and Interbuilding Backbones.

T1—Layout of complete building per floor—Building Area/Serving Zone Boundaries, Backbone Systems, and Horizontal Pathways.

T2—Serving Zones/Building Area Drawings—Drop Locations and Cable IDs.

T3—Telecommunications Rooms—Plan Views—Racks and Walls Elevations.

T4—Typical Detail Drawings—Faceplate Labeling, Firestopping, Safety, etc.

T5—Schedules—(spreadsheets) to show information for cut-overs and cable plant management.

This annex also gives suggestions for standard line types, symbols, abbreviations, etc.

Work Orders

Work orders record all moves, adds, and changes. They form a history of the cabling system's life. An equally important practice is to update the records each time a work order is executed.

Reports

A report is a group of records organized in a specific manner. A computer database for a cabling system can allow you to generate specific reports of selected parts of records. An example is a cable report showing all cables running from a wiring closet. The report can include the cable identifier, pathway, termination position in the closet, termination position in the work area, cable length, and application.

TIA/EIA-606-A gives an extensive set of requirements for a cable management system. Unlike the other cabling standards, however, it is often only partially implemented. There are few, if any, products on the market that provide all the functionality required in TIA-606-A, and few users who really want that much functionality. From the user's perspective, there are three types of information that the cable-management system can keep track of:

- Information about the cables, connecting hardware, and associated interconnections. This information is essential in any cabling management system. It allows you to trace existing connections, find new circuits, and so forth. This is the primary purpose for which cable management systems are usually used.
- Information about pathways and spaces. This is the information about the physical placement of the cables and connecting hardware. For example, what conduit is a cable in? Some users want their cable management system to keep track of this type of information, and others keep this information in other formats, such as a set of CAD drawings.
- Information about the applications carried by the cabling and their associated users. This consists of a directory that associates use with particular cable runs and applications (voice, Ethernet, Fast Ethernet, etc.) they are running on that cable. If desired, this can be kept in the cable management system, but many users have other telecommunications administration where this information already exists, such as PBX and LAN administration software.

TIA-606-A attempts to customize the administration system to the needs of various users by providing four classes of administration. However, it is important to consider what information and functionality are needed for a particular site and install and configure a cable management system to meet the users' needs without requiring them to track a lot of other information they don't care about. The cable management features that are needed should be implemented in a way that is compliant with TIA-606-A, but there is no need to implement features of little value just because they are included in TIA-606-A.

What's in a Record?

TIA-606-A defines several types of records and associates a unique indentifier with each. Both mandatory and optional information are defined for each type of record. For example, the information in a horizontal link record is summarized in Figure 10–1.

Different types of records are required for each of the four classes of administration, as shown in Figure 10–2.

Type	Information	Description
Required	Identifier	Primary indexing identifier, such as 1A-A47
	Cable type	Type and category of cable
	Location of TO	Room, office, or grid location
	TO type	8-position modular, category, etc.
	Cable length	
	Cross-connect hardware	Description of cross-connect hardware in TR
	Service record	When link was installed, tested, etc.
Optional	Location of test results	File names, etc.
	Location of outlet within room	
	Color of connector or icon	
	Other connectors at same location	Usually lists other connectors in the same faceplate
	Any other desired information	

Figure 10–1. TIA-606-A Horizontal link record

Color Coding

Color coding of circuits can be useful for identifying circuits by type or function. Some confusion exists regarding standards-based color coding. TIA/EIA-606-A recommends color coding only for cross connects within the cabling system. Figure 10–3 summarizes the color coding. Proper color coding is a valuable aid in making moves, adds, and changes and in troubleshooting the cabling plant. Color coding allows technicians to identify cables quickly and to double-check work readily. Color coding reduces wiring errors.

Your color-coding scheme does not have to be limited to those recommended in TIA-606-A. Some users also color code by level of performance so that a Category 5e connection is easily distinguished from a Category 6 connection. While color coding in TIA-606-A is intended for cross connects, it can be extended to other parts of the building. For example, the blue recommended for horizontal cabling is for the telecommunications closet, not for the outlet in the work area. Color coding an outlet, however, for telephones and low- and high-speed network connections can help identify ports as well as icons and words.

Cable Management Software

Record keeping is essential to managing a structured cabling system. And nothing excels at keeping records flexibly and powerfully like a computer. Not surprisingly, there are many programs designed specifically to help you manage a cabling plant. They cover a wide range of features, capabilities, power, ease of use, and price. But if you're attempting to manage a network of more than a handful of outlets, they are indispensable.

Identifier	Class of admin.			
	1	2	3	4
Telecommunications space	R	R	R	R
Horizontal link	R	R	R	R
Telecommunications main grounding busbar	R	R	R	R
Intrabuilding backbone cable		R	R	R
Intrabuilding backbone pair or strand		R	R	R
Telecommunications grounding busbar		R	R	R
Firestop location		R	R	R
Interbuilding backbone cable			R	R
Interbuilding backbone pair or strand			R	R
Building			R	R
Campus or site				R
Horizontal or intrabuilding backbone pathway element		O	O	O
Horizontal or intrabuilding backbone pathway between two TRs		O	O	O
Outside plant pathway element			O	O
Interbuilding pathway or element			O	O
Wide area network link				O
Private network link				O

R= Required, O= Optional

Figure 10–2. TIA-606-A identifiers

Cable management software typically contains a large database with complete information about all the cables, connecting hardware, patch cables, networking equipment, and so forth. The initial population of this database with complete and accurate information is a critical step. It can also be very labor intensive. When a cable management system is being used with an existing cabling installation, it is usually necessary to do a physical inventory of the entire system to make sure that all the records are entered accurately. It is much easier to compile accurate data when documenting a cabling system as it is installed, although entering the data into the database can still be a time-consuming task. However, if the initial data entry is not done completely and accurately, the cable management software will never be fully useful.

As part of the life cycle of the cabling plant, moves, adds, and changes must be accommodated by any administrative system. Failure to document all changes and update all related records in a timely, orderly manner can result, over time, in a cabling plant whose physical structure no longer matches the records and drawings that supposedly documented it. Inaccuracies increase the likelihood of errors occurring during moves, adds, and changes. Network reliability degrades, downtime increases, and users are peeved.

Cable management software typically offers a graphical interface that eases the chores of learning and using the program. While the essential "backend" of the program is one or

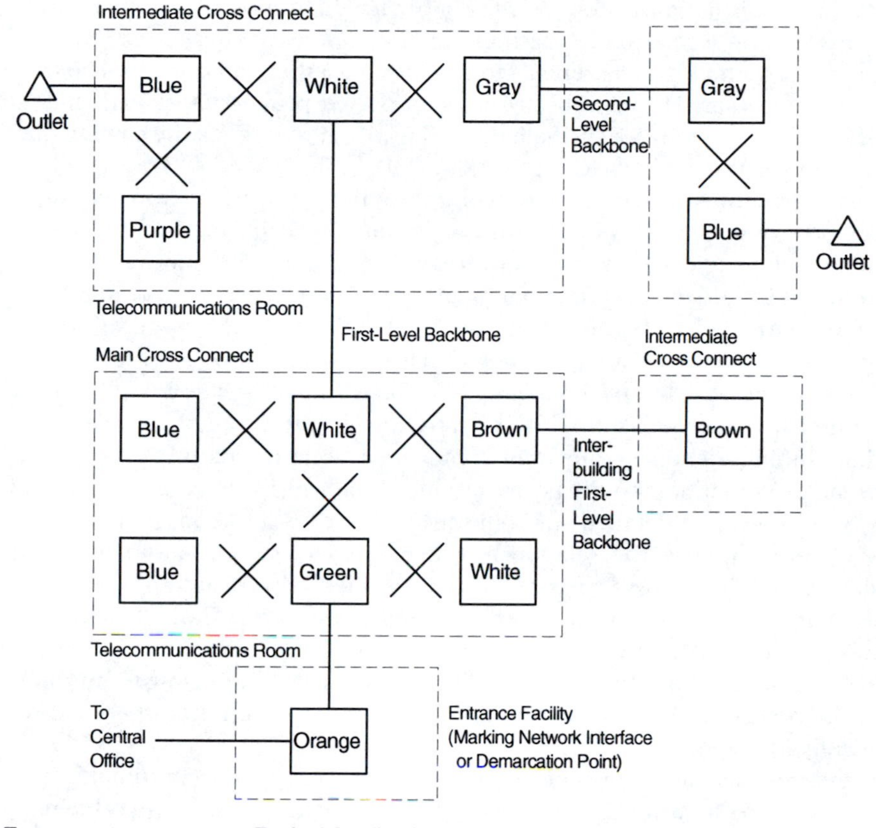

Color	Type	Typical Application	Pantone
Orange	Demarcation point	termination of cable from telephone company's central office	150 C
Green	Networks:	network connections or auxiliary circuit terminations	353 C
Purple	Common equipment:	switching and data equipment (PBXs, hubs, etc.)	264 C
White	First-level backbone:	connections between main and intermediate cross connects	
Gray	Second-level backbone:	connections between intermediate and horizontal cross connects	422 C
Blue	Horizontal	horizontal cable terminations in TRs	291 C
Brown	Interbuilding Backbone	campus cables	465 C
Yellow	Other	alarms, security, environmental controls, etc.	101 C
Red	Key telephone systems		184 C

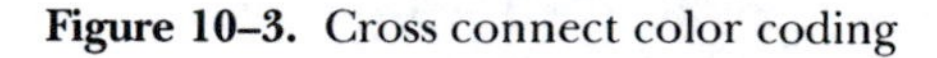

Figure 10–3. Cross connect color coding

more databases storing all records of the system, the graphical front end simplifies using the software.

While simply viewing the cabling plant on multiple levels might be beneficial, the power of cable management systems is in helping you maintain the system. The program can automatically validate data paths pair by pair or for an entire subsystem. This doesn't mean that the program can identify installation errors where the cable was not run according to the plan. Cable management programs do not connect directly to the wiring, so they can only validate that the plan is correct, that the cabling is "theoretically" interconnected properly. Installers must still physically verify the installation with testers. Still, the ability to logically check the cabling plant quickly and automatically beats manual checking.

Cable management software will allow you to use floor layouts for your building. To ease creating such drawings, many will allow you to import CAD drawings from other programs (such as the industry-standard .dxf or .dwg formats) so that existing layouts can be used. The cabling system can be laid out and viewed in many ways: from a floor plan point of view or from a connectivity point of view. Floor layouts are invaluable. As you may have noticed, typical naming conventions for cables and connecting hardware are not overly intuitive even if they are logical and systematic. By quickly being able to visually associate components with their physical location, you save yourself a lot of headaches. You will be able to trace a cable from end to end, including identifying any interconnections along the path.

The visual presentation should be hierarchical, allowing you to move from an overall system-level view down to a detailed view of individual circuits and components. For example, the high-level view will show the relationships between wiring closets and the general cable runs to office areas. The low-level view might be a detailed view of a wiring outlet in an office, showing how each conductor is terminated (remember that T568A and T568B are wired differently) and providing all related identifications of connected cables.

Changes are handled automatically. If a PC moves from one room to another, you can simply drag the object on the screen from the old location to the new. All related records will be automatically updated.

One of the most useful features of a cable management system is automatic circuit generation. If it is desired to move a PC or phone from one room to another, the cable management software can automatically find a path through the cabling system and determine what cables to use and what connections to make at the patch panels. This can greatly facilitate moves, adds, and changes while keeping the database up-to-date.

Cable management software can also help you to specify applications running over a specific cable. It will prevent you from assigning Fast Ethernet to a Category 3 voice cable.

A built-in library of components also makes it easier to manage the system and prevent errors. Real-world components, rather than generic representations, allow capacities—both used and spare—to be seen easily. Capacity planning becomes easier because it is more convenient. Do you have spare patch panels or cross connects? Is there room in the telecommunications closet for more equipment? Is there spare cable behind the walls? Is Category 6 or Category 3 being used for the telephone equipment? Is the unused outlet available for a modem? Or is it connected to a network hub?

The software should also be able to accommodate records for attached active devices, such as switches, hubs, telephones, and PCs. By extension, it will be able to document what application is running over each circuit. This becomes invaluable over time as repeated adds, moves, and changes must be made. Being able to associate network equipment with the cabling system is extremely beneficial in that is provides an end-to-end view of the total system.

The software can also help you manage the system by generating work orders and tracking their progress and completion. Technicians can be dispatched more easily because locations and all related identification are added to the order. On completion, all records are automatically updated. Work can be arranged to make the most efficient use of manpower and achieve higher productivity. When the work is completed, the paperwork can be filed (an important task in any large corporation!) and the software can be instructed to update the information automatically based on the work order.

In short, high-end cabling management software is capable of doing just about anything you can think of—and several additional useful things you haven't thought of yet. About the only things the software won't do is install the system and certify it through hands-on testing.

Smart Patching

In recent years, a number of vendors have developed systems that incorporate some intelligence into the patch panel hardware so it can be linked to the cable administration database. This is often referred to as *smart patching* or *intelligent patching*. This integration of the patch panel hardware and the cable administration software helps to minimize cross connection errors, and keeps the database in sync with the actual configuration of the cross connect panel.

A typical smart patching system adds sensors and indicators to the patching hardware so that the location of patch cords can be monitored. It also provides a display capability (typically an LCD screen) that can be accessed by field personnel who are working at the cross connect location. Figure 10–4 shows a typical smart patching system.

Smart patching systems can support both fiber and copper networks, and provide a number of useful features, such as:

- Guidance for installing patch cords—when a new connection is entered into the cable administration database, a work order is displayed on the LCD screen and LEDs are illuminated at each jack where the patch cord should be installed.

Figure 10–4. Typical smart patching system. Note the buttons and LEDs above each modular jack and the LCD display screen at the top of the unit *(SYSTIMAX Solutions graphics are courtesy of SYSTIMAX Solutions™, a CommScope company)*

- Prompting for incorrect patch cords—if a patch cord is installed in the wrong location, the installer receives an error message on the display, as shown in Figure 10–5.
- Circuit trace—if the button above a jack is pushed, LEDs are illuminated at each end of the patchcord, as shown in Figure 10–6. This greatly simplifies finding the other end of a patchcord in a crowded cross connect field.
- Enhanced security—if an authorized cross connect change is made, an alarm can be issued. This can help prevent evesdropping and other types of tampering with the network.
- Record of changes—the administrative database retains a permanent record of all changes that are made, which can be very useful in troubleshooting any problems that arise.

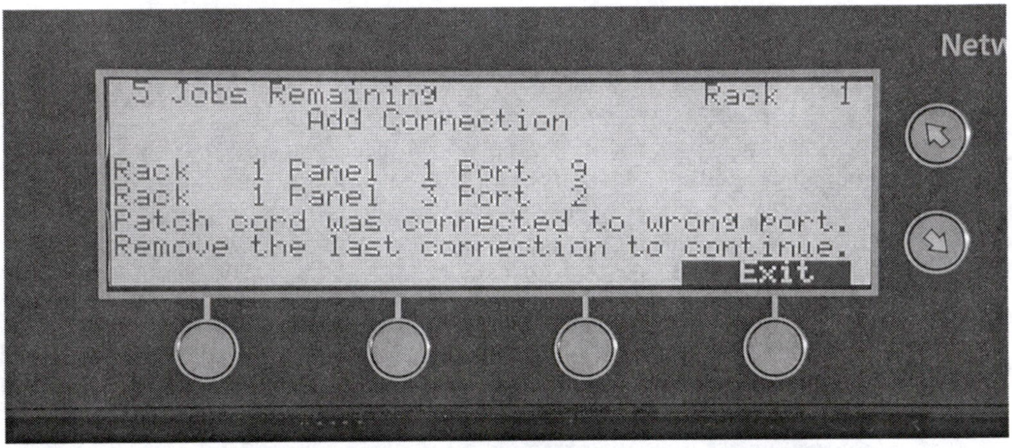

Figure 10–5. Smart patching system—error message for incorrectly installed patch cord *(SYSTIMAX Solutions graphics are courtesy of SYSTIMAX Solutions™, a CommScope company)*

Figure 10–6. Smart patching system—circuit trace operation *(SYSTIMAX Solutions graphics are courtesy of SYSTIMAX Solutions™, a CommScope company)*

Smart patching systems have the potential to make cable administration more efficient and cost effective by minimizing cross connect errors, eliminating paper work orders, and ensuring that the database stays synchronized with the actual state of the cross connect panel. They can also increase the security of the network by detecting and reporting unauthorized or incorrect changes to the cross connects.

SUMMARY

- TIA/EIA-606-A is the standard for documenting and administering a cabling system.
- TIA/EIA-606-A defines the contents of a record.
- Record information includes required information, optional information, required linkages, and optional linkages.
- TIA-606-A provides four classes of administration that can be used for implementations of different complexity.
- Smart patching systems provide integration between the cross connect hardware and the cable management software.
- Cable management software is an invaluable aid for documenting and maintaining the cabling system.

REVIEW QUESTIONS

1. What U.S. standard covers administration of cabling systems?
2. What four types of information are covered in the standard?
3. What color is used for connections between the first-level backbone?
4. What color is used for station-level interconnections?
5. What is the advantage of hierarchical views in cable management software?
6. What are the major benefits of using cable management software?
7. What is the biggest challenge in implementing a cable management software system?
8. From the user's perspective, what three types of information does the cable management software keep track of?
9. What feature of cable management software is most useful for moves, adds, and changes?
10. What are the three methods of implementing a cabling administration system?
11. What is the purpose of the four classes of administration?
12. What class of administration would be used for a typical office building with several floors?
13. What four requirements must labels meet?
14. What is the primary difference between a smart patching system and a regular patch panel?
15. List four advantages of smart patching systems.

CHAPTER 11

Residential Cabling Systems

Objectives

After reading this chapter, you should be able to:

- List five trends that help make a residential cabling system an attractive feature in a home.
- List the two main types of cable used in residential cabling systems.
- Describe typical types of home networks.
- Define the similarities and differences between residential and commercial structured cabling systems.
- Give examples of broadband access to the home.

Structured cabling systems are finding use in homes and will soon become standard equipment in many new homes. The reasons are similar to those for office buildings: a structured system is more flexible, adaptable, and capable of handling new high-speed services. Until recently, a great deal of flexibility and adaptability were not needed in a home. You had basic phone service over simple telephone wiring—much home wiring isn't even up to Category 1 standards. And you probably had cable television with a few outlets scattered around the house.

Residential Trends

Entertainment

The needs and wants of consumers are changing for many reasons. Television is becoming more sophisticated (in capabilities, not necessarily content), with cable offering hundreds of channels. Direct broadcast satellite is a strong competitor to cable. What's more, the viewing expectations of American households are changing, with home entertainment systems with enhanced surround sound system, with VCRs and DVD players, and with the arrival of digital and high-definition TV. Families want to be able to distribute video throughout the house. Parents watch a TV show in the family room where the DVD player is located. The DVD output is redirected to a television in another room for the kids.

The Internet

Another influence is the Internet. In Chapter 1, we mentioned that Internet traffic now surpasses telephone traffic on the public telephone systems. In the United States, the most common way to access the Internet is via dial-up lines with 56 kb/s analog modems. In 2005, well over half of the homes in the U.S. still use dial-up Internet access. There are a variety of broadband access techniques available, including cable modems, digital subscriber line (DSL), satellite, Wi-Fi, and others. But guess what? Your antiquated telephone wiring or cable TV wiring might not be up to the task. Your access might be dumbed down to slower speeds because that's all your house can handle.

Many families now have more than one computer, one or more for the adults and one or more for the kids, to prevent arguments over access and prevent inadvertent deletion of files (such as the financial records by an inexperienced user.). Sharing Internet access is easier with a high-speed link. You don't want to share a low-speed modem link. And you don't want arguments on whose turn it is for Internet access. Many families are installing a second telephone line just for Internet access and other computer communications.

Broadband access also holds the possibilities of other services, like video on demand (VOD). VOD is like watching a tape, but it is provided to your home by an outside service similar to cable television. VOD differs from pay-per-view in that you can choose from hundreds or thousands of movies or shows, can start them at your convenience, and can even "rewind" or pause a program.

Another potential "killer application" that is getting a lot of attention is voice over Internet protocol, or VoIP. VoIP involves digitizing and packetizing voice signals and transmitting them over the Internet. Because Internet service providers typically charge a fixed monthly fee rather than charging per call as long-distance voice carriers do, VoIP service is potentially much cheaper for people who make lots of long-distance calls. However, there can be issues with voice quality, especially on relatively low-speed broadband access links. There are also a number of regulatory issues that are being addressed, such as whether to require VoIP to provide E911 service, etc.

Home Offices/Home Networks

A third influence is the growing number of home offices, telecommuters who work at home either all or part of the time. The home office is a popular trend with new homes, with over a third of new houses having a home office built in.

With multiple computers comes an interest in a home network for linking computers together. A network allows the family to share resources like printers, scanners, and Internet access. Multiplayer games allow people to go head-to-head with one another over the network or over the Internet. A trend everybody in the communications industry likes to discuss is

Cable TV	Hundreds of channels, including pay-per-view. High definition programming is becoming increasingly available.
Video on demand (VOD)	Movies or other video when you want them. You order the movie you want to see from hundreds or thousands immediately available. Watch the *Blues Brothers* every night or every episode of *Gilligan's Island*. A related, but less expansive version, is near video on demand (NVOD), which allows you to pick from dozens of movies that begin at regular intervals. VOD differs from pay-per-view in two respects: First, you select what you want to see rather than watching what is offered. Second, VOD can act like a VCR, allowing you to pause, rewind, and otherwise control the show.
Internet access	Surf the Internet at high speeds. By this we mean at speeds faster than what is possible with an analog voice modem. Speeds can range from 640 Kbps to several megabits per second. By contrast, voice modems operate at 28.8, 33.6, or 56 Kbps. Most of the following services can be obtained over the Internet, although they can be provided by other means by the service provider.
VoIP	Make and receive inexpensive long-distance phone calls over the Internet.
Information	Get stock quotes, sports scores, weather, or nearly any kind of regular information.
Home shopping	Order pizza from the corner pizzeria or a new computer from an international company. Such buying with online electronic commerce is becoming an important avenue of commerce.
Videoconferencing	Talk to and see others through a video camera connected to a computer or other device.
Distance learning	Get educated by connecting with a remote location, say a college across the state or across the globe. The ability to have live, interactive video is helpful to obtain maximum benefit.
Work at home	So-called teleworking is a growing trend. The ability to have high-speed connection for voice, video, and data to the office and customers make working easier and more productive.

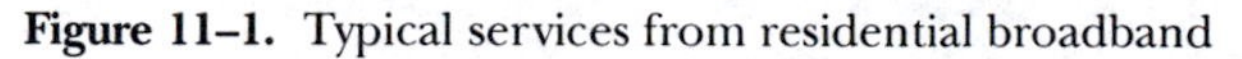

Figure 11–1. Typical services from residential broadband

convergence—the merging of telephony and computers into a single entity. As telephony becomes digital end to end, it increasingly can make use of the computer and the powerful capabilities technology offers for video telephony, call management, and other needs.

High-speed Internet access has also made teleworking (also called telecommuting) much more popular. Figure 11–1 summarizes some of the major residential Internet applications.

Home Automation

Home automation and security are related to structured cabling systems. Many families want the convenience and energy savings that can be had by controlling heating and cooling, turning lights on and off automatically, and even watering the lawn on a fixed schedule. They even want a sprinkler system than doesn't turn on at the appointed time if the lawn is wet from rain.

Or consider safety. Intruder alarms. Remote monitoring by a security service. Or a front door camera that you can watch from any television in the house. Or a camera in the baby's room so the mother can monitor a napping baby from the kitchen. While home automation isn't strictly part of a structured cabling system as we generally discuss in this book, automation systems fit nicely in with cabling systems. Many vendors of residential cabling systems also offer various types of home automation equipment.

Smart Appliances

As microprocessor chips become cheaper and more powerful, they allow more and more intelligence to be built into appliances. Most appliances today—from microwave ovens to washers—have one or more microprocessors built in. The next logical step is to provide an interface to the outside world so that the appliance can communicate and be controlled externally. Figure 11–2 shows an interesting, if somewhat extreme, example of a refrigerator with a

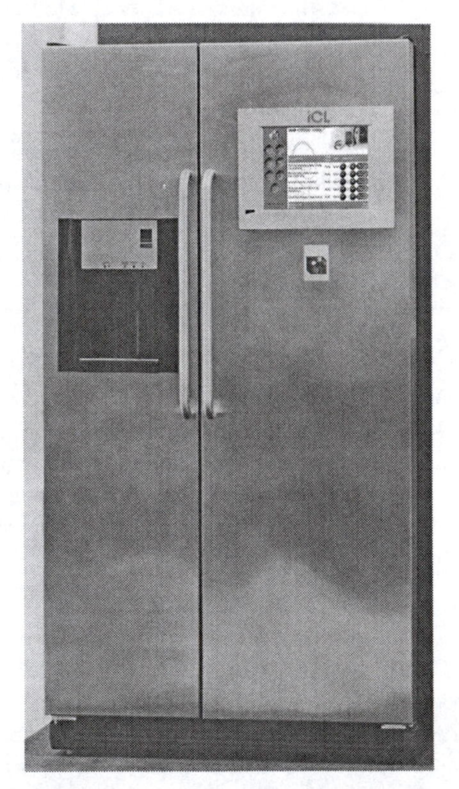

Figure 11–2. Smart refrigerator with built-in communications center *(Courtesy of Frigidaire Home Products)*

built-in communications center. Refrigerators in many homes serve to hold communications—notes, appointments, children's achievements fastened with magnets. But the refrigerator in the figure takes things a giant step further by including a Web browser, display, and touch-sensitive keys. You can leave notes for the family, maintain a calendar, send and receive e-mail, keep grocery lists, and so forth. The refrigerator can connect directly to the Internet over a phone line or through a home network.

If the smart refrigerator seems extreme, consider that it may well be the future as appliances use built-in intelligence to interface to the world. It may be awhile before your toaster is equally smart, but the convenience of smart appliances will make them more popular.

Two Notes about Home Cabling

You should realize two things about home cabling systems. While there are standards containing recommendations for a home cabling system, they do not as yet have quite as much influence as a generic building standard like TIA-568B. While a company might specify that their cabling system must meet the requirements of TIA-568B, few homeowners will specify that their system should meet TIA-570B. Indeed, how many homeowners have even heard of the standard? However, builders are increasingly aware that a structured cabling system will help sell homes. And providers of such systems pattern their offerings around the cabling standards.

The second thing is that structured cabling systems are not just for high-end expensive homes. A basic system adds only a few hundred dollars to the cost of the home and is in the reach of most families. Even mid-price homes and apartments will be equipped with structured systems. Widespread adoption of structured cabling systems depends on two things.

- Greater acceptance by builders that a structured cabling system is a desirable feature that can help sell houses.
- Greater demand from consumers as they become aware of the advantages of structured cabling systems in better entertainment, Internet access, and home automation. Plus, someday having a structured cabling system will even help make your home more salable when you move on.

Structured cabling is becoming a standard feature in new homes. As of 2005, roughly 60% of the new homes built in the United States included structured cabling.

TIA/EIA-570B: The Residential Cabling Standard

Yes, there is a residential counterpart to TIA-568B. TIA-570B is a standard for residential cabling systems. In many respects it is patterned after 568B, with a reliance on UTP and a home run architecture. A typical system uses cross connects from a central distribution point and individual cables running to each outlet throughout the house. Wiring rules are similar: 90-meter maximum runs for UTP, T568A wiring of outlets, and so forth.

The main difference is the use of coaxial cable for video applications like cable TV or direct satellite broadcast TV. The recommended cable is Series 6 coaxial cable, which is quad shielded and offers a bandwidth over 1.5 GHz. A traditional cable TV system has a bandwidth of 500 MHz, and newer systems run up to about 750 MHz. The quad shielding—two layers of

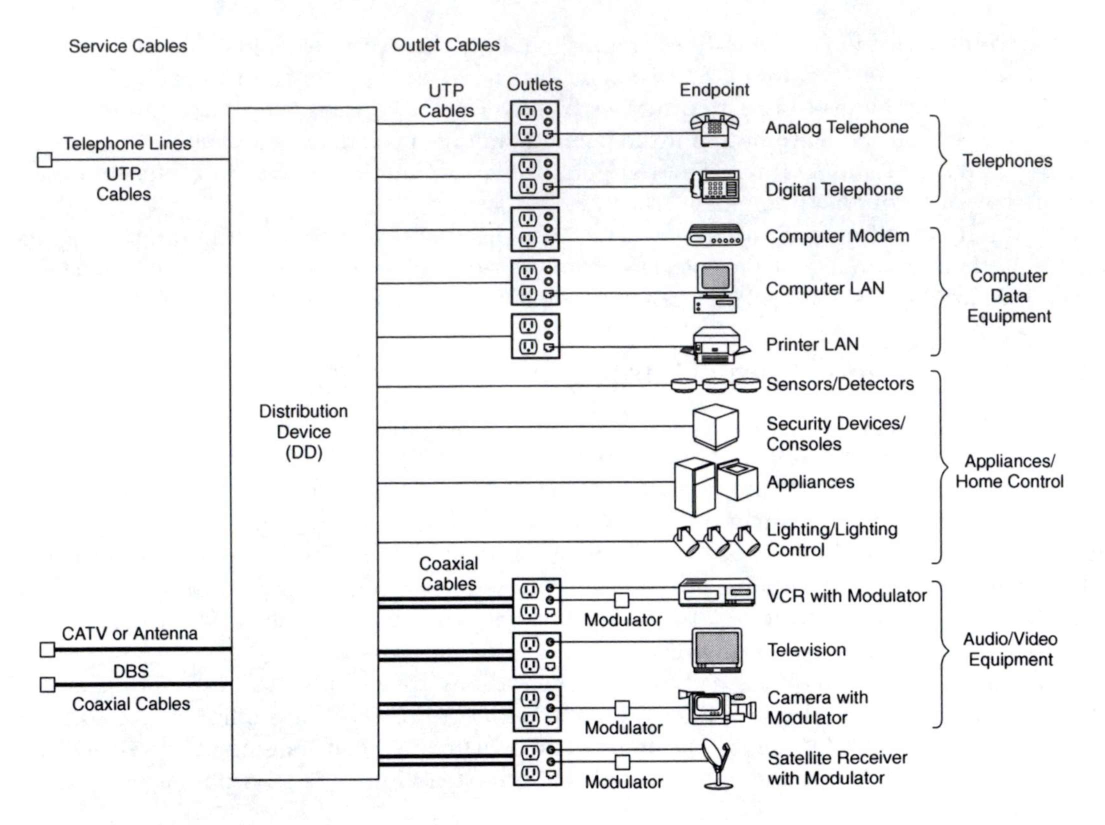

Figure 11–3. Typical residential structured cabling system

braid and two layers of foil—gives much better shielding than the RG-59 cable typically used with cable television and provides greater bandwidth.

Figure 11–3 shows a typical wring scheme for a home. A central distribution device (DD) serves as a consolidation point for services coming into the home, including telephony, cable or satellite TV, and broadband high-speed services. The DD is sometimes called a residential gateway (because it serves as the entry for outside services into the home) or a service center (because it is the center from which services are distributed). From the DD, home run cabling is run to outlets throughout the home.

Figure 11–4 shows a typical DD. Some systems are passive, essentially serving as a cross connect to distribute services through the home with 110-style punch-down blocks. Other systems contain a variety of modules for different purposes. Typical modules might include:

Video modules for splitting the video signal. The module can be passive or active. An active module amplifies the video signal. A typical video module might accept one incoming cable TV signal and provide six or eight output signals for distribution throughout the house.

Telephone modules for splitting telephone signals. Most systems will accommodate two or more telephone lines. Multiple telephone lines are becoming increasingly popular, either to give teenagers their own phone or to provide Internet access without tying up the telephone.

Figure 11–4. Distribution device with voice, data, and video modules installed *(Courtesy of UStec)*

Audio modules distribute stereo audio signals throughout the house.

Network modules for building a home network. The module typically is a Fast Ethernet hub to provide a 100 Mbps shared-media network.

Home automation controller to provide the central intelligence for a home automation system.

Many distribution gateways are modular and expandable. Modularity is desirable for many reasons. For the builder, it allows several levels of service to be provided. Buyers can choose from basic, advanced, or deluxe systems. But modularity also allows a pay-as-you-go flexibility so that services can be added as needed. There is one small drawback to the expandability: some forethought is required for the cabling. The cabling ideally should be in place so that only modules need to be added. Need to add a network? If the cabling and outlets are in place, then it is simple to add a hub and connect the cables. Some homes use plastic tubing to distribute cables throughout the house. The tubes can greatly simplify pulling new cables if needed.

But, as with an office building, the prudent homeowner will insist on installing extra cables even if the services are not yet needed. While this may seem a waste, consider cable TV. In the early days of cable, a home typically had only one or two outlets. The cable companies charged extra for each outlet. If you wanted to add an outlet, you paid the cable company first to come and install the cable and then a monthly fee for use of the outlet. Today, you pay only for the service, regardless of the number of outlets and TVs. So many homes are now built with a cable outlet in each room. It's neat and convenient. Want to add a TV to a room? Just plug it into the cable outlet. The same is true for telephone service. What is new is the desire to also accommodate data services, including the Internet, e-mail, and connections to the office.

Although TIA-570B makes provisions for two different grades of cabling (Grades 1 and 2), there is little useful distinction between them. Both Grade 1 and 2 require at least Category 5e UTP and recommend Category 6 UTP.

Residential Cabling Design Considerations

Location of TOs

It is important to have a sufficient number of TOs in each room so that equipment can be connected without using long extension cords. The only requirement in TIA-570B for TO placement is that at least one TO shall be placed in each of the following rooms: Kitchen; bedroom; family/great room; and den/study. TIA-570B also recommends that additional TOs be placed:

- On any unbroken wall space of 12 feet or more in length.
- So that no point along the floor line is more that 25 feet (measured along the floor line) from the nearest TO.

This is the absolute minimum. While the growth of wireless devices (including cordless phones, cell phones, and Wi-Fi) has reduced the need for TOs somewhat, what is most important is that there is a TO available where you need it. In most houses, many additional TOs beyond those required by TIA-570B will be provided.

To ensure that the house contains an adequate number of TOs for the voice and data network, the following guidelines are recommended.

1. Provide at least one TO in every room, including bathrooms, large entry halls, and other rooms not covered by the TIA-570B requirement.
2. Provide at least one TO in each basement and garage area.
3. In rooms that are likely to have extensive data or entertainment networking, provide at least one TO for every 12 linear feet of wall. In such areas, consideration should be given to installing dual-TOs.
4. Provide at least one TO on all unbroken wall spaces where it is likely that telecommunications equipment will be needed, even if the wall space is less than 12 feet in length. This insures that telecommunications cords will never have to be run across a doorway.
5. In bedrooms, provide at least two TOs on opposite walls so that a telecommunications cord never has to be run across the room. Most people will want a telephone near their bed, so place a TO near where the bed(s) will likely be.
6. Rooms that may be used as a home office should follow recommendation 3 above.

The intent of these guidelines it to ensure that sufficient TOs are installed so that the house can accommodate a wide variety of equipment configurations, at locations desired by the homeowner, without running long telecommunications cords in areas where they present tripping hazards.

Cable

UTP cabling is universally used for voice and data networks. TIA-570B requires Category 5e or better UTP cable and recommends Category 6 UTP. Due to the large improvement in performance at a relatively small additional cost, Category 6 UTP should always be used whenever possible. Only low-end installations where minimizing the cost is critical should use Category 5e cabling.

As an intermediate-cost solution, it sometimes makes sense to pull Category 6 cable and install Category 5e connecting hardware. This only gives Category 5e performance, but the cabling system can be easily upgraded to Category 6 performance in the future by replacing the connecting hardware.

Wireless Access Points

Wireless networks (such as Wi-Fi) have become extremely popular. A TO should be placed at locations where WAPs are likely to be placed.

Telecommunication Outlets

A variety of outlets are available to suit different needs, from telephony to TV to networking. Most outlets include several interfaces, including modular jacks for telephony/data, F-connectors for cable TV, and even receptacles for power. More commonly, telephone and data jacks are provided separately for convenience. The standard recommends an outlet for every 12 feet of wall space so that there is no point along a wall that is more than 25 feet from an outlet. This lessens the chance of unsightly extension cords or the need to place televisions or computers because of the location of an outlet. Do you want your entertainment center against this wall or that wall? A lack of outlets may answer the question faster than your decorating needs. If you have an outlet on both walls, you obviously have more flexibility. Most people find that convenience is served by generous, rather than parsimonious, application of outlets.

Different rooms will have different combinations of outlets. A kitchen may have only a phone and TV. A home office may have two phone lines, two data lines, and TV.

What about Fiber?

The TIA-570B standard recognizes 50/125-µm and 62.5/125-µm multimode fiber for horizontal runs. A family home is essentially only a horizontal cabling system. (Single-mode fiber is allowed for multifamily buildings, such as apartments, where runs between floors form a backbone.) The need for fiber is being hotly debated, as is the type of fiber appropriate. Again, the pro-fiber camp sees fiber as the best way to future-proof the home and ensure it is prepared for whatever services are available in the future. Opposing this group are those who see fiber as not yet appropriate and probably not needed. Fiber *in* the home seems silly until fiber *to* the home is provided. At the same time, many builders are not comfortable with fiber.

Of all the various types of fiber, plastic optical fiber (POF) is the one that seems most suited to residential use. While POF has traditionally offered very high losses and very low data rates, a newer type of POF offers a graded-index core that will allow gigabit data rates over 300 meters. Plastic fibers are very rugged, have large cores (250-μm diameter is typical for the newer graded index types), and require simpler and less expensive connectors.

Proponents of POF point to lower costs all around—for cable and connectors, for associated electronics, and for installation. POF is the cost-effective way to put more fiber in your home. Detractors point to POF's lack of market acceptance in data networking applications and the widespread acceptance of 50/125-μm and 62.5/125-μm fibers. Whether POF finds a niche in the residential market remains to be seen.

A Closer Look at Broadband Services

At the beginning of this chapter, we discussed some of the applications now available or coming soon to homes. We also said that broadband services will allow more flexibility and more convenience by providing a high-speed link to the outside world. In this section, we will look at some of the services.

An important aspect in providing high-speed services to the home is the ability to use the existing cabling plant, either telephone cables or coaxial TV cable. The so-called *local loop* or *last mile* to the home is already wired and few companies want to commit to the expensive task of rewiring it. That is one reason why fiber-to-the-home is not coming anytime soon—especially if the existing telephone cables can be used. While fiber and better copper cables may find application in new developments, they are less likely in older ones.

Analog Modem

If you have a computer, you probably have an analog modem operating at 56 kbps. For our purposes, this is a slow-speed connection, the type of connection that, while easily available in every home, is too slow for the wonderful world of broadband. Still, most of us use an analog modem for one of three reasons:

1. It came with the computer and represents the lowest cost Internet access.
2. It's the only way to communicate: high-speed access is not available.
3. Broadband services are available, but not cost-effective.

DSL

DSL stands for digital subscriber line, and is the means to bring high-speed digital services to the home. More importantly, it will bring these services to the home over the copper twisted-pair cabling the telephone companies already have installed. One of the great dilemmas of the recent communications revolution is how to provide high-speed access without rewiring all the local telephone wiring to homes. DSL is one answer.

There are many variations of DSL services supplying data rates between 1 and 8 Mbps. The two most important to home users are ADSL and G.Lite.

ADSL is *asymmetric* DSL, offering different speeds depending on the direction of transmission: downstream from the service provider or upstream from the home. Downstream rates range from 1 to 8 Mbps, while upstream rates top out at 768 kbps. Asymmetric transmission makes more sense than may appear at first. Consider the Internet. Most upstream transmissions are requests for new pages, e-mail, and similar tasks that require modest bandwidth. But transmission downstream to the home requires greater bandwidth to accommodate graphics-heavy pages, streaming audio and video, and large files. As far as asymmetric transmission, remember that 56k modems are also asymmetric—capable of speeds of 56 kbps downstream and 33.6 kbps upstream. What's more, 56k modems do not reach 56 kbps speeds for both regulatory and practical reasons (mainly line conditions).

The exact speed attained with ADSL depends on the transmission distance from home to service provider, the gauge and overall quality of the cables, and on the level of service the provider wants to provide. Some service providers provide tiered services, charging different amounts for different speeds. One variation of ADSL is called rate-adaptive ADSL, which automatically adjusts to achieve the best mix of speed and quality. This simplifies the system for the service provider, who doesn't have to worry as much about users having problems because of distances or line conditions.

G.Lite is a slower DSL variation also called Universal ADSL or ADSL Lite. G.Lite tops out at 1.5 Mbps downstream and 386 kbps upstream. G.Lite is designed to be easily installed by the consumer. ADSL, in contrast, typically requires a truck rollout—the telco has to send out a technician to perform the installation of a POTS splitter to separate the voice signal from the data signal. G.Lite is a splitterless technology and can be easily installed.

While it is easy to think that more bandwidth is always better, this is not always the case. How much bandwidth do you need to maximize the Internet and get the fastest response? No matter how fast your connection, Internet users typically don't realize speeds much above about 400 kbps. While the Internet backbone operates very fast at gigabits per second, other bottlenecks limit the practical speeds you can realistically achieve at home. These bottlenecks include:

- Service provider used (some are better than others at managing traffic and supplying bandwidth)
- Traffic on the 'net (causes slowdowns)
- Time of day (traffic varies with time of day)
- Internet protocols (the TCP/IP has inherent delays)
- Sites visited (some sites are busier than others)
- Servers (servers vary in performance and in the ability to handle heavier loads)

If you do much surfing from home, you probably know that performance can vary considerably. While faster access services will definitely improve things, there are limits—at least for now.

One nice feature shared by DSL and cable modems is called *always-on Internet,* meaning that when your computer is turned on, you are connected to the Internet. There is no need to dial up a service provider and wait for a connection. Your computer can check for e-mail as soon as you turn it on. Plus, you won't tie up a phone line using the 'net. Phone calls can still come over the DSL line while you are connected to the 'net. Some observers believe that this always-on feature will help sell DSL as much as higher speeds will. It simply makes the Internet an easier and more pleasant experience.

Broadband services like DSL also allow multiple users to access the Internet at the same time. Multiple computers in the home create multiple users who may want to be on-line at the same time. High-speed access not only permits more than one user, it will not degrade the experience of each user.

Cable Modems

Cable modems allow broadband services to be carried on cable TV networks. At the residence, the incoming coaxial cable goes through a splitter. Television sets receive the video signals just as though the broadband data was not there. A cable modem, which packs broadband data into a previously unused frequency band, is connected to a PC or data hub in the home.

Traditionally, cable TV systems used frequencies from 50 to 550 MHz to send TV signals, using 6 MHz of bandwidth for each TV channel. To support broadband data, frequencies from 5 to 40 MHz are used for uplink signaling, and frequencies above 550 MHz are used for sending data downlink. As with ADSL, there is a lot more bandwidth available in the downlink direction.

This is conceptually similar to the way the phone network supports broadband data with a DSL modem but there are a couple of key differences. First, the coaxial cable is designed to provide hundreds of MHz of bandwidth, while the phone lines were designed for low frequencies. Second, the traditional cable TV network is inherently a shared, unidirectional resource. The traditional cable TV system starts at a headend, or hub, and goes through a series of trunks and amplifiers that branch out and finally have a drop at each house. To support broadband data, transmission in the uplink direction must also be supported.

A modern cable TV system typically uses a Hybrid Fiber-Coax (HFC) architecture where fiber is run from the headend to a location near the homes to be served, and the lines to the houses are coax. Due to the large bandwidth of the coax, cable TV systems can offer higher bandwidth to the subscriber than DSL systems. A typical cable system might offer a subscriber downlink data rates of about 3 Mb/s and uplink data rates of 256 kb/s. Keep in mind that this is the peak rate and will be shared among a number of subscribers, so the performance of cable modem systems tends to decline somewhat as the number of users increases. In particular, if there is a server in your neighborhood generating a lot of traffic, the uplink delays can become noticeable.

Technical specifications for cable modems are developed by Cable Labs, Inc. They are referred to as DOCSIS (Data-Over-Cable Service Interface Specifications). The DOCSIS documents are available for downloading at http://www.cablemodem.com/specifications/.

Cable modem Internet access is generally available in any area with cable TV service.

Satellite

Broadband access via satellite has one huge advantage—it is available anywhere in the country. This is in contrast to cable modem and DSL access services, which are widely deployed in urban and surburban settings but are not always available outside of densely populated areas. Maximum downstream data rates for satellite access are about 400 to 500 kb/s. Upstream data rates are much slower, from 50 to 100 Kb/s.

Satellite networks use geostationary satellites that orbit the earth at an altitude of about 22,300 miles. Because of the large distance to the satellite, a fairly sophisticated antenna and transceiver are required, which can cost a few hundred dollars to install. The antenna is usually about 2×3 feet in size and must have an unimpeded view towards the equator (i.e., to the south for antennas located in the northern hemisphere).

Also, because of the long transmission distance, there is about a half-second round trip delay in sending a signal over the satellite link and getting a response. While this is not a problem for some applications, it is very noticeable for VoIP.

Wi-Fi

Wi-Fi networks are sometimes used as a means of broadband access. For example, from a cable or DSL modem, Wi-Fi can be used to extend access throughout a home, campus, apartment building, or coffee shop. The Wi-Fi data rates of 11 or 54 Mb/s are more than adequate to handle the data delivered by a cable modem, DSL modem, or satellite interface.

New Access Technologies

In addition to the technologies mentioned above, there are several new access technologies currently being rolled out, including fiber to the home (FTTH), broadband over power line (BPL), and WiMax. It remains to be seen which of these technologies will become widely deployed.

Fiber to the Home (FTTH)

Fiber to the home, as the name implies, consists of a network that extends fiber connectivity all the way to the residence. FTTH is often regarded as the "Holy Grail" of access technologies. It can deliver far more bandwidth to the home than any other access technology—potentially hundreds of Mb/s or more. Unfortunately, FTTH networks are expensive to deploy, often costing several thousand dollars per house. Although the installed base of FTTH is growing rapidly, it is currently available only in limited areas.

IEEE 802.16 (WiMax)

The IEEE 802.16 standards specify a fixed wireless network that can provide broadband access over a metropolitan area. It is often referred to as WirelessMAN and is also called WiMax. It is promoted by the WiMax Forum, which is an association of more than 60 companies that provides compatibility testing and certification of WiMax equipment.

WiMax provides high bandwidth channels in both the uplink and downlink directions. Throughput of up to 100 Mb/s can be achieved, although the standard does not specify how much bandwidth should be provided to each subscriber. The range is 30 miles or more. Due to the relatively short round-trip path, delay is not a factor like it is with satellites.

Although there are a number of open issues regarding WiMax, it is a very attractive technology. It provides substantial bandwidth without requiring a large infrastructure investment at every potential subscriber location.

Broadband over Power Line (BPL)

The concept of broadband over power line (BPL) relies on one obvious fact—virtually every home is connected to the power grid by copper wires. So why not just send broadband data over them instead of building a separate access network? This would be especially beneficial in rural areas where the cost of providing DSL or cable modem service would be prohibitive. This idea has been around for years, but has been beset with technical and regulatory problems.

A number of different BPL architectures have been proposed. Estimates of the bit rates that can be achieved over the power lines vary quite a bit. DSL-like figures of a few hundred kilobits to 1 Mb/s or so are often mentioned, although some estimates are higher. The actual bit rates will depend on the type of technology and architecture that is ultimately deployed.

BPL technology is still in its early stages. Although BPL is supported by the FCC and power companies, there are still many technical and regulatory issues, such as safety, interference, etc.

Home Networking

Home networking is a fast-growing segment of the network market. As families put in more than one computer, they want to share resources, including peripherals and Internet access. Networks also allow game players to face off against one another and can help enable easier home automation.

Every type of wiring can be used for home networking:

- Existing telephone wiring
- Powerline wiring
- UTP cabling
- Coaxial cable
- Wireless, including both radio frequencies and infrared

The ingenuity that creators of home network systems use in adapting every sort of wiring to the networks stems from several factors. Not too many years ago, a simple home network could cost over $500 for a hub and NIC cards. Today, you can outfit a network for under $100. In addition, you needed new cabling to run standard 10-Mbps Ethernet. Now you can build it into your residential cabling system. Because of the costs and difficulty in wiring existing homes, systems have been devised to run on powerlines, over existing telephone cabling, and through the air over wireless links.

A structured cabling simplifies home networking, both by providing up-to-date cabling, a home run structure, and even the hub. Some structured wiring systems offer hubs as an option in the service center. It's also easy to add a hub outside the service center and connect it to a cross connect inside the center.

Here is a brief introduction to some of the major types of home networking technology available today:

IEEE 802.3 (Ethernet)

While it was originally developed for commercial applications, the price of Ethernet hubs and NICs has dropped to the point where they are often used in residential applications. In fact, many manufacturers make Ethernet equipment specifically for home use. Some even offer hubs directly integrated into the service center of a structured cabling system. For more details about Ethernet, refer to Chapter 5.

Residential Ethernet

In July 2004, the IEEE 802.3 committee approved the formation of a Residential Ethernet study group to investigate the "additional application of Ethernet to time sensitive consumer Audio/Video Devices." In other words, to add isochronous capabilities to Ethernet. By October 2004, the study group had agreed on a set of draft objectives that look very much like a description of an IEEE 1394 network (see the next section for a description of IEEE 1394). The objectives include:

- At least 75% of aggregate bandwidth is available for isochronous traffic.
- At least 10% of aggregate bandwidth is reserved for best-effort (i.e., asynchronous) traffic.
- Isochronous traffic will only be supported on full-duplex links of 100 Mb/s or greater.
- Isochronous bridging to IEEE 1394 and IEEE 802.11 networks will be supported.
- No new physical layer interfaces will be specified.

It is far too early to predict how this specification will end up, but one possible outcome is essentially a merging of Ethernet and FireWire (IEEE 802 and 1394). The standard is not expected to be completed until at least 2006.

IEEE 1394 (FireWire)

FireWire was invented by Apple Computer in 1986. It was originally designed as a serial bus for interconnection of high-performance computer peripheral equipment and has since expanded into a full-fledged LAN. It has been standardized as IEEE 1394. There are three versions of the specification: IEEE 1394-1995, IEEE 1394a-2000, and IEEE 1394b-2002.[1] The terms "FireWire" and "IEEE 1394" are often used interchangeably.

FireWire has a couple of unique properties that make it ideal for residential networks:

- It supports both local clusters of endpoints as well as long-distance (up to 100m) connections for links between clusters.
- It provides integral support for isochronous data with reserved bandwidth and guaranteed latency and jitter. This makes it ideal for transmitting real-time data such as streaming audio and video.

[1] *For more details on the various IEEE 1394 specifications, see Residential Networks (Baxter, 2005) or FireWire System Architecture (Anderson, 2002).*

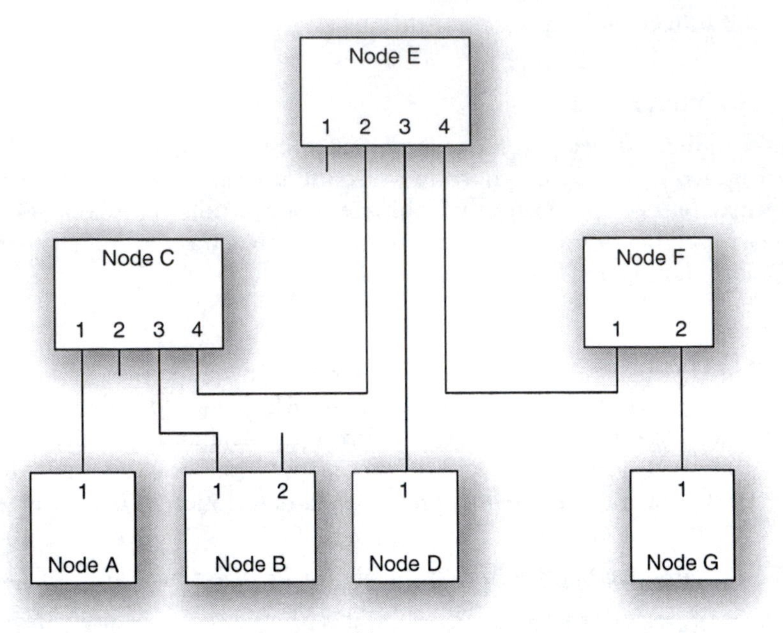

Figure 11–5. IEEE 1394 bus, node and port architecture

Architecture

A FireWire network consists of a collection of nodes tied together by a serial bus. Each node may optionally have multiple ports. Ports on the various nodes are hooked together with cables or cords to form a bus, as shown in Figure 11–5.

The figure shows seven nodes (indicated by the letters A–G), having a total of 15 ports (numbered from 1 on each node). The exact configuration of interconnections is not critical. If, for example, the cable from Node F port 1 were removed from Node E port 4 and connected to Node C port 2, the network would operate the same as before. When a signal is received at one port on a node, it is repeated on all the other ports so that the signal propagates throughout the entire network.

FireWire uses a memory-mapped architecture with 64-bit addressing. Six bits of the address are used to specify the node, which means that a maximum of 63 nodes (plus a broadcast address) can be on a single serial bus. Ten additional bits are allocated to indicate the bus ID, so up to 1,023 serial busses can be connected in a single IEEE 1394 network. Figure 11–6 shows how bridges are used to create a large IEEE 1394 network.

FireWire runs at speeds starting at 98.304 Mb/s (referred to as S100) and multiples thereof, including S200, S400, and S800. S1600 and S3200 have been documented but not implemented yet. The FireWire bus has two fundamental modes of operation—asynchronous and isochronous. Up to 80% of the bus bandwidth can be reserved for isochronous streams, and the remaining 20% is used for asynchronous data.

This dual-mode operation allows video to be sent isochronously (with guaranteed bandwidth, latency, and jitter) while control information, on-screen display information, etc. are sent asynchronously.

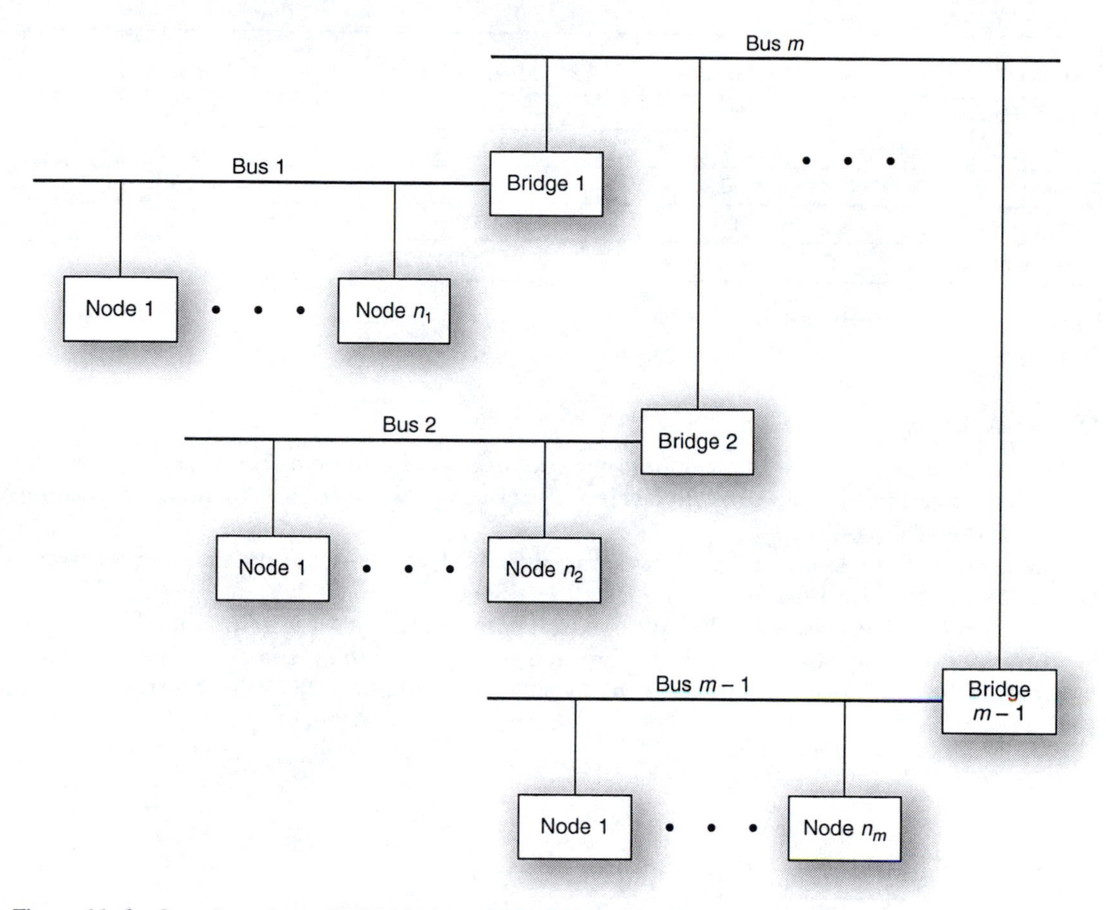

Figure 11–6. Creating a large IEEE 1394 network with bridges

Cabling Media

IEEE 1394b, the most recent FireWire specification, supports five different types of media, as indicated in Figure 11–7.

The three types of fiber are not widely used. Short-haul copper uses specially designed FireWire cords for communication within a local cluster. The UTP interface can be used to connect to another cluster up to 100m away.

As of late 2005, two additional types of physical layer interfaces are under development.

- IEEE 1394c, which should go out for industry ballot before the end of 2005, uses a 1000BASE-T interface to transmit a FireWire S800 signal up to 100m over UTP. This takes advantage of the availability of existing Gigabit Ethernet interface chips.
- The extension of the short-haul copper interface to support S400 signals for distances of 50 to 100m over Cat 6 UTP.

Media	Max. dist. (m)	Max. speed	Connector
Short-haul copper	4.5	S1600	Beta
UTP (Cat 5 or better)	100	S100	RJ-45
1000-μm POF	50	S200	PN
225-μm HPCF	100	S200	PN
50-μm MMF	100	S1600	LC

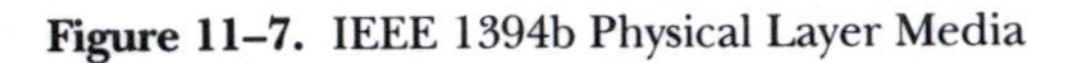

Figure 11–7. IEEE 1394b Physical Layer Media

Structured Cabling

FireWire networks fit very nicely into a structured cabling system. Figure 11–8 shows an example of a residential network with a video cluster in the family room and a data processing cluster in the home office.

Each cluster uses short-haul copper connection. UTP cables run from each cluster to the DD. A 1394 repeater can be used at the DD if necessary.

One of the nice features of FireWire is that computing and entertainment devices can be combined on the same network. In Figure 11–8, for example, it is possible to access a movie or video clip stored on the PC in the home office and play it back on the HDTV in the family room.

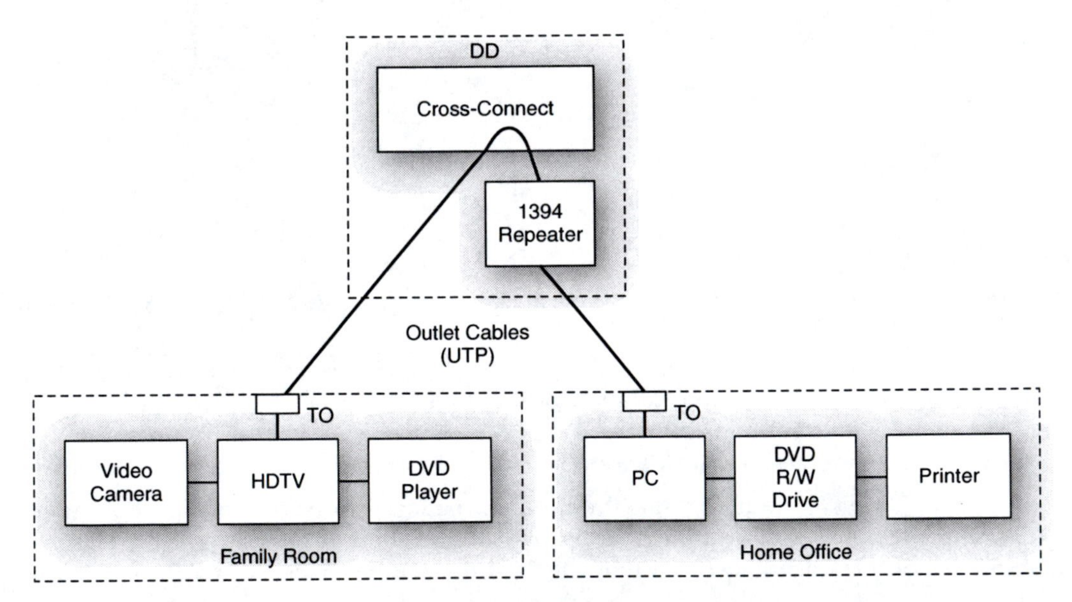

Figure 11–8. IEEE 1394 network with structured cabling

IEEE 802.11 (Wi-Fi)

The IEEE 802.11 wireless LAN specifications (which were discussed in Chapter 7), while originally developed for commercial applications, have found widespread use in the home. The three types of Wi-Fi networks most commonly found in homes are:

- IEEE 802.11b, which operates at data rates up to 11 Mb/s
- IEEE 802.11g, which operates at data rates up to 54 Mb/s
- IEEE 802.11n, which will operate at data rates of 100 Mb/s or more. As of late 2005, the specification has not yet been finalized, but "pre-11.n" equipment is available.

Wi-Fi networks are an excellent addition to a home data network. They can provide access throughout the house, patio, etc. However, they typically do not have enough bandwidth (or QoS) to handle entertainment networks.

Security is also an issue with wireless networks in the home. Because the signals propagate outside the home, it is important to ensure that the Wi-Fi security features are enabled to prevent evesdropping and unauthorized access.

Wired networks (such as Ethernet and FireWire) present a different set of tradeoffs than wireless networks. The ideal wireless application would be a small, battery-powered portable device where mobility is essential, data rate requirements are modest, and phantom power is not required. The best residential networks combine wired and wireless technologies in a way that emphasizes the strengths and advantages of each. As a rough rule of thumb, if it moves – go wireless. If not, wire it.

No New Wires (NNW) Networks

For new construction, the networks previously discussed are the best choices. But for application in existing homes that do not have structured cabling installed, several types of networks have been developed that do not require the installation of new wiring. In general, the NNW networks have lower performance (in terms of data rates, distance, etc.) than networks such as Ethernet and FireWire.

HomePNA

The Home Phoneline Networking Alliance (HomePNA) was founded in June 1998 by 11 companies in the computer and networking fields. The alliance has grown to include more than 100 member companies. The purpose of the group was to define a home networking protocol that was inexpensive to implement on the existing wire found in most homes. The first specification (HomePNA 1.0) was released in late 1998. It was based on technology developed by Tut Systems, Inc. and operated at 1 Mbps. More recently, the HomePNA 2.0 specification was released. HomePNA 2.0 operates at 10 Mbps with full backward compatibility with HomePNA 1.0. A number of companies make chipsets to efficiently implement this specification. Functionally, the network operates similarly to an Ethernet network, but uses a different encoding technique at the physical layer. HomePNA networks can use Ethernet software drivers. While HomePNA was designed to work over the installed base of existing wiring, of course, it works very well over new structured cabling installations as well.

HomeRF

The HomeRF Working Group was established in March 1998 to develop a set of specifications for a low-cost home wireless network. It currently has more than 90 members. The Shared Wireless Application Protocol (SWAP) uses frequency-hopping spread spectrum to achieve a data rate of up to 2 Mbps over a 50-meter maximum range. It supports both circuit and packet switching to efficiently carry voice and data. HomeRF has been almost completely replaced in the marketplace by Wi-Fi.

MOCA

MOCA (Multimedia Over Coax Alliance) is a private consortium devoted to the reuse of existing coaxial cable in homes. MOCA supports data rates of 270 Mb/s or more and provides HDTV-quality video. Because there is a limited amount of coax in most homes, Wi-Fi or PLC extensions are used.

Powerline Carrier (PLC)

The concept of using the power wiring in the house to carry data has been around for a long time. Vintage technologies such as CE-Bus and X-10 have had PLC options for many years. In those networks, PLC was generally used for low-bandwidth applications such as lighting control. Early PLC implementations had problems with noise and reliability, which has given the technology a bad reputation with many people.

PLC systems have historically supported very low bit rates. X-10, for example, supports only about 60 bits/sec over the powerline. A newer type of PLC technology called the Universal Powerline Bus (UPB™) supports 480 bits/sec. UPB is a proprietary technology (developed by Powerline Control Systems) that is used mainly for lighting controls.

The electrical powerlines in a house are a very harsh environment for the transmission of signals. Some specific problems include:

- The transmission path between any two points is very complicated and includes multiple paths and many stubs terminated in different impedances. This causes great variation in both amplitude and phase response with frequency.
- The transfer function is variable over time as appliances are plugged and unplugged, or turned on and off.
- There can be significant levels of electrical noise caused by motors, fluorescent lamps, power supplies, lamp dimmers, etc. Radiated interference may also be picked up by the electrical wiring.
- Signals placed on the wiring in one home can easily be transmitted to other homes which share the same power transformer (typically six or so homes in the U.S.)

HomePlug

Recently, there has been renewed interest in using the power wiring for higher-bandwidth data communication as well as low-bandwidth control functions. The HomePlug Powerline Alliance (www.homeplug.org) was formed in 2000 to standardize support for powerline communication. The HomePlug 1.0 specification supports data rates up to 14 Mb/s, which is roughly comparable to 10 Mb/s Ethernet. HomePlug 1.0 speeds are adequate for Internet access, but not for entertainment networks. In January 2004, a powerline device operating at speeds in excess of 100 Mb/s was demonstrated at the Consumer Electronics Show. This provides sufficient bandwidth even for HDTV signals. The HomePlug AV specification, which will provide data rates of 100 to 200 Mb/s, is expected to be completed in 2005. HomePlug appears to be a significant advance from early PLC systems, but it is not clear that all the problems related to noise and interference have been solved. This is definitely a technology worth watching in the future.

Legacy Home Networking Technologies

CEBus

The Consumer Electronics Bus (CEBus, pronounced *see-bus*) is an example of home automation that is standards-based and offers compatibility among equipment from different manufacturers. The CEBus standard (EIA 600) mandates standardized distribution of control and data, and provides a common language for interproduct operation. Standardized physical product interfaces allow users to install new devices without complicated wiring. CEBus-compliant products can share data from their internal sensors, such as temperature, light status, channel selection, or room occupancy, as well as using data from a central point, such as date/time and programming instructions. Centralization of redundant functions and the reduction of user interfaces make it easy to install new products without learning new ways to operate them, and allow products to work together so that, for example, when the doorbell rings, the TV automatically switches to monitor the front door camera. CEBus can run over a variety of physical-layer media, including:

- UTP cable (phone wire)
- Powerline
- Infrared
- Coaxial cable
- RF Wireless
- Fiber-optic cable

The CEBus standard has been around for quite a while—the CEBus Industry council was formed in 1984, and draft versions of the standard were available in the late 1980s. While CEBus-compliant equipment is widely available, it has not become the dominant home automation technology like it once appeared it would.

X10

X10 is a home automation protocol that has been around for many years. It is often considered to be low-tech and low-performance; however, it is the lowest cost and largest installed base of any home automation technology. It supports a variety of physical media, including phone wire and infrared, but is most commonly implemented over powerline.

LONWorks

LONWorks was developed by the Echelon Corporation and is standardized under EIA-709. LONWorks uses carrier-sense multiple access (like Ethernet) at the physical layer and is relatively independent of the transmission media. Like most home automation protocols, it can be used over Cat. 3 or Cat. 5 UTP, powerline, and wireless. The LONWorks protocol has been built into the "Neuron" chip available from Cypress Semiconductor. These chips are incorporated into home automation equipment by many manufacturers. LONWorks and X10 are the most widely implemented home automation technologies.

Summary

- A residential cabling system is based on Category 5e or better UTP for telephone and data and Series 6 coaxial cable for video.
- A residential cabling system uses a star-wired homerun topology as specified in TIA-570B.
- IEEE 1394 networks (also known as FireWire) have ideal properties for a home network.
- The most commonly used broadband access technologies in the United States are cable modems and DSL.
- Residential cabling systems will help families take advantage of broadband services to the home, including Internet and new video options.
- A variety of home networks systems are available for the home; some use structured cabling, while others use alternative techniques such as wireless.

Review Questions

1. What is the main difference in the types of cabling between building and residential cabling systems?
2. What is asymmetric transmission?
3. Give three examples of asymmetric transmission.
4. What topology does a residential cabling system use?
5. What is a principal advantage of HomePNA?
6. Name five media that CEBus runs on.
7. Name the two main uses of home automation.
8. Research which broadband access services are available for your home: DSL and cable. What is the basic price for the service (with Internet access)?
9. What is the main advantage of G.Lite?
10. Name at least 3 types of "no new wires" technologies for broadband access.
11. What two properties of IEEE 1394 (FireWire) networks makes them particularly well-suited to home networks?
12. How many nodes can be accommodated by a single FireWire bus? By a maximum-size system with bridges?
13. What are the three most common types of Wi-Fi network and what data rates do they support?
14. Name three new access technologies that are just starting to be deployed.
15. TIA-570B requires at least one telecommunications outlet in what types of room?

Appendix A
PREMISES CABLING SPECIFICATIONS

This appendix summarizes selected specifications of most interest to testing and certifying premises cabling according to the requirements of TIA/EIA-568B. It does not include other characteristics of concern to manufacturers in designing and categorizing components.

The information is simply to allow you to do brief comparisons of different types of cables and connecting hardware and to see the differences among components, links, and channels. The TIA/EIA-568B standards contain equations that specify the normative requirements for compliance. The tables below contain sample values that have been calculated from these equations.

UNSHIELDED TWISTED-PAIR CABLE

NEXT Loss

NEXT Loss for Horizontal Cables

Freq.	Pair-to-Pair NEXT Loss (dB)			Power-sum NEXT Loss (dB)	
(MHz)	Cat 3	Cat 5e	Cat 6	Cat 5e	Cat 6
0.772	43.0	67.0	76.0	64.0	74.0
1	41.3	65.3	74.3	62.3	72.3
10	26.3	50.3	59.3	47.3	57.3
16	23.2	47.2	56.2	44.2	54.2
25		44.3	53.3	41.3	51.3
31.25		42.9	51.9	39.9	49.9
62.5		38.4	47.4	35.4	45.4
100		35.3	44.3	32.3	42.3
200			39.8		37.8
250			38.3		36.3

NEXT Loss for Connecting Hardware

Freq.	Pair-to-Pair NEXT Loss (dB)		
(MHz)	Cat 3	Cat 5e	Cat 6
1	58.0	65.0	75.0
10	38.0	63.0	74.0
16	33.9	58.9	69.9
25		55.0	66.0
31.25		53.1	64.1
62.5		47.1	58.1
100		43.0	54.0
200			48.0
250			46.0

NEXT Loss for Permanent Link

Freq. (MHz)	Pair-to-Pair NEXT Loss (dB)			Power-sum NEXT Loss (dB)	
	Cat 3	Cat 5e	Cat 6	Cat 5e	Cat 6
1	40.1	>60	65.0	>57	62.0
10	24.3	48.5	57.8	45.5	55.5
16	21.0	45.2	54.6	42.2	52.2
25		42.1	51.5	39.1	49.1
31.25		40.5	50.0	37.5	47.5
62.5		35.7	45.1	32.7	42.7
100		32.3	41.8	29.3	39.3
200			36.9		34.3
250			35.3		32.7

NEXT Loss for Channel

Freq. (MHz)	Pair-to-Pair NEXT Loss (dB)			Power-sum NEXT Loss (dB)	
	Cat 3	Cat 5e	Cat 6	Cat 5e	Cat 6
1	39.1	>60	65.0	>57	62.0
10	22.7	47.0	56.6	44.0	54.0
16	19.3	43.6	53.2	40.6	50.6
25		40.3	50.0	37.3	47.3
31.25		38.7	48.4	35.7	45.7
62.5		33.6	43.4	30.6	40.6
100		30.1	39.9	27.1	37.1
200			34.8		31.9
250			33.1		30.2

Equal-Level FEXT loss

ELFEXT Loss for Horizontal Cables

Freq. (MHz)	Pair-to-Pair ELFEXT Loss (dB)		Power-sum ELFEXT Loss (dB)	
	Cat 5e	Cat 6	Cat 5e	Cat 6
1	63.8	67.8	60.8	64.8
10	43.8	47.8	40.8	44.8
16	39.7	43.7	36.7	40.7
25	35.8	39.8	32.8	36.8
31.25	33.9	37.9	30.9	34.9
62.5	27.9	31.9	24.9	28.9
100	23.8	27.8	20.8	24.8
200		21.8		18.8
250		19.8		16.8

FEXT Loss for Connecting Hardware

Freq. (MHz)	Pair-to-Pair FEXT Loss (dB)	
	Cat 5e	Cat 6
1	65.0	75.0
10	55.1	63.1
16	51.0	59.0
25	47.1	55.1
31.25	45.2	53.2
62.5	39.2	47.2
100	35.1	43.1
200		37.1
250		35.1

ELFEXT Loss for Permanent Link

Freq. (MHz)	Pair-to-Pair ELFEXT Loss (dB)		Power-sum ELFEXT Loss (dB)	
	Cat 5e	Cat 6	Cat 5e	Cat 6
1	58.6	64.2	55.5	61.2
10	38.6	44.2	35.6	41.2
16	34.5	40.1	31.5	37.1
25	30.7	36.2	27.7	33.2
31.25	28.7	34.3	25.7	31.3
62.5	22.7	28.3	19.7	25.3
100	18.6	24.2	15.6	21.2
200		18.2		15.2
250		16.2		13.2

ELFEXT Loss for Channel

Freq. (MHz)	Pair-to-Pair ELFEXT Loss (dB)		Power-sum ELFEXT Loss (dB)	
	Cat 5e	Cat 6	Cat 5e	Cat 6
1	57.4	63.3	54.4	60.3
10	37.4	43.3	34.4	40.3
16	33.3	39.2	30.3	36.2
25	29.4	35.3	26.4	32.3
31.25	27.5	33.4	24.5	30.4
62.5	21.5	27.3	18.5	24.3
100	17.4	23.2	14.4	20.3
200		17.2		14.2
250		15.3		12.3

Insertion Loss

Insertion Loss for Horizontal Cables

Freq. (MHz)	Insertion Loss (dB)		
	Cat 3	Cat 5e	Cat 6
1	2.6	2.0	2.0
10	9.7	6.5	6.0
16	13.1	8.2	7.6
25		10.4	9.5
31.25		11.7	10.7
62.5		17.0	15.4
100		22.0	19.8
200			29.0
250			32.8

Insertion Loss for Connecting Hardware

Freq. (MHz)	Insertion Loss (dB)		
	Cat 3	Cat 5e	Cat 6
1	0.4	0.1	0.10
10	0.4	0.1	0.10
16	0.4	0.1	0.10
25	0.4	0.2	0.10
31.25		0.2	0.11
62.5		0.3	0.16
100		0.4	0.20
200			0.28
250			0.32

Insertion Loss for Permanent Link and Channel

Freq. (MHz)	Permanent Link Insertion Loss (dB)			Channel Insertion Loss (dB)		
	Cat 3	Cat 5e	Cat 6	Cat 3	Cat 5e	Cat 6
1	3.5	2.1	1.9	4.2	2.2	2.2
10	9.9	6.2	5.6	11.5	7.1	6.3
16	13.0	7.9	7.0	14.9	9.1	8.0
25		10.0	8.9		11.4	10.1
31.25		11.2	10.0		12.9	11.4
62.5		16.2	14.4		18.6	16.5
100		21.0	18.6		24.0	21.3
200			27.4			31.5
250			31.1			35.9

Return Loss for Permanent Link and Channel

Freq. (MHz)	Permanent Link Return Loss (dB)		Channel Return Loss (dB)	
	Cat 5e	Cat 6	Cat 5e	Cat 6
1	17	19.1	17	19.0
10	17	21.0	17	19.0
16	17	20.0	17	18.0
25	16.3	19.0	16	17.0
31.25	15.6	18.5	14.9	16.5
62.5	13.5	16.0	12	14.0
100	12.1	14.0	10	12.0
200		11.0		9.0
250		10.0		8.0

Propagation Delay and Delay Skew for Horizontal Cables and Connecting Hardware

Freq. (MHz)	Cable (3, 5e, 6)		Connecting Hardware	
	Propagation Delay (ns/100m)	Delay Skew (ns/100m)	Propagation Delay (ns)	Delay Skew (ns)
1	570	45	2.5	1.25
10	545	45	2.5	1.25
100	538	45	2.5	1.25
250	536	45	2.5	1.25

Propagation Delay and Delay Skew for Permanent Links and Channels

Permanent Link Propagation Delay (all categories): 498 ns measured at 10 MHz

Permanent Link Delay Skew (all categories): 44 ns

Channel Propagation Delay (all categories): 555 ns measured at 10 MHz

Channel Delay Skew (all categories): 50 ns

Characteristic Impedance

Characteristic impedance: 100 ohms, ± 15%

AUGMENTED CATEGORY 6 (6A) UTP SPECIFICATIONS

The following augmented category 6 specifications for permanent links and channels were taken from Draft 2.0 of Addendum 10 to TIA-568-B.2. These values are provided for information only and may have changed slightly during the ballot process. Refer to the TIA specification for the official values.

Extensions of Category 6 specifications for IL, NEXT, FEXT, and RL

In general, for parameters that are defined for Category 6 (i.e., everything except alien crosstalk), the Category 6A specifications are identical to Category 6 for frequencies up to 250 MHz.

Category 6A Insertion Loss (dB)

Freq. (MHz)	Permanent Link	Channel
≤250	Same as Category 6	
300	32.6	37.3
400	38.4	43.6
500	43.8	49.3

Category 6A NEXT Loss, worst pair-to-pair (dB)

Freq. (MHz)	Permanent Link	Channel	Field Channel
≤250	Same as Category 6		
300	34.0	31.7	31.7
400	29.9	28.7	28.3
500	26.7	26.1	24.9

Because of the presence of an "extra" modular connector at each of the channel when field testing (Figure 6-2), a small allowance is made at frequencies above 300 MHz when making field measurements of either NEXT or PSNEXT.

Category 6A PSNEXT Loss (dB)

Freq. (MHz)	Permanent Link	Channel	Field Channel
≤250	Same as Category 6		
300	31.4	28.8	28.8
400	27.1	25.8	25.3
500	23.8	23.2	21.8

Category 6A ELFEXT, worst pair-to-pair (dB)

Freq. (MHz)	Permanent Link	Channel
≤250	Same as Category 6	
300	14.6	13.7
400	12.1	11.2
500	10.2	9.3

Category 6A PSELFEXT (dB)

Freq. (MHz)	Permanent Link	Channel
≤250	Same as Category 6	
300	11.6	10.7
400	9.1	8.2
500	7.2	6.3

Category 6A Return Loss (dB)

Freq. (MHz)	Permanent Link	Channel
≤250	Same as Category 6	
300	8.4	7.2
400	6.0	6.0
500	6.0	6.0

Category 6A Propagation Delay and Delay Skew

Specifications for propagation delay and delay skew of both permanent links and channels are TBD.

Alien Crosstalk Specifications

The augmented category 6 specification gives requirements for power sum alien near-end crosstalk (PSANEXT) and power sum alien equal-level far-end crosstalk (PSAELFEXT). Since the alien crosstalk requirements have no precedent in Category 6, sample values will be given across the entire frequency band.

In addition to the normal power-sum specifications, there are also requirements for average power-sum alien crosstalk. These specifications represent the average of the four power-sum calculations for each pair in the cable. The intent of this requirement is to ensure that there is not a large variation in alien crosstalk performance for different pairs.

Category 6A PSANEXT Loss (dB)

Freq. (MHz)	PSANEXT Loss		Average PSANEXT Loss	
	Permanent Link	Channel	Permanent Link	Channel
1	TBD	75.0	TBD	81.0
10	TBD	70.0	TBD	71.0
16	TBD	68.0	TBD	69.0
25	TBD	66.0	TBD	67.0
31.25	TBD	65.1	TBD	66.1
62.5	TBD	62.0	TBD	63.0
100	TBD	60.0	TBD	61.0
200	TBD	55.5	TBD	56.5
250	TBD	54.0	TBD	55.0
300	TBD	52.8	TBD	53.8
400	TBD	51.0	TBD	52.0
500	TBD	49.5	TBD	50.5

Category 6A PSAELFEXT (dB)

Freq. (MHz)	PSAELFEXT		Average PSAELFEXT	
	Permanent Link	Channel	Permanent Link	Channel
1	TBD	77.0 (TBD)	TBD	81.0
10	TBD	57.0 (TBD)	TBD	61.0
16	TBD	52.9 (TBD)	TBD	56.9
25	TBD	49.0 (TBD)	TBD	53.0
31.25	TBD	47.1 (TBD)	TBD	51.1
62.5	TBD	41.1 (TBD)	TBD	45.1
100	TBD	37.0 (TBD)	TBD	41.0
200	TBD	31.0 (TBD)	TBD	35.0
250	TBD	29.0 (TBD)	TBD	33.0
300	TBD	27.5 (TBD)	TBD	31.5
400	TBD	25.0 (TBD)	TBD	29.0
500	TBD	23.0 (TBD)	TBD	27.0

OPTICAL FIBERS

Fiber-Optic Cable

Property	**Multimode Fiber**		**Single-Mode Fiber**
	850 nm	**1300 nm**	**1310 nm and 1550 nm**
Bandwidth (MHz-km)	160	500	—
Attenuation (dB/km)	3.5	1.5	1.0 (indoor cable)
			0.5 (outdoor cable)

Fiber-Optic Connecting Hardware

Connector Insertion Loss (dB)		0.75 maximum
Splice Insertion Loss (dB)		0.3 maximum
Durability		500 mating cycles
Return Loss (dB)	Multimode	20
	Single Mode	26

Appendix B
ABBREVIATIONS AND ACRONYMS

ε	dielectric constant (epsilon)
λ	wavelength
Ω	ohms (omega)
μm	micrometer (a millionth of a meter)
ACR	attenuation-to-crosstalk ratio
ADO	auxiliary disconnect outlet
ADSL	asymmetric digital subscriber line
AFEXT	alien far-end crosstalk
AHJ	authority having jurisdiction
ALF	assisted living facility
ANEXT	alien near-end crosstalk
ASCII	American Standard Code for Information Interchange
ATM	asynchronous transfer mode
ATSC	Advanced Television Systems Committee
AUI	attachment unit interface
AWG	American wire gauge
AXT	alien crosstalk
BAS	building automation system
BER	bit error rate
BICSI	Building Industry Consulting Service International
BL	blue
BNC	Bayonet Neil-Concelman (coaxial cable connector)
BPL	broadband over power line
BR	brown
BSS	basic service set
c	speed of light (electromagnetic energy) in free space
CAP	carrierless amplitude and phase
CATV	community antenna television
CBX	computerized branch exchange
CD	compact disk
CEA	Consumer Electronics Association
CEDIA	Custom Electronic Design & Installation Association
CO	central office
CRC	cyclic redundancy check
CSMA/CD	carrier sense, multiple access, with collision detection
dB	decibel
dB_m	decibel referenced to a milliwatt
DBS	direct broadcast satellite
DD	distribution device
DOCSIS	Data-Over-Cable Service Interface Specifications
DS	data-strobe
DSL	digital subscriber line
DSP	digital dignal processing
DWDM	dense wavelength division multiplexing
EBCDIC	Extended Binary Coded Decimal Information Code
ECL	emitter-coupled logic

EF	entrance facility
EIA	Electronics Industries Association
ELFEXT	equal-level far-end crosstalk
EMC	electromagnetic compatibility
EMD	equilibrium mode distribution
EMI	electromagnetic interference
E-O	electrical-to-optical
ER	equipment room (telecommunications)
ESS	extended service set
FC	fibre channel
FCC	Federal Communications Commission
FDDI	fiber distributed data interface
FEXT	far-end crosstalk
FFT	fast Fourier transform
FOTP	fiber-optic test procedure
FTP	foil twisted pair
FTTH	fiber to the home
G	green
Gbps	gigabit per second
GHz	gigahertz
GI-POF	graded index plastic optical fiber
GOF	glass optical fiber
HCP	horizontal connection point
HDTV	high definition TV
HomePNA	home phoneline networking alliance
HPCF	hard polymer-coated fiber
HVAC	heating, ventilation, and air-conditioning
IBSS	independent basic service set
IC	intermediate cross connect
IDC	insulation displacement contact/connector
IDF	intermediate distribution frame
IEC	International Electrotechnical Committee
IEEE	Institute of Electrical and Electronic Engineers
IEEE-SA	IEEE Standards Association
IETF	Internet Engineering Task Force
IL	insertion loss
IP	internet protocol
ISP	internet service provider
ISO	International Standards Organization
ITU	International Telecommunications Union
kHz	kilohertz
km	kilometer
KVM	keyboard/video/mouse
LAN	local area network
LCL	longitudinal conversion loss
LCTL	longitudinal conversion transfer loss
LED	light-emitting diode
m	meter
MAC	media access control
Mbps	megabit per second

MC main cross connect

MC microprocessor-based controller

MDF main distribution frame

MHz megahertz

MIC medium interface connector

MICE mechanical, ingress, climatic, and electromagnetic

MLT-3 multilevel transmission (three level)

MMF multimode fiber

MOCA multimedia over coax alliance

NA numerical aperture

NEC® *National Electrical Code*®

NEXT near-end crosstalk

NFPA National Fire Protection Association

NIC network interface card

NID network interface device

nm nanometer

NMS network management system, network management station

NNW no new wires

NRZ nonreturn to zero

NRZI nonreturn to zero inverted

NTSC National Television Steering Committee

NVP nominal velocity of propagation

O orange

OC outlet cable

ODVA Open DeviceNet Vendors Association

O-E optical-to-electrical

OFSTP optical fiber standard test procedure

OLTS optical loss test set

OSI open systems interconnection

OSP outside plant

OTDR optical time domain reflectometer

PABX private automatic branch exchange

PAL phase alternate line

PBX private branch exchange

PC personal computer

PD Powered device

PF-POF perfluorinated plastic optical fiber

PHY physical signaling sublayer

PLC powerline communication

PMD physical media dependent

PoE power over ethernet

POF plastic optical fiber

PSAELFEXT power-sum alien equal-level far-end crosstalk

PSANEXT power-sum alien near-end crosstalk

PSE power sourcing equipment

PSFEXT power-sum far-end crosstalk

PSELFEXT power-sum equal-level far-end crosstalk

PSNEXT power-sum near-end crosstalk

QoS quality of service

RCLED resonant cavity LED

RF radio frequency

RJ	registered jack
RL	return loss
RZ	return to zero
SAN	storage area network
SCS	structured cabling system
ScTP	screened twisted pair
SDSL	symmetric digital subscriber line
SDTV	standard definition TV
SECAM	sequential couleur a memoire
SMF	singlemode fiber
SMF-PMD	single-mode fiber physical media dependent
SNMP	simple network management protocol
SRL	structural return loss
STB	set-top box
STP	shielded twisted-pair cable
sUTP	screened unshielded twisted-pair cable
TBD	to be determined
Tbps	terabit per second
TCL	transverse conversion loss
TCP	transmission control protocol
TCTL	transverse conversion transfer loss
TE	telecommunications enclosure
TIA	telecommunications industry association
TO	telecommunications outlet
TP-PMD	twisted pair physical media dependent
TR	telecommunications room
TTL	transistor-transistor logic
USOC	universal service order code
UTP	unshielded twisted-pair cable
UWB	ultra wideband
V	volt
VCSEL	vertical cavity surface-emitting laser
VFL	visual fault locator
VoIP	voice over internet protocol
W	white
WAP	wireless access point
WDM	wavelength division multiplexing
Wi-Fi	wireless fidelity
WLAN	wireless local area network
WPAN	wireless personal area network
xDSL	refers to any type of digital subscriber line

Appendix C
GLOSSARY

10BASE-2 Ethernet over thin coaxial cable at 10 Mbps.

10BASE-5 Ethernet over thick coaxial cable at 10 Mbps.

10BASE-F Ethernet over fiber-optic cable at 10 Mbps.

10BASE-T Ethernet over unshielded twisted-pair cable at 10 Mbps.

100BASE-FX Fast Ethernet over fiber-optic cable at 100 Mbps.

100BASE-T4 Fast Ethernet over Category 3 unshielded twisted-pair cable at 100 Mbps and using four cable pairs.

100BASE-TX Fast Ethernet over Category 5 unshielded twisted-pair cable at 100 Mbps and using two cable pairs.

1000BASE-LX Gigabit Ethernet over fiber with a long-wavelength (1300-nm) source.

100BASE-SX Gigabit Ethernet over fiber with a short-wavelength (850-nm) source.

1000BASE-T Gigabit Ethernet over Category 5 unshielded twisted-pair cable at 1000 Gbps.

abandoned cable Communications cable which is not either terminated at both ends (on a connector of other equipment) or labeled for future use.

access method The rules by which a network device gains the right to transmit a communication on the network. Common methods include carrier sense/multiple access, with collision detection, and token passing.

access network The facilities that connect a building or residence to the long-haul networks (telephone and internet.)

aerial cable Telecommunications cable installed above ground on supporting structures such as poles, sides of buildings, etc.

alien crosstalk (AXT) Crosstalk between conductors in different cables.

american wire gauge (AWG) A nominal indication of the diameter of non-ferrous conductors, most commonly used for communication and power cables.

asynchronous Not coordinated in time, used to describe events that happen at random or non-periodic times.

asynchronous transfer mode (ATM) A cell switching network using short, fixed-length cells operating at a basic speed of 155 Mbps, with other speeds including 25 and 51 Mbps to the desktop and 622 Mbps in the backbone.

attenuation A decrease in power from one point to another. See insertion loss.

attenuation-to-crosstalk ratio (ACR) The ratio of the attenuation on a received pair to the crosstalk on the same pair, giving an indication of the margin between the noise level and signal level. The ACR typically decreases with increasing frequency.

authority having jurisdiction (AHJ) The building official, electrical inspector, fire marshal, or other individuals or entities responsible for interpretation and enforcement of local building and electrical codes.

backbone cabling Cable running between buildings or between telecommunications closets. In star networks, the backbone cable interconnects hubs, switches, and similar devices, as opposed to cables running between hub and station. In bus networks, the bus cable.

balanced transmission A mode of signal transmission in which each conductor carries the signal of equal magnitude but opposite polarity. A 5-volt signal, for example, appears as +2.5 volts on one conductor and –2.5 volts on the other.

balun An impedance-matching device that also provides conversion between balanced and unbalanced modes of transmission. The term derives from balanced/unbalanced.

bandwidth The capacity of a communication channel to carry information.

baud The number of signaling elements (i.e., changes of state of the transmission line) transmitted per second. The baud rate is equal to the bit rate *only* if each signaling element represents exactly one bit of information.

bit A contraction of the phrase "binary digit." A bit is the smallest unit of information that can be transmitted, typically represented by a binary value of 1 or 0.

bonding The permanent joining of metallic parts to form an electrically conductive path that will ensure electrical continuity and the capacity to conduct safely any current likely to be imposed.

bridge A device which connects two or more LANs at the data link layer.

buffer The outer coating on an optical fiber that protects the core and cladding from damage.

buried cable Telecommunications cable installed under the ground in direct contact with the soil.

cable modem A device which allows a PC or home network to attach to the cable TV network and receive broadband data.

campus subsystem Cabling for interconnecting buildings which are located relatively close together and have continuous legal connectivity (i.e., without right-of-way issues.)

carrier sense/multiple access, with collision detection (CSMA/CD) A networks access method used by Ethernet in which a station listens for traffic before transmitting. If two stations transmit simultaneously, a collision is detected and both stations wait a brief time before attempting to transmit again.

cell The area in which a wireless access point (WAP) is located.

central office In the telephone network, a location which houses local switching equipment.

channel The end-to-end transmission path between two points at which application-specific equipment is connected.

characteristic impedance The impedance presented at the input of an infinitely long transmission line.

cladding The outer concentric layer of an optical fiber, with a lower refractive index so that light is guided through the core by total internal reflection.

coaxial cable A cable in which the center, signal-carrying conductor is concentrically centered within an outer shield and separated from the conductor by a dielectric. The structure, from inside to outside, is center conductor, dielectric, shield, outer jacket.

common-mode transmission A (usually undesirable) method of signal transmission where each conductor in a pair carries a signal which is equal in both amplitude and polarity.

conduit A raceway of circular cross section. Also used more generally to indicate an enclosed cable pathway.

convergence The merging of telephony, computing, and television into a single digital network.

core The central light-carrying portion of an optical fiber.

coverage area The area served by the BAS devices attached to a horizontal cable.

crimp connector A connector that is attached to a conductor by mechanical force.

crosstalk The unwanted transfer of energy from one circuit to another. Crosstalk can be measured at the same (near) end or far end with respect to the source of energy.

cutoff wavelength For a single-mode fiber, the wavelength above which the operation switches from multimode to single-mode propagation.

data center A building or portion of a building whose primary function is to house a computer room and its support areas

data connector A four-position connector for 150-ohm STP and used primarily in Token Ring networks.

DC loop resistance The total DC resistance of a cable. For a twisted-pair cable, it includes the round-trip resistance, down one wire of the pair and back up the other wire.

decibel (dB) The logarithmic ratio of two powers, voltages, or currents, often used to indicate either gain and loss in a circuit or to compare the power in two different circuits. $dB = 10 \log_{10} (P_1/P_2) = 20 \log10 (V_1/V_2) = 20 \log_{10} (I_1/I_2)$. Power is most often used in premises cabling.

delay skew The difference in propagation time for signals on the different pairs in a cable.

dense wavelength-division multiplexing (DWDM) A form of optical multiplexing in which different wavelengths (signals) are transmitted simultaneously throgh a fiber.

differential signal Another name for balanced signal.

differential-mode transmission A method of signal transmission where each conductor in a pair carries a signal that is equal in amplitude and opposite in polarity so that the net current at any instant of time is zero.

digital subscriber line A technology which uses the existing copper loop plant to provide broadband access by using frequencies above the voice band.

direct sequence spread spectrum (DSSS) A spread-spectrum technique in which the data to be transmitted is chopped up into small pieces and spread out across a wide frequency band.

dispersion A general term for those phenomena that cause a broadening or spreading of light as it propagates through an optical fiber. The two main types are modal and material.

dwell time In a FHSS system, the amount of time spent in each channel before hopping to another channel.

equal-level far-end crosstalk (ELFEXT) Crosstalk that is measured on the quiet line at the far end from the energy source on the active line relative to the received signal level. ELFEXT = FEXT – attenuation.

ethernet The most widely installed LAN, specified in IEEE 802.3

far-end crosstalk (FEXT) Crosstalk that is measured on the quiet line at the far end from the energy source on the active line.

fiber distributed data interface (FDDI) A token-passing ring network originally designed for fiber optics and having a data transmission rate of 100 Mbps.

fiber-optic cable An assembly of multiple optical fibers under a common outer jacket.

FireWire another name for an IEEE 1394 network

foil twisted pair (FTP) Cable with 4 twisted pairs surrounded by an overall foil screen. Sometimes called Screened Twisted Pair (ScTP).

frequency hopping spread spectrum (FHSS) A spread-spectrum technique in which the transmitter and receiver "hop" from one frequency channel to another according to a specified algorithm. At each channel, a short burst of narrowband data is transmitted.

geostationary satellite A satellite orbiting directly over the equator at an altitude of about 22,300 miles, where the orbital period is 24 hours. The satellite appears to be in a fixed position from any point on the earth's surface.

graded-index fiber An optical fiber whose core has a nonuniform index of refraction. The core is composed of concentric rings of glass whose refractive indices decrease from the center axis. The purpose is to reduce modal dispersion and thereby increase fiber bandwidth. The 62.5/125-μm fiber recommended for premises cabling has a graded index.

grounding Establishing a conducting connection between an electrical circuit or equipment and the earth.

home automation system (HAS) Equipment and infrastructure which supports automatic control of home services, such as climate and lighting control, security and fire alarms, etc.

homeRF A specification for a wireless home networking technology issued by an industry consortium. It has been largely rendered moot by the success of IEEE 802.11.

horizontal cabling That portion of the premises cabling that runs between the telecommunications outlet of the work area and the horizontal cross connect in the telecommunications closet.

horizontal cross connect The cross connect that connects horizontal cable to the backbone or equipment.

hub A device at the center of a star-wired network. Named in analogy to the center of a wheel where the spokes come together.

hybrid cable An assembly consisting of two or more cables of the same or different types under one overall sheath.

hybrid circuit An electronic circuit that separates out the signals going in different directions on a bidirectional transmission line.

hybrid cord A fiber cord with different types of connectors on the two ends.

IEEE 802.3 The standards committee defining Ethernet networks; networks meeting these standards.

index of refraction The ratio of the speed of light in a medium to the speed of light in a vacuum.

industrial cabling Cabling installed in factories and similar environments which are used for manufacturing, processing, refining, etc.

insertion loss (IL) The loss of power that results from inserting a component into a circuit. For example, a connector causes insertion loss across the interconnection (in comparison to a continuous cable with no interconnection). See attenuation.

insulation displacement connector (IDC) A connector which pierces the insulation and attaches to the copper conductor without stripping back the insulation.

intermediate cross connect A cross connect between the first and second levels of backbone cabling.

internet telephony Another name for Voice over IP (VoIP.)

isochronous Uniform in time and recurring at regular intervals.

jacket The outer covering of a cable. Sometimes called a sheath.

jumper A cable without connectors, typically used at the cross connect to join circuits. Similar to a patch cable (which has connectors).

KVM switch A keyboard/video/mouse switch allows a single keyboard, monitor, and mouse to be connected to any of several computers. KVM switches are often used in data centers.

laser A light source producing, through stimulated emissions, coherent, near-monochromatic light—an intense, narrow beam at a single frequency. Lasers in fiber optics are solid-state semiconductors.

LC connector A small-form-factor fiber-optic connector using a 125-mm ceramic ferrule. Two connectors gang together to form a duplex connector.

legacy Technology, applications, and/or data that have been inherited from a previous generation. For IEEE 1394b, 1394a and 1394-1995 represent legacy systems. For 1000BASE-T, 100BASE-T and earlier versions of Ethernet represent legacy systems.

light-emitting diode (LED) A semiconductor that spontaneously emits light when current passes through it.

link power budget An accounting of all the losses and margins along an optical path.

local area network A group of computers and other devices that share a communications network (wired or wireless) within a small geographic area such as an office building or home.

logarithm The logarithm of a number is the power to which the base (10 for dB calculations) must be raised to equal the number. For example, if $y = 10^x$, then $x = \log(y)$.

longitudinal conversion loss (LCL) One of the parameters describing the balance of a UTP cable, consisting of the ratio of the differential voltage at the transmitter to the longitudinal voltage at the transmitter.

longitudinal conversion transfer loss (LCTL) One of the parameters describing the balance of a UTP cable, consisting of the ratio of the differential voltage at the receiver to the longitudinal voltage at the transmitter.

longitudinal Another name for unbalanced or common-mode transmission.

main cross connect A first-level cross connect for backbone cable, entrance cable, and equipment. The main cross connect is the top level of the premises cabling tree.

material dispersion Dispersion that results from each wavelength travelling at a different speed through an optical fiber. Also called chromatic dispersion.

MICE Levels Characterization of an industrial cabling environment according to its Mechanical, Ingress, Climatic, and Electromagnetic characteristics.

modal dispersion Dispersion that results from the different transit lengths of different propagating modes in a multimode optical fiber.

mode field diameter The diameter of optical energy in a single-mode fiber. Because the mode-field diameter is larger than the core diameter, it replaces core diameter as a practical parameter.

modular jack The equipment-mounted half of a modular interconnection.

modular plug The cable-mounted half of a modular interconnection. While available in 4, 6, and 8 positions, the 8-position version is specified for premises cabling.

MT-RJ connector A small-form-factor fiber-optic connector offering a duplex interface in a MT-style ferrule.

multimode fiber (MMF) An optical fiber that supports more than one propagating mode.

near-end crosstalk (NEXT) Crosstalk that is measured on the quiet line at the same end as the source of energy on the active line.

nominal velocity of propagation (NVP) The velocity of propagation in a medium relative to the velocity of propagation in a vacuum.

numerical aperture (NA) The "light-gathering ability" of an optical fiber, defining the maximum angle to the fiber axis at which light will be accepted and propagated.

OptiJack connector A small form-factor fiber connector, originally designed by Panduit.

orthogonal frequency division multiplexing (OFDM) A spread spectrum technique in which the frequency band is divided up into multiple overlapping channels.

outside plant (OSP) Telecommunications infrastructure (cable, pathways, and spaces) that is installed exterior to buildings.

open standard interconnect (OSI) A seven-layer ISO model for defining a communications network. Premises cabling deals with the lowest layer, the physical layer.

optical fiber A thin glass or plastic filament used that propagates light. The signal-carrying part of a fiber-optic cable.

optical time-domain reflectometry A method for evaluating optical fiber based on detecting and measuring backscattered (reflected) light. Used to measure fiber length and attenuation, evaluate splice and connector joints, locate faults, and certify cabling systems.

passive link An optical fiber test configuration consisting of a duplex run of fiber and the pair of connectors on each end.

patch cable A cable terminated with connectors and used to perform cross connects. Similar to a jumper cable (which has no connectors).

patch cord A patch cable.

patch panel A passive device, typically an arrangement of feed-through connectors, to allow circuit connections and rearrangements by plugging and unplugging patch cords.

permanent link A test configuration for a link that excludes test cords and patch cords.

phase alternate Line (PAL) The analog television format used in most of Europe, Asia, Australia, and Africa.

plastic optical fiber (POF) An optical fiber made of plastic rather than glass.

plenum The air-handling space between walls, under structural floors, and above drop ceilings used to circulate and otherwise handle air in a building. Such spaces are considered plenums only if they are used for air handling.

plenum cable A flame- and smoke-retardant cable that can be run in plenums without being enclosed in a conduit.

power sum In NEXT and FEXT testing, a measurement of crosstalk that takes into account the sum of crosstalk from all other conductors coupling noise onto a quiet line.

power sum equal-level far-end crosstalk (PSELFEXT) A computation of the unwanted signal power coupling from multiple transmitters at the near-end into a pair measured at the far-end and normalized to the received signal level.

power sum near-end crosstalk (PSNEXT) A computation of the unwanted signal power coupling from multiple transmitters at the near-end into a pair measured at the near-end.

propagation delay The time for a signal to travel from one end of a transmission path to the other end.

protocol adaptation layer (PAL) A protocol layer that interfaces between a higher layer designed for one protocol and a lower layer that provides a different set of services.

punch-down To terminate a cable on an IDC connector.

quad cable A cable consisting of 2 untwisted pairs. Not compliant with TIA cabling standards.

quality of service (QoS) The concept of providing guaranteed bandwidth and low delay in asynchronous networks.

raceway Any channel designed to hold cables.

repeater A device that receives, amplifies (and sometimes reshapes), and retransmits a signal. It is used to boost signal levels and extend the distance a signal can be transmitted. It can physically extend the distance of a LAN or connect two LAN segments.

return loss (RL) The ratio of the outgoing signal power to the reflected signal power.

ring network A network topology in which terminals are connected in a point-to-point serial fashion in an unbroken circular configuration. Many logical rings are wired as a star for greater reliability.

riser A backbone cable running vertically between floors.

router A device which connects two or more LANs at the network layer.

SC connector A fiber-optic connector having a 2.5-mm ferrule, push-pull latching mechanism, and the ability to be snapped together to form duplex and multifiber connectors.

SC-DC connector A small form-factor fiber connector, originally designed by IBM and Siecor.

scrambling The process of converting a data stream into a pseudo-random stream with more desirable statistical properties before transmitting.

screened twisted pair (ScTP) see Foil Twisted Pair (FTP).

sequential couleur a memoire (SECAM) An analog television format developed in France.

sheath: see Jacket.

shielded twisted pair (STP) A cable consisting of 2 twisted pairs, each with an individual shield, covered by an overall screen and a jacket.

signal-to-noise ratio (SNR) A metric of signal quality calculated by dividing the (desired) signal power by the (undesired) noise power.

single-mode fiber (SMF) An optical fiber that supports only one mode of light propagation above the cutoff wavelength. Single-mode fibers have extremely high bandwidths and allow very long transmission distances.

SONET Synchronous optical network, an international standard for fiber-optic-based digital telephony.

splice: A permanent joining of fiber conductors.

spread spectrum A form of wireless communication which uses a wide-bandwidth, low-power signal.

ST connector A popular fiber-optic connector having a 2.5-mm ferrule and a bayonet latching mechanism. ST is a trademark of Lucent Technologies.

star network A network in which all stations are connected through a single point. Star configurations tend to be reliable.

step-index fiber An optical fiber, either multimode or single mode, in which the core's refractive index is uniform throughout so that a sharp demarcation or step occurs at the core-to-cladding interface. Step-index multimode fibers typically have lower bandwidths than graded-index multimode fibers.

stranded conductor A conductor composed of several small wires. Typically used in cords for better flexibility.

strength member That part of the fiber-optic cable that increases the cable's tensile strength and serves as a load-bearing component. Usually made of Kevlar aramid yarn, fiberglass filaments, or steel strands.

structured cabling system (SCS) A cabling system that is designed to support a range of applications with standard interfaces and specified transmission performance.

structural return loss (SRL) A measure of the impedance uniformity of a cable. It measures energy reflected due to structural variations in the cable. A higher SRL number indicates better performance (more uniformity and lower reflections).

telecommunications enclosure (TE) A case or housing for telecommunications equipment, cable terminations, and cross-connect cabling.

telecommunications closet Obsolete (though still frequently used) terminology—replaced by 'telecommunications room.'

telecommunications outlet (TO) A connecting device (usually a modular jack or coaxial connector) in the home on which the outlet cable terminates.

telecommunications room (TR) An enclosed space for housing telecommunications equipment, cable terminations, and cross-connect cabling—the location of the horizontal cross-connect. Formerly known as the 'telecommunications closet.'

teleworking Doing work outside the traditional workplace. This could be a traditional telecommuting environment where the employee works from home instead of traveling to an office, or it could involve working on the road while traveling, working at a customer's site, etc.

thicknet A popular term for an IEEE 802.3 10BASE-5 network.

thinnet A popular term for an IEEE 802.3 10BASE-2 network.

time-domain reflectometry A method for evaluating cables based on detecting and measuring reflected energy from impedance mismatches. Used to measure cable length and attenuation, evaluate splice and connector joints, locate faults, and certify cabling systems. See *optical time-domain reflectometry.*

tiny TR Another name for telecommunications enclosure

token passing A network access method in which a station must wait to receive a special token frame before transmitting.

token ring A token-passing ring network, such as 802.5 or FDDI. Often capitalized to refer specifically to IEEE 802.5 networks.

total internal reflection The principle by which light travels through an optical fiber. If the light strikes the core-cladding interface at an angle greater than the critical angle, it will be totally reflected and will propagate down the fiber.

transverse conversion loss (TCL) One of the parameters describing the balance of a UTP cable, consisting of the ratio of the longitudinal voltage at the transmitter to the differential voltage at the transmitter.

transverse conversion transfer loss (TCTL) One of the parameters describing the balance of a UTP cable, consisting of the ratio of the longitudinal voltage at the receiver to the differential voltage at the transmitter.

twisted-pair cable A cable having the conductors of individual pairs twisted around one another to increase noise immunity. Twisted-pair cables operate in the balanced mode.

ultra wideband (UWB) A wireless technology that uses a very wide frequency band to transmit at low power and high data rates for short distances.

underground cable A telecommunications cable installed under the ground in a trough or duct that isolates the cable from direct contact with the soil.

unshielded twisted pair cable Cable consisting of 8 insulated copper conductors twisted together into 4 pairs without any screens or shields.

velocity of propagation The speed of electromagnetic energy in a medium (including copper and optical cables) in relationship to its speed in free space. Usually given in terms of a percentage. Test devices use velocity of propagation to measure a signal's transit time and thereby calculate the cable's length.

vertical-cavity surface-emitting laser (VCSEL) A low-cost laser that emits light from its surface rather than from its edge and well suited to operation with multimode fibers.

voice over IP (VoIP) A technology that allows voice calls to be digitized and transmitted over the internet using the IP protocol. Also referred to as internet telephony.

volition connector A ferruleless small form-factor fiber connector, originally designed by 3M.

wavelength division multiplexing (WDM) Transmitting multiple wavelengths of light through a fiber to increase the bandwidth. See CWDM and DWDM.

Wi-Fi protected access (WPA) An encryption system for wireless networks which is supposed to be more secure and easier to use than WEP.

WiFi Short for "wireless fidelity"—another name for an IEEE 802.11 network.

wireless access point (WAP) In a WLAN, a station which transmits to and receives from other stations to provide connectivity between them. The WAP may also serve as a interconnection point between the WLAN and other networks.

wireless equivalent privacy (WEP) A protocol specified in IEEE 802.11b to provide wireless networks with a degree of privacy and security equivalent to what is generally expected with a wired network.

wireless local area network (WLAN) A local area network that uses wireless transmission to communicate between nodes.

wireless personal area network (WPAN) A wireless network that allows the interconnection of information technology devices within the range of an individual person (typically about 10 meters.)

wiring closet A general term for any room housing equipment or cross connects. See telecommunications room.

work area That area of the premises cabling where users are located; the area from the communications outlet to the equipment connected to the premises cabling. Loosely, an office, cubicle, and so forth.

Appendix D
CABLE COLOR CODES

CABLE COLOR CODES

4-Pair UTP

The preferred color code is for both backbone and horizontal four-pair cable. The alternative is for stranded patch cable only. The chart also shows the termination positions for modular plugs.

Pair	Preferred		Alternative (Stranded Patch Cable)		Connector Pins	
	Color	**Abbreviation**	**Color**	**Abbreviation**	**T568A**	**T568B**
Pair 1	White-Blue	W-BL	Green	G	5	5
	Blue	Blue	Red	R	4	4
Pair 2	White-Orange	W-O	Black	BK	3	1
	Orange	O	Yellow	Y	6	2
Pair 3	White-Green	W-G	Blue	BL	1	3
	Green	G	Orange	O	2	6
Pair 4	White-Brown	W-BR	Brown	BR	7	7
	Brown	BR	Slate	S	8	8

25-Pair UTP

TIA/EIA-568B color coding of 25-pair cable is according to ANSI/ICEA S-80-576, which is based on the industry-standard Western Electric scheme. Each group of five pairs has a subgroup color that is matched to a standard set of five colors. The table also shows the standard terminating 25-pair cable in miniature ribbon connectors and wiring blocks.

Subgroup Color	Pair	Color	Abbreviation	Connector Pins	
				Miniature Ribbon Connector	Wiring Block
White	Pair 1	White-Blue Stripe	W-BL	1	1
		Blue-White Stripe	BL-W	26	2
	Pair 2	White-Orange Stripe	W-O	2	3
		Orange-White Stripe	O-W	27	4
	Pair 3	White-Green Stripe	W-G	3	5
		Green-White Stripe	G-W	28	6
	Pair 4	White-Brown Stripe	W-BR	4	7
		Brown-White Stripe	BR-W	29	8
	Pair 5	White-Slate Stripe	W-S	5	9
		Slate-White Stripe	S-W	30	10
Red	Pair 6	Red-Blue Stripe	R-BL	6	11
		Blue-Red Stripe	BL-R	31	12
	Pair 7	Red-Orange Stripe	R-O	7	13
		Orange-Red Stripe	O-R	32	14
	Pair 8	Red-Green Stripe	R-G	8	15
		Green-Red Stripe	G-R	33	16
	Pair 9	Red-Brown Stripe	R-BR	9	17
		Brown-Red Stripe	BR-R	34	18
	Pair 10	Red-Slate Stripe	R-S	10	19
		Slate-Red Stripe	S-R	35	20
Black	Pair 11	Black-Blue Stripe	BK-BL	11	21
		Blue-Black Stripe	BL-BK	36	22
	Pair 12	Black-Orange Stripe	BK-O	12	23
		Orange-Black Stripe	O-BK	37	24
	Pair 13	Black-Green Stripe	BK-G	13	25
		Green-Black Stripe	G-BK	38	26
	Pair 14	Black-Brown Stripe	BK-BR	14	27
		Brown-Black Stripe	BR-BK	39	28
	Pair 15	Black-Slate Stripe	BK-S	15	29
		Slate-Black Stripe	S-BK	40	30

Subgroup Color	Pair	Color	Abbreviation	Connector Pins	
				Miniature Ribbon Connector	Wiring Block
Yellow	Pair 16	Yellow-Blue Stripe	Y-BL	16	31
		Blue-Yellow Stripe	BL-Y	41	32
	Pair 17	Yellow-Orange Stripe	Y-O	17	33
		Orange-Yellow Stripe	O-Y	42	34
	Pair 18	Yellow-Green Stripe	Y-G	18	35
		Green-Yellow Stripe	G-Y	43	36
	Pair 19	Yellow-Brown Stripe	Y-BR	19	37
		Brown-Yellow Stripe	BR-Y	44	38
	Pair 20	Yellow-Slate Stripe	Y-S	20	39
		Slate-Yellow Stripe	S-Y	45	40
Violet	Pair 21	Violet-Blue Stripe	V-BL	21	41
		Blue-Violet Stripe	BL-V	46	42
	Pair 22	Violet-Orange Stripe	V-O	22	43
		Orange-Violet Stripe	O-V	47	44
	Pair 23	Violet-Green Stripe	V-G	23	45
		Green-Violet Stripe	G-V	48	46
	Pair 24	Violet-Brown Stripe	V-BR	24	47
		Brown-Violet Stripe	BR-V	49	48
	Pair 25	Violet-Slate Stripe	V-S	25	49
		Slate-Violet Stripe	S-V	50	50

Multifiber Cables

TIA/EIA-598B defines color coding for multifiber cables. Either the fiber or the cable tube can be colored.

Fiber	Fiber/Tube Color	Fiber	Fiber/Tube Color
1	Blue	13	Blue-Black Stripe
2	Orange	14	Orange-Black Stripe
3	Green	15	Green-Black Stripe
4	Brown	16	Brown-Black Stripe
5	Slate	17	Slate-Black Stripe
6	White	18	White-Black Stripe
7	Red	19	Red-Black Stripe
8	Black	20	Black-Black Stripe
9	Yellow	21	Yellow-Black Stripe
10	Violet	22	Violet-Black Stripe
11	Rose	23	Rose-Black Stripe
12	Aqua	24	Aqua-Black Stripe

STP

The following is the color code information for 150-ohm shielded twisted-pair cable.

Pair	Color Code
Pair 1	Red
	Green
Pair 2	Orange
	Black

Appendix E
FURTHER READING

What follows is a brief list of books that expand on many issues covered in this book. The list leans toward application-oriented material.

Equally important are the standards that drive premises cabling. Standards can be ordered from

American National Standards Institute (ANSI)
25 West 43rd St.
New York, NY 10036
212-642-4900
www.ansi.org

Global Engineering Documents
15 Inverness Way East
Englewood, CO 80112
800-854-7179
303-397-7956
www.global.ihs.com

STANDARDS

TIA/EIA-568B.1: Commercial Building Telecommunications Cabling Standard Part 1: General Requirements

TIA/EIA-568-B.2: Commercial Building Telecommunications Cabling Standard Part 2: Balanced Twisted-Pair Cabling Components

TIA/EIA-568-B.2-1: Commercial Building Telecommunications Cabling Standard—Part 2: Balanced Twisted Pair Components—Addendum 1—Transmission Performance Specifications for 4-Pair 100 Ohm Category 6 Cabling

TIA/EIA-568B-B.3: Commercial Building Telecommunications Cabling Standard Part 3: Optical Fiber Cabling Components Standard

TIA/EIA-568-B.2-10: Transmission Performance Specifications for 4-pair 100Ω Augmented Category 6 Cabling

TIA/EIA-569A: Commercial Building Standard for Telecommunications Pathways and Spaces

TIA/EIA-570B: Residential Telecommunications Cabling Standard

TIA/EIA-606A: Administration Standard for Commercial Telecommunications Infrastructure

J-STD-607-A: Commercial Building Grounding (Earthing) and Bonding Requirements for Telecommunications

TIA/EIA-758A: Customer-Owned Outside Plant Telecommunications Infrastructure Standard

ISO/IEC 11801 Revision 02: Generic Cabling for Customer Premises

TIA TSB-140: Additional Guidelines for Field-Testing Length, Loss, and Polarity of Optical Fiber Cabling Systems

BOOKS

Bartnikas, R. & Srivastava, K. D. eds. *Power and Communication Cables: Theory and Applications.* (New York: IEEE Press, 2000). ISBN 0-7803-1196-5.

Baxter, Les (2005). *Residential Networks.* Albany, NY: Delmar. ISBN 1401862675.

BICSI (2002). *Residential Network Cabling.* New York: McGraw-Hill. ISBN 0-07-138211-9.

BICSI (2004). *Information Transport Systems Installation Manual* (4th ed.) New York: McGraw-Hill. ISBN 1-928886-25-6.

Elliott, Barry J. (2000). *Cable Engineering for Local Area Networks,* Cambridge, England: Woodhead Publishing Ltd. ISBN 0-8247-052504.

Gardiol, Fred E. (1987) *Lossy Transmission Lines.* Norwood, MA: Artech House. ISBN 0-89006-198-X.

Gast, Matthew S. (2002). *802.11 Wireless Networks: The Definitive Guide.* Sebastopol, CA: O'Reilly & Associates. ISBN 0596001835.

Hecht, Jeff (2001). *Understanding Fiber Optics* (4th ed.) Upper Saddle River, NJ: Prentice-Hall. ISBN 0130278289.

Hughes, Harold. *Telecommunications Cables: Design, Manufacture, and Installation.* New York: John Wiley & Sons, 1997. ISBN 0-471-97410-2.

Kartalopoulos, Stamatios V. *Introduction to DWDM Technology: Data in a Rainbow.* New York: IEEE Press, 2000. ISBN 0-7803-5399-4.

Kipp, Scott (2003). *Broadband Entertainment: Digital Audio, Video and Gaming in Your Home.* Westminister, CO, All Digital. ISBN 0974141011.

LaRocca, James & LaRocca, Ruth (2002). *802.11 Demystified.* New York: McGraw-Hill. ISBN 0-07-138528-2.

National Fire Protection Association, Inc. (2005). National Electrical Code, 2005 Edition. Quincy, MA: National Fire Protection Association, Inc. ISBN 0-87765-625-8.

Spurgeon, Charles E. (2000). *Ethernet – The Definitive Guide.* Sebastopol, CA: O'Reilly & Associates. ISBN 1-56592-660-9.

Sterling, Donald J. Jr. (2004). *Technician's Guide to Fiber Optics* (4th ed.) Albany, NY: Delmar. ISBN 1401812708.

Sterling, Donald J. Jr. and Wissler, Steven P. (2003). *The Industrial Ethernet Networking Guide.* Albany, NY: Delmar. ISBN 0-7668-4210-X.

Tanenbaum, Andrew S. (2002). *Computer Networks* (4th ed.) Upper Saddle River, NJ: Prentice-Hall. ISBN 0130661023.

Appendix F
THE DECIBEL

Objectives

The decibel (dB) is referred to in many places in this book. After reading this chapter, you will be able to:

- Define a decibel (dB).
- Understand when it is appropriate to use decibels.
- Do simple calculations with decibels.

Introduction

The decibel (dB) is commonly used, but not always well understood. This chapter will review the history and purpose of decibels and the mathematics of how decibels are used.

The decibel, as the name implies, is 1/10 of a Bel, which is named for Alexander Graham Bell[1] who first studied the logarithmic nature of human hearing. The Bel is too large a unit for most purposes, so the decibel is almost universally used instead.

Unlike most other units, the decibel is not an absolute measurement, but represents a ratio of two power measurements[2]. Decibels use a logarithmic scale, which means that a large signal range can be expressed with relatively small numbers. Decibels were originally defined for use with audio power measurements, because the human ear responds to sound intensity in a roughly logarithmic manner. Their use has since been extended to many other fields. The decibel is used extensively in communication systems for various purposes such as quantifying optical intensity, amplification, attenuation, crosstalk, etc.

What is a Decibel?

A dB is defined as a logarithmic ration of two power measurements, as indicated in the following equation:

$$dB = 10 \log \left(\frac{P_1}{P_2}\right)$$

If the dB value is known, the power ratio can be determined using the equation:

$$\left(\frac{P_1}{P_2}\right) = 10^{\left(\frac{dB}{10}\right)}$$

[1] *The term "Bel" is capitalized because it comes from a proper name. Similarly dB and deciBel should also be capitalized, although the latter is often written without the capital B (decibel).*

[2] *The decibel can also be used for some quantities other than power, such as voltage and current, with certain restrictions, such as the voltage or current measurements must be made relative to the same impedance.*

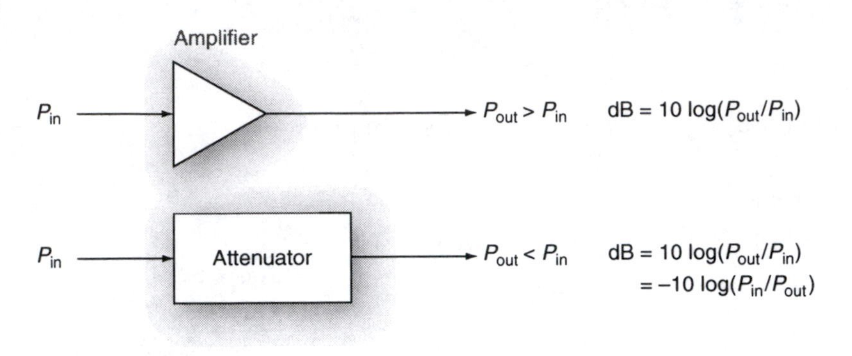

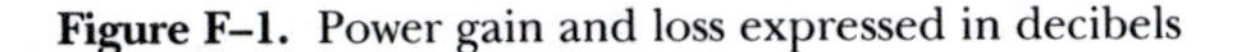

Figure F–1. Power gain and loss expressed in decibels

A common use of decibels is to express the gain or attenuation of an electrical circuit, as shown in Figure F-1. Note that if the ratio P_1/P_2 is less than 1 (i.e., an attenuation rather than a gain) the logarithm becomes negative[3].

Let's look at a couple of typical examples to get a feel for what a figure expressed in dB means. If the ratio $P_1/P_2 = 2$, that corresponds to a dB value of 10log2 which equals 3.01, or roughly 3 dB. Similarly, if $P_1/P_2 = 1/2$, then the dB value is $10\log(1/2) = -10\log(2) = -3$ dB. So a doubling of power corresponds to 3 dB while a halving of power corresponds to −3 dB. Similarly, a power gain of 10 corresponds to 10 dB, while a gain of 100 corresponds to 20 dB, etc.

Why use Decibels?

Decibels have several properties that make them very useful.

- First, they can cover a very large range with relatively small, easy use numbers. In other words, decibels allow a quantity with a large dynamic range to be described by a relatively small number.
- Second, they are excellent for comparing two different quantities, regardless of how large or small the quantities are.
- Finally, they make it very easy to do mathematics when the performance of each component in a system is described in decibels. The total gain or loss of components in series can be easily calculated by adding the decibel value of each component.

[3]*Attenuation can also be expressed as a positive number if this is obvious from the context. For example, if is proper to say "The NEXT loss is 40 dB" because the negative sign is indicated by the term "loss."*

If several amplifiers and/or attenuators are placed in series with each other, the power gain or loss of the entire group is simply the arithmetic sum of the dB rating of each element. For example, if amplifiers with gains of 2, 10, and 20 are connected together, the resultant circuit has a gain of 3 + 10 + 13 = 26 dB. Adding the decibel values is usually easier than multiplying the gains, which results in a value of 400.

Expressing power levels in decibels can also give some insight into when a change in level is significant. For audible and visual signals, a change of 3 dB (or a doubling of signal power) is usually needed in order to be perceived as a significant change. Consider the example of a 3-way light bulb, which is found in many table lamps. Such lamps often have wattage values of 50, 100, and 150 for the 3 positions of the lamp switch. When going from 50 to 100 watts, the change is very obvious. However, when going 100 to 150 watts, the change is barely noticeable. If you entered the room and were asked which wattage the lamp was set for, it is very easy to tell if it is 50 watts. However, most people cannot reliably distinguish between 100 and 150 watts. The reason for this is apparent if we consider the values in decibels. Even though the difference is 50 watts in either case, going from 50 to 100 is a change of 3 dB, while going from 100 to 150 is a change of only 1.8 dB. Due to the logarithmic response of our eyes, this latter change is barely noticeable.

Relative Decibels

Instead of using a ratio of two variable power levels, such as the input and output of an amplifier, a decibel ratio can be calculated using a fixed power level in the denominator of the equation. For communication signals, a fixed value of 1 mW is often used. Decibels calculated in this way are referred to as "decibels relative to a milliwatt", or dB_m. A dB_m is defined by the equation $10\log(P_1/1\text{ mW})$. A power level of 2 mW would be referred to as 3 dB_m. Similarly, 1 watt (1000 mW) can be referred to as 30 dB_m.

There are many other types of relative dB used in different fields, such as dB_{rnc} in telecommunications and dB_V and dB_u in audio engineering. A more recent use of the decibel concept is dB_{FS}, or decibels compared to full scale, which is used in digital audio systems.

When measuring the loudness of audible sounds, the audio power level is compared to the faintest sound which the human ear can detect (known as the threshold of hearing, or TOH) and sound intensity decibels are calculated using the ratio of the sound waves to the TOH, which is about 10^{-12} W/m^2. At the high end of the audio range, the threshold of pain is around 130 dB, or about 10^{13} times as large.

Common dB values

The most commonly encountered decibel values are 3dB and 10 dB (which represent factors of 2 and 10, respectively.) Many other decibel values can be easily calculated. 6 dB, for example, corresponds to a power ratio of 4 and 13 dB corresponds to a power ratio of 20 (because they are 3 + 3 dB and 10 + 3 dB, respectively.)

The tables below show some common decibel values and the power ratios they correspond to. Figure F-2 shows power ratios greater than 1 which correspond to positive decibel values and Figure F-3 shows power ratios less than 1 which correspond to negative decibel values.

dB	Power Ratio	dB	Power Ratio
0	1.0	15	31.6
1	1.3	16	39.8
2	1.6	17	50.1
3	2.0	18	63.1
4	2.5	19	79.4
5	3.2	20	100.0
6	4.0	25	316.2
7	5.0	30	1,000
8	6.3	40	10,000
9	7.9	50	100,000
10	10.0	60	1,000,000
11	12.6	70	10,000,000
12	15.8	80	100,000,000
13	20.0	90	1,000,000,000
14	25.1	100	10,000,000,000

Figure F–2. Positive dB values

dB	Power Ratio	dB	Power Ratio
0	1.0	−15	0.0316
−1	0.794	−16	0.0251
−2	0.631	−17	0.0200
−3	0.501	−18	0.0158
−4	0.398	−19	0.0126
−5	0.316	−20	0.0100
−6	0.251	−25	0.0032
−7	0.200	−30	0.0010
−8	0.158	−40	0.0001
−9	0.126	−50	0.00001
−10	0.100	−60	0.000001
−11	0.079	−70	0.0000001
−12	0.063	−80	0.00000001
−13	0.050	−90	0.000000001
−14	0.040	−100	0.0000000001

Figure F–3. Negative dB values

Use of Decibels in Communication Systems

The properties of decibels are ideally suited for representing many different quantities in communications systems.

Decibels in Fiber Optic Systems

In fiber optic systems, decibels are used extensively for specifying optical power levels, often in dB_m. The attenuation, or optical loss, of various components such as optical fibers, splitters, connectors, splices, and other passive components are usually specified in decibels. The gain of optical amplifiers and other active components is also specified in decibels. Finally, the optical link budget (Figure 3-20), which takes all of the above items into account, is expressed in decibels.

Decibels in UTP Systems

In UTP systems, like fiber optic systems, many of the transmission parameters are expressed in decibels. For components (such as cables, cords, patch panels, plugs and jacks, etc.) as well as the permanent link and channel configurations, the UTP parameters which are specified in decibels include:

- Insertion loss or attenuation (usually specified in dB/100m for cables)
- Near-end crosstalk (NEXT and PSNEXT)
- Far-end crosstalk (FEXT, ELFEXT, and PSELFEXT)
- Alien crosstalk (ANEXT, PSANEXT, AFEXT, PSAFEXT, and PSAELFEXT)
- Structural return loss (SRL)
- Return loss (RL)
- Attenuation to Crosstalk Ratio (ACR)

For UTP components and systems, these parameters are specified as a function of frequency.

Summary

This annex has reviewed the basics of decibels and how they are used in communication systems. The major points covered in this annex are:

- The decibel (dB) is a logarithmic unit which represents the ratio of two power measurements.
- Decibels have several useful properties, including the ability to represent a quantity with a large dynamic range by a relatively small number.
- Decibels can also be used to specify a power level relative to a fixed quantity, such as a milliwatt.
- Decibels are used extensively in communication systems for representing many different quantities.

Review questions

1. What is a logarithm?
2. What is a decibel?
3. Given an amplifier with input power P_{in} and output power P_{out}, what is the equation for the gain in decibels?
4. For the amplifier in question 2, if P_{in}= 4 mW and P_{out} = 60 mW, what is the gain in dB?
5. For the amplifier in question 3, what is the power output in dBm?
6. List at least 3 advantages of using decibels.
7. What value of decibels is often called the "half-power point"?
8. What power ratio is represented by a gain of 27 dB?
9. The sound intensity during a normal conversation is about 60 dB greater than the threshold of hearing (TOH). What does this correspond to in terms of power levels?
10. The NEXT signal in a UTP cable is attenuated by 40 dB and the insertion loss of the channel is 10 dB. If a single transmitter at each end is sending the same power level, what is the ratio of signal power to crosstalk power at the receivers in decibels? What does that correspond to in terms of a power ratio?
11. A Category 6 UTP permanent link has NEXT loss measurements of 54.6 dB at 16 MHz and 35.2 dB at 250 MHz. What is the ratio of noise power at 250 MHz compared to the noise power at 16 MHz?
12. The PSNEXT loss for a Category 5e and 6 UTP channels at 100 MHz are 32.3 dB and 41.8 dB, respectively. What does that difference correspond to in terms of power levels?
13. The insertion loss of a Category 6 channel is 1.9 dB at 1 MHz and 31.1 dB at 250 MHz (remember that lower decibels figures are better for insertion loss.) If 1 watt of signal power is launched into the channel at each frequency, how much power at each frequency arrives at the receiver?
14. A fiber optic link has a loss of 20 dB. What percentage of the transmitted photons arrives at the receiver?
15. A Category 6 UTP link has a return loss of 11 dB at 200 MHz. What percentage of the transmitted signal power is reflected back?

Index

CPSIA information can be obtained
at www.ICGtesting.com
Printed in the USA
FFOW02n1659050516
23854FF